Nimrod's Genesis

RAF Maritime Patrol Projects and Weapons Since 1945

Chris Gibson

HIKOKI
PUBLICATIONS

'Cush became the father of Nimrod; he was the first on Earth to be a mighty man. He was a mighty hunter before the Lord.' Genesis 10:8-12

'I sincerely trust that you will be successful, but if not, I can hardly suppose we shall have anyone but ourselves to blame. It is now almost exactly four years since OR350, itself preceded by protracted discussions, was issued on 18 July 1960. This has been supplanted by two further requirements, consideration of the problem has been continually postponed in the supposed interest of other projects, and various paper studies have been made. As far as I can see, however, there has been no significant progress towards a solution.'
Air Marshal A D Selway, C-in-C Coastal Command, 9 July 1964

To Harry Fraser-Mitchell

First published in 2015 by
Hikoki Publications Ltd
1a Ringway Trading Est
Shadowmoss Rd
Manchester
M22 5LH
England

Email: enquiries@crecy.co.uk
www.crecy.co.uk

© Chris Gibson

Colour drawings: © Adrian Mann
Line drawings: © Chris Gibson
Front cover and original photography:
© Graham Wheatley

Layout by Russell Strong

ISBN 9 781902 109473

Printed in China

Front cover Illustrations:

Top: Nimrod MR.2P XV244 climbs out on another operation. XV244 is preserved at Kinloss Barracks, Morayshire. ©*Graham Wheatley*

Bottom: An Avro Vulcan MR.3 banks away from a Soviet Navy *Yankee*-Class SSBN in the Norwegian Sea. *Adrian Mann*

Back cover illustrations:

Top: Saunders Roe P.104 flying boat as tendered for R.2/48, the Sunderland replacement specification. *Via RAF Museum*

Centre left: A pair of VC10MR on a *Tapestry* patrol over the Beryl Field in 1986. *Adrian Mann*

Centre middle: Dornier Do 340 6/7 flying boat design study for the NBMR.2 requirement. *via Dornier Museum Friedrichshafen*

Centre right: Avro Shackleton MR.2 WL754 of 42 Sqn RAF on patrol. *via Tony Buttler*

Bottom: A putative HS.1011 at low level on Windermere. *Adrian Mann*

Rear flap: Lockheed Neptune MR.1 banks away, showing its AN/APS.20 radar and gun turrets. *Terry Panopalis Collection*

Half-title page: Hawker Siddeley's HS.1011 variable-geometry supersonic maritime patrol aircraft performs a low-level pass at a Battle of the Atlantic commemoration ceremony at the former Shorts factory on Windermere. *Adrian Mann*

Title page, top: Nimrod XV226 strikes a pose on a test flight above the clouds. *Author's Collection*

Title page, bottom: Nimrod MR.1 XV243 of 206 Squadron taxies out Kinloss, a base that hosted land-based maritime patrol operations for sixty years. *Terry Panopalis Collection*

Contents

Introduction 4

Acknowledgements 5

1 *Voenno-morskoj flot SSSR* and the Threat From Below 10

2 Hunter/Killer: Sensors and Weapons for Aircraft 19

3 Old Men and Airships: The Last Flying Boats 47

4 The Kipper Fleet: Short and Medium Range MR 71

5 Avro's Bespoke Maritime Landplane 81

6 Sideshows: NATO and the Trinity 96

7 Following Trinity: OR.350 114

8 *Plus Ça Change*: OR.357 136

9 Not So Interim MR: OR.381 156

10 Replacing the Maritime Comet 175

Conclusion: A 21st Century Swordfish 197

Postscript 199

Glossary 200

Selected Bibliography 201

Appendix 1: Requirements & Specifications 202

Appendix 2: Rescue and Tapestry 205

Appendix 3: The Blind Alley 218

Index 223

Is this the future of the RAF's maritime patrol? A Boeing P-8A Poseidon of the US Navy VX-1 'Pioneers'. The type operated in 2012 from RAF Leuchars during exercise Joint Warrior and is the RAF's preferred type if the maritime patrol role is resurrected. *US Navy*

Acknowledgements

Producing *Nimrod's Genesis* has drawn together information from a myriad of sources and archives around the UK. These are maintained by volunteers, mainly retired employees of the various companies and as such are experts in particular fields. The benevolence of BAE Systems PLC provides much of the funding for archives at Woodford, Brooklands, Brough and Warton and therefore BAe and its former employees deserve better recognition as the keepers of Britain's aerospace heritage. When it comes to the professional archivists, the staff at the National Archives at Kew, that 'Burgess Shale' of aircraft projects, are always helpful and patient with a rummager like me. The RAF Museum at Hendon, always willing to help, holds a treasure trove of archive material and I would like to thank Peter Elliott, Gordon Leith and Belinda Day for their help. Continuing the geological metaphor, if Kew is the Burgess Shale there is also a Klondike: the Secret Projects Forum created by Paul Martell-Mead, built by its thousands of contributors and like the artisanal gold mines of the Klondike, nuggets can be found therein.

Firstly, I would like to thank my dear wife Kirsten for her considerable patience and continuing encouragement. When it comes to artwork, there is no finer exponent of 'what-if' artwork than Adrian Mann who can bring my drawings and sketches to life in living colour. Tony Buttler and Phil Butler must be thanked for their continued guidance on matters of research and provision of images. A book like *Nimrod's Genesis* requires photographs of a variety of subjects from aircraft to submarines and Terry Panopalis must be thanked for allowing access to his collection and supplying many of the images for this book. Thanks are also due to David Cook and aviation photographer Graham Wheatley who supplied images *Nimrod's Genesis*.

Another group whose ongoing help and advice contributed so much to *Nimrod's Genesis* are the aforementioned volunteers at the various archives who always provide a warm welcome and friendly assistance. When it comes to maritime patrol aircraft, Brooklands Museum and Avro Heritage hold much of the material on the RAF's occasionally fraught efforts to acquire maritime patrol aircraft. George Jenks at Woodford sourced much of the material on Hawker Siddeley studies, while Chris Farara, Albert Kitchenside and Jack Fuller are to be thanked for their continuing work at Brooklands and providing information on the Vickers/BAC studies into Maritime patrol types. Albert Kitchenside continues to amaze with the material on the Vickers VC10, the extent of its variants providing many a surprise. Last, but by no means least, I must thank Tony Wilson and the staff at North West Heritage Group at Warton, the last place I expected to find material on maritime patrol aircraft. Having always associated Warton with 'fast-movers' it came as a surprise when Tony turned up a key document on BAC's later proposals for OR.357. Ray Williams very kindly supplied the information on the fascinating AW.681MR, another of those types whose maritime application came as a surprise. The information on Dornier's submissions to the NBMR.2 competition came from the Dornier Museum and thanks are due to Ingo Weidig and Wolfgang Muehlbauer for helping in this. Engines were a major influence on maritime patrol studies and my questions on the obscurest of engine variants were answered by Dave Birch of the Rolls-Royce Heritage Trust. Lastly, I would like to thank Duncan Greenman of Bristol *AiRchive* for information and drawings of the Maritime Britannia projects.

Thanks are also due to a number of people who have provided invaluable help in its creation. Gillian Richardson and Jeremy Pratt at Crécy Publishing for their continuing support and are, as always, a pleasure to work with. The layout and the overall look of *Nimrod's Genesis* was the work of Russ Strong while Neil Lewis kept me on the straight and narrow. The research for *Nimrod's Genesis* could not have been carried out in the first place without the help of Robert Thornton and Bobbie Alexander, while I must also thank Neil Armstrong, Andy Paterson and my colleagues in the Beryl and Forties drilling teams.

To mark the 70th anniversary of the Battle of the Atlantic, a series of commemorative events were held in London and Liverpool. The longest continuous campaign of the Second World War, the battle saw the application of the latest and most secret technologies yet developed by either side. The Allies deployed radar, radio intercept and cryptography against the *Kriegsmarine* U-boats while they countered with equipment such as the *schnörkel*, radar warning receivers and infrared suppression. On 26th May 2013 the crowds gathered at Liverpool watched a flypast by Fleet Air Arm and Royal Air Force aircraft. It would be churlish to point out that while the Battle of Britain Memorial Flight flies the flag for the RAF at many events, perhaps a Catalina would have been more apt for this event. The Battle of Britain might have saved Britain, but the Battle of the Atlantic allowed Britain and its Empire to fight on and ultimately, with the additional blood and treasure of the USA and USSR, prevail against the Third Reich.

Coastal Command, since its establishment in 1936, had been denied the long-range aircraft its commanders requested while Fighter and Bomber Commands, perceived as the keys to victory, were re-equipped in the period of pre-war re-armament. Working closely with the Admiralty and tasked with protecting the trade routes around the UK and Mediterranean, as well as British shipping across the Empire, Coastal Command struggled to fulfil this task as its equipment, Avro Ansons and Lockheed Hudsons, was found wanting. This was particularly true where the protection of Atlantic routes was concerned, an undertaking that mounted in importance after mid-1940 as the U-boat 'Wolfpacks' wreaked havoc in the Atlantic. The long-range missions, out in the Atlantic and conducted by Shorts Sunderlands and Consolidated Catalinas were effective, but there were never enough long-endurance aircraft. The Great War had shown that the presence of *any* aircraft restricted U-boat operations but aircraft that could operate in the

The last thing many U-boat *Kapitans* saw. PBY-5A Catalina, G-PBYA, the last flying link with Coastal Command's operations during the Battle of the Atlantic. The RAF operated more than 600 Catalinas with only eleven examples being amphibians.

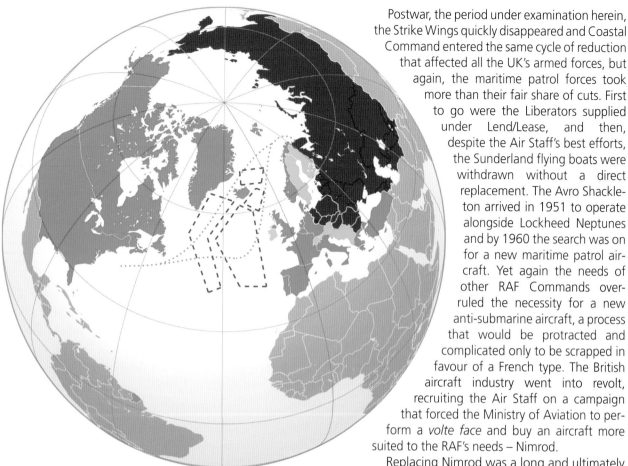

Key to illustration:

▨ NATO Nations 1989
▨ Warsaw Pact 1989
▨ Neutral/Non-Aligned 1989

– – SOSUS Network (Illustrative)
···· Soviet Submarine Access Routes

The world view in 1970, in the era of Flexible Response. If Warsaw Pact formations pushed into Western Europe, the United States would need to reinforce NATO by sea and air. Soviet naval forces, particularly the submarine units, would attempt to interdict the supply ships. The map also shows the SOSUS arrays that were used to track Soviet submarines.

Postwar, the period under examination herein, the Strike Wings quickly disappeared and Coastal Command entered the same cycle of reduction that affected all the UK's armed forces, but again, the maritime patrol forces took more than their fair share of cuts. First to go were the Liberators supplied under Lend/Lease, and then, despite the Air Staff's best efforts, the Sunderland flying boats were withdrawn without a direct replacement. The Avro Shackleton arrived in 1951 to operate alongside Lockheed Neptunes and by 1960 the search was on for a new maritime patrol aircraft. Yet again the needs of other RAF Commands overruled the necessity for a new anti-submarine aircraft, a process that would be protracted and complicated only to be scrapped in favour of a French type. The British aircraft industry went into revolt, recruiting the Air Staff on a campaign that forced the Ministry of Aviation to perform a *volte face* and buy an aircraft more suited to the RAF's needs – Nimrod.

Replacing Nimrod was a long and ultimately futile saga that spanned almost twenty years. In the late Nineties, save for a Russian amphibian, there were no brand new maritime aircraft, only revamped designs from the Fifties. Typically, the MoD opted for a revamped design from the Forties in the shape of the Nimrod MRA.4, whose genesis was protracted and ended with 'reduction to produce' by the scrappie's gas-axe. At the time of writing, summer 2014, the United Kingdom has no true maritime patrol capability and *Nimrod's Genesis* concludes by examining the options for the future.

Nimrod's Genesis will examine the maritime patrol aircraft, design studies, weapons, sensors and subsidiary roles of what was originally Coastal Command in 1945 until the end of specific RAF maritime patrol operations in 2010. The subject matter examines the Air Staff's operational requirements, the project studies that attempted to fulfil these requirements, political, financial and technical reasons why some proposals, such as the flying boat and the types to meet OR.350 and OR.357, failed to prosper. Look elsewhere for detailed information on anti-submarine warfare tactics, strategy or operations, nor are unit and aircraft histories covered in *Nimrod's Genesis*. It does, however, answer the question of why the RAF acquired

mid-Atlantic 'Black Hole' were in short supply, especially the four-engined bombers that Bomber Command had first call on.

The Atlantic was the theatre where Coastal Command became best known, but the Command itself had two facets; the offensive operations of the anti-ship units, such as the Banff Strike Wing, and the anti-submarine patrol units, both becoming highly efficient after 1943 when the Air Staff and the Ministries realised that the opening of the Second Front could only be possible by securing the Atlantic for shipping. By combining the technologies mentioned above and supplying Coastal Command with the long-range aircraft they had been asking for, defeating the Wolfpacks was possible. And so, supplied with long-range aircraft in sufficient numbers and equipped with the latest technology, the U-boat menace diminished and the Battle of the Atlantic was won.

an aircraft that was essentially a jet-powered Shackleton, a type with which the RAF had a love/hate relationship. The attempts to replace the Sunderland flying boat that figures highly in the early postwar years are examined, as are the reasons for the disappearance of flying boats from the order of battle in all air arms.

Notes on Nomenclature and Units

The Breguet Br.1150 appears repeatedly throughout this book and for consistency the Anglicised name adopted by the Avro and Hawker Siddeley companies, Atlantic, has been used. However, when the French name Atlantique appears in a quote it appears as per the original document.

Nautical miles (nm) have been used throughout this book. The RAF used statute miles until the 1960s and where these are included in quoted text from original documents, the conversions to nautical miles and kilometres are appended. In the main text, nautical miles are used throughout to allow comparison with post 1960 data. Speeds are quoted in knots (kt) and kilometres per hour (km/hr), mass/weight is quoted in pounds (lb), or kilogrammes (kg) while heavier weights are expressed as long tons (2,240 lb) and metric tonnes (mt). Metric equivalents of all units are shown for clarity. Throughout this book all-up-weight (AUW) is quoted and should be taken as the maximum gross take-off weight which includes the aircraft's structure, engines, fuel and oil, crew and payload. It is quoted in pounds (lb), with its metric equivalent in kilograms (kg) alongside.

Where the costs of items or programmes is concerned the British abbreviations for millions (m) and billions (bn) are used throughout.

Chris Gibson
Washington, October 2014

Personal Prologue

On 11th December 2011, the storm had been lashing the North Sea for three days and the supply boats had sought shelter in the Moray Firth. The Forties Delta production platform had been 'waiting on weather' (WOW) until the weather improved to a sea state that allowed the supply boats to be worked. The rig crew, having completed the usual WOW activities of maintenance and 'scrubbing', had been set to 'chipping and painting' in sheltered areas around the rig. One hundred miles west of the Forties Field the supply boats rode out the storm in the waters of the Firth with some very unusual bed-fellows; a battlegroup of the Russian Navy. Comprising a destroyer, two frigates, an oiler, an ocean-going tug and the flagship of the Russian Navy, the aircraft carrier *Admiral Kuznetzov*, this naval squadron was on a deployment to the Mediterranean when the storm blew up and so had sought shelter. As the rig crew chipped and painted, 600nm (1,200km) to the south, the Type 42 destroyer HMS *York*, set out from Portsmouth and headed north through rough seas to keep an eye on the Russians. An odd situation made even odder by the fact that barely eighteen months before, the *Admiral Kuznetzov* and its escorts would have been under continual observation by a reconnaissance asset that was based less than 100nm (185km) away, twenty minutes flying time, from the carrier's position. That asset was the Hawker Siddeley Nimrod MR.2P, Britain's 'Mighty Hunter' and as of May 2010 these had been retired and the United Kingdom had no maritime patrol aircraft to search the seas for submarines, aid ships in distress, co-ordinate search and rescue (SAR) operations or keep an eye on trespassers.

How did a maritime nation come to have no maritime patrol capability, save for Reims Cessna F406s operated as fishery protection platforms and RAF C-130J Hercules transports operating as interim maritime reconnaissance platforms in the search and rescue role? A further question is – interim to what?

9

1 *Voenno-morskoj flot SSSR* and the Threat from Below

'There is no-one to fight at sea.'
Clement Attlee, January 1946

Before delving into the evolution of the United Kingdom's maritime patrol and anti-submarine aircraft and weapons, a brief look at the driving force behind their development would be useful. The *raison d'être* for maritime patrol aircraft is to locate, track and, if necessary, destroy enemy submarines. In the case of RAF Coastal Command, in a period of less than five years it had traded depth-charging U-boats for locating and tracking Soviet submarines and these, as will be shown below, were many and varied.

Two countries came under sustained blockade by submarine during World War Two. One, Japan, suffered complete and utter destruction of its merchant marine at the hands of the United States Navy. The second, Great Britain, had stared into the abyss of defeat by the *Kriegsmarine's* U-boat wolfpacks in 1942, but had by 1943 applied an all-arms approach to counter-ing the submarine. Success lay in technology and courage. The technology lay in areas as diverse as programmable computing machines designed and built in rural England by the most talented of academics and technicians, to arc welders drinking their milk and sparking their rods in American shipyards. The courage came from those who risked their lives minute-by-minute on the ships carrying the materiel that would ultimately win the war. What was proved was the submarine's efficacy as a weapon, and with developments in propulsion and weapons technologies, these would become the capital ships of the late 20th century.

Postwar, the United States and Royal Navies went unchallenged, having destroyed the Imperial Japanese Navy, the only navy that had posed a serious challenge on the surface, and the *Kriegsmarine's* U-boat force. Attlee was correct when less than six months after the war's end he said 'There was no-one to fight at sea.' However, as the Forties drew to a close, that would change.

A *Victor III* attack submarine showing its streamlined hull. Also visible is the housing for the towed sonar (the streamlined pod on the tip of the rudder). The *Victor I* and *II* had much improved sound-proofing, but the *Victor III* (also known, unofficially, to the Americans as the *Walker*-Class) was a quantum leap in sound suppression in Russian submarines. *US Navy*

The Soviet Union sought to build a 'blue water navy' in the form of the *Voenno-morskoj flot SSSR* (Military Maritime Fleet of the USSR) – an ocean-going fleet on a par with the Royal and US Navies – and, like their former Allies, America and Britain, the Soviets had learned much from the German U-boat campaign. Initially the Soviets built a fleet of submarines to counter surface vessels but by the late Fifties the advent of the nuclear-powered boat carrying submarine-launched ballistic missiles had upped the stakes. No longer were ships and a few hundred crewmen at risk from submarines: entire cities and their populace were now at the mercy of submarines. Therefore it is useful to have a very brief outline of what was the main target for the RAF's maritime patrol operations throughout the Cold War – the Soviet Navy's submarine fleet.

In the Great Patriotic War, 1941-45, Russian submarines did not have a very good war. If tonnage per submarine lost is used as a benchmark for success, the Soviet Navy fared badly with 3,692 tons per boat lost against the Royal Navy's 20,266 tons per boat lost or US Navy with 101,923 tons per boat lost. Comparing the Royal and Soviet navies may seem unfair, but the majority of the Soviet submarine operations were in the Baltic and Black Seas, which like the RN's major operating area in the Mediterranean, restricted submarine operations and made detection and destruction by aircraft easier. The Soviet Navy did cause the greatest loss of life in a single maritime incident when submarine *S13* torpedoed the German liner *Wilhelm Gustloff* in January 1945. The death toll is unknown, but is thought to be up to 9,000 souls.

Postwar, the Soviets learned much from their occupation of the German shipyards on the Baltic coast and like the Western allies, were very impressed by the Type XXI U-boat. They acquired four Type XXIs and these were commissioned into the Soviet Navy. Its most impressive aspect was its ability to operate submerged for extended periods, up to three days, and the use of the *schnörkel* to charge the batteries. The streamlined casing made for higher speeds while sonar and noise reductions systems made the Type XXI a formidable weapon that made the work of the World War Two era anti-submarine aircraft very difficult.

With imitation being the sincerest form of flattery, the Soviets set out to copy the Type XXI under Project 614, but what ultimately emerged was Project 613, a Soviet design from the early Forties that was latterly influenced by

the Type XXI and known in the West as the *Whiskey*-class. Not that the Soviets were alone in this as the US Navy's *Gato*, *Balao* and *Tench* class submarines owed much to the Type XXI. Further developments of the *Whiskey* included the *Cylinder* and *Long Bin* series that were fitted with cylindrical housings for SS-N-3 *Shaddock* inertially-guided missiles. Such developments mirrored those of the US Navy's Chance-Vought Regulus I and II or Avro's proposed system for the Royal Navy that used the Avro W.112 missile. The *Whiskey Long Bin* was succeeded by the Project 675, *Echo*-class, a cruise missile carrier with three missile launchers arranged along the boat's sides, aft of the conning tower. Originally intended to carry anti-ship missiles, the lack of a suitable long-range missile guidance system saw the *Echo*-class boats converted to carry SS-N-3 *Shaddock* cruise missiles for attacks on land targets. The *Echo*-class was in turn replaced by the Project 651, *Juliett*-class, that was designed to carry four *Shaddock* or SS-N-12 *Sandbox* land-attack missiles that used satellite targeting information. Such submarines gave the Soviet Navy a credible deterrent when operating off the eastern seaboard of the United States, but a new technology would raise the stakes in the field of anti-submarine warfare.

Caught! This *Foxtrot*-Class submarine was photographed running on the surface during a NATO naval exercise. Running on the surface at night was standard procedure for diesel/electric submarines, making them visible on radar. *Author's collection*

The Russian (and Soviet) Navy fielded a number of surface attack missiles. The SS-N-3 *Shaddock* and SS-N-12 *Sandbox* that were launched from 'bins' on the deck on *Whiskey* classes or built into the hull of the *Juliett* class or attack submarines such as the *Victor* and *Akula*.

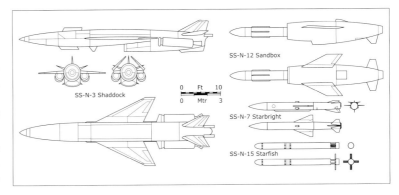

SS-N-3 Shaddock

SS-N-12 Sandbox

SS-N-7 Starbright

SS-N-15 Starfish

0 Ft 10
0 Mtr 3

Soviet missile submarines, such as this *Whiskey Twin-Cylinder* carried their missiles in large launch tubes attached to the deck. This *Whiskey* has its tubes mounted aft of the conning tower whereas the *Whiskey Long Bin* had its launchers mounted for'ard. On surfacing the launch cylinders were elevated and the watertight hatches opened. *Author's collection*

The experience of developing the diesel/electric *Whiskey*-class led to the larger *Zulu*-class that became the first submarine to launch a R-11FM SS-N-1 *Scud* ballistic missile on 16th September 1955. By this act alone the submarine threat changed from being one that could sink ships to one that could annihilate entire cities. Later versions of the *Zulu* could carry two *Scuds*, installed in an extension of the conning tower and although the submarine had to surface and the missile raised clear of the conning tower prior to

launch, the *Zulu* showed where the future of submarine warfare lay. Despite its range being less than 100nm (1,185km) that was enough to make the coastal cities of the USA vulnerable to a *Scud* with a nuclear warhead.

The *Zulu* subsequently formed the basis of the *Foxtrot*-class, which was intended to replace the *Zulu* from 1958, but was not as successful as the *Zulu*, being much noisier than its contemporaries. *Foxtrots* were the last of the line from the Type XXI and most submarines designed after the *Foxtrot* featured the streamlined ichthyoid shape that is familiar today. The *Foxtrot* formed the basis of the much larger *Golf*-class that was the first Soviet submarine designed from the start to carry and launch ballistic missiles. Like the *Zulu*, the initial *Golf*-class submarines carried their three *Scuds* in an extended conning tower, but these missiles were soon replaced by the R13 SS-N-4 *Sark*, the Soviet Union's first purpose-designed SLBM (Submarine Launched Ballistic Missile). The liquid-fuelled *Sark* had a range of 324nm (600km) with a 1-2Mt nuclear warhead and as before, *Sark* had to be launched on the surface.

The next development in the *Golf*-class was the deployment of the R-21 SS-N-5 *Serb* (also referred to as *Sark* in some literature) that used a 'cold launch' technique to eject the missile from its launch tube before the main engines fired. This meant that the *Serb* could be launched while the submarine was submerged and with a range of 700nm (1,300km), the *Serb* was a much more formidable weapon. With this extended strike range the scene was set for the ballistic missile submarine to reach its acme though the application of nuclear propulsion.

Another game changer was the submarine launched ballistic missile (SLBM). Shown here is an early surface test launch of a Polaris A1, but the definitive A3 with more than twice the range of the A1 served with the US and Royal Navies. The ability to launch such a weapon from a submerged nuclear-powered submarine made the task of the maritime patrol aircraft much tougher. *Author's collection*

The Soviet Union's first nuclear-powered submarine was the Project 627, known to NATO as the *November*-class. This was an attack submarine intended to tackle surface units with torpedoes rather than hunting other submarines and the first *November* entered service in 1958. Despite its sleek hull and lack of the myriad of holes and vents typical of earlier submarines, the *November*-class was particularly noisy and easy for NATO forces to detect and track. By adding the missile launch tubes from the *Golf*-class to the *November*, the first Soviet nuclear-powered ballistic missile submarine (in NATO parlance an SSBN – Ship, Submersible, Ballistic missile, Nuclear-powered) was developed as Project 658, designated the *Hotel*-class by NATO and entered service in 1959. The *Hotel*-class also carried three *Serb* missiles and the deployment of the *Hotel*-class came at a particularly tense period of the Cold War, essentially making many of the sensors the RAF used to find submarines, such as ASV radar and Autolycus exhaust detectors, obsolete in the submarine-hunting role.

The *Hotels* shared the *November's* noisy powerplant and hull design but development of a replacement was soon under way and this scheme, Project 667, reflected the current thinking in the West by being streamlined and carrying its missiles within the hull rather than behind the conning tower. In fact the resultant submarine shared so many attributes of the American '41 for Freedom' Polaris-armed Fleet Ballistic Missile Submarines that NATO designated it the *Yankee*-class. These were much quieter than the *Hotels* and carried sixteen *Serb* missiles within the hull and the new SLBM's improved range allowed the Soviet Navy to establish deterrent patrols in the Atlantic to the east of Bermuda and between Iceland and Scotland. The *Yankee*-class entered service in 1968 and served until the end of the Cold War. A further development of Project 667 was the *Delta*-class, which comprised four variants that were intended to take advantage of the new long-range missiles that were under development and the four variants, *Delta* I, II, II and IV reflected the deployment of new and improved SLBMs.

The basic premise of the *Delta*-class was to avoid the need to enter the Atlantic and thus pass through the NATO surveillance network that included the Sound Surveillance System (SOSUS), ever-present maritime patrol aircraft and hunter/killer submarines. SOSUS involved an array of sonar sensors on the seabed that could detect and track submarines. These were laid on the Atlantic seabed with the British terminals at RAF Brawdy in Pembrokeshire and

A stern-on view of a *Yankee*-Class showing the housing for the 16 SS-N-6 *Serb* missiles. The dive planes mounted on the conning tower, a feature that it shares with the *Delta*-Class and, of course, American '41 for Freedom' classes. These were the boats that patrolled the Atlantic and Pacific seaboards of the USA in the late 1960s and 1970s. *Author's collection*

Scatsta in the Shetlands. The SOSUS network was installed and calibrated using NATO submarines and maritime aircraft under the code-name *Backscratch* and could even detect Soviet Tupolev Tu-95 *Bear* aircraft in flight. Earlier Soviet missiles were relatively short-ranged so the launch submarine was required to close with the American coastline to reach their targets. The development of the initial R29 (SS-N-8) *Sawfly* with a range of 4,160nm (7,700km) allowed the USA to be threatened from afar, even from under the Arctic ice. The *Delta*-class was designed with this in mind, their patrols being conducted in the Norwegian and Barents Seas, carrying up to twelve *Sawfly* missiles in launch tubes within the hull aft of the conning tower. It was the prospect of a boat like the *Delta*-class and long-range SLBMs deployed away from the SOSUS network in the waters around Spitzbergen that drove the Air Staff's requirement for maritime patrol aircraft to

The size of the missile bay housing on this *Delta IV* alludes to the increased size, and subsequently, capability of the SS-N-23 *Skiff* SLBMs carried within. The *Delta* class submarine with long-ranged SLBM such as SS-N-8 *Sawfly* allowed the Soviet deterrent to remain in Russian waters thus avoiding crossing the SOSUS arrays. *US Navy*

operate at 1,000nm (1,852km) and employ high transit speeds to get there. The concept of threatening the USA from high latitudes was taken to its logical conclusion by the next class of SSBN under consideration in this short canter through Soviet submarines – the *Typhoon*.

The largest submarines yet built, the Project 941 boats are known to NATO as the *Typhoon*-class and are massive SSBNs with a submerged displacement of 47,240 tons (48,000 tonnes). Typhoons can carry twenty R39 (SS-N-20) *Sturgeon* ballistic missiles with a range of 4,482nm (8,300km), with each missile carrying up to ten

MIRVs (Multiple Independently-targeted Re-entry Vehicles). This makes the R39 a formidable ballistic missile with intercontinental capability and thus can threaten the USA from the seas around Russia, if not from their home ports. The *Typhoon* was designed to cruise under the Arctic ice cap and to fire its missiles from areas of open water known as *polynyas,* which are ice-free areas, formed where warmer water reaches the surface or winds cause the pack-ice to break up and drift away from islands. As with the *Delta*-class, there was no need for the *Typhoons* to enter the North Atlantic through the Greenland-Iceland-UK (GIUK) Gap and run the gauntlet of the ocean surveillance systems.

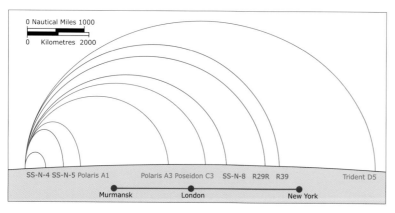

Having developed the SLBM, both sides sought to extend range and increase both the throw-weight of the missiles and the fitting of multiple re-entry vehicles also increased the deterrent value of the SLBM. The longer range of the American Trident D5 was probably driven by the need to reach targets in the vastness of the Soviet Union.

The continual improvements in Soviet SLBM range allowed the submarines that would launch them to operate well away from NATO anti-submarine assets. This map shows where the various generations of Soviet SSBNs operated and why the Air Staff wanted high transit speeds and long endurance to counter the later *Delta* and *Typhoon* classes.

Having described the SSBNs, the attack and hunter/killer submarines that would have attacked NATO surface units such a carrier battle groups and convoys, or sought out NATO's submarines, need to be examined. The *Foxtrot*-class was replaced by the diesel/electric *Tango*-class, fitted with better propulsion systems that allowed it to operate on electrical power for almost a week before using the *schnörkel* and running its three diesel engines to charge the batteries. The *Tango*-class boats were SSKs (Ship, Submersible, conventionally-powered, hunter/Killer) that specialised in attacking NATO nuclear submarines in areas such as the approaches to bases where the likelihood of detecting a NATO submarine was much improved. First commissioned in 1980, the *Kilo*-class SSK took over from the *Tango* and exhibited some interesting features such as being fitted with a SAM system. The *Kilo*–class has been quite widely exported, particularly to China and India.

Then, in 1959 the attack submarines went nuclear, with the appearance of the *November*-class SSN (Ship, Submersible, Nuclear-powered, hunter/killer), the first nuclear-powered submarines in the Soviet Navy. Developed as the Project 627, the *November*-class was tasked with attacking NATO convoys and surface warships with nuclear-armed torpedoes. As noted above, the *Novembers* were noisier than the earlier diesel-powered submarines, but the *Novembers* were fast, faster than most US Navy submarines, a trait that aided their task of attacking surface units. As the Sixties progressed, the anti-submarine screens around convoys and carrier task forces made torpedo attacks more difficult, prompting the development of a long-range anti-ship missile that could be launched from submarine torpedo tubes, such as the SS-N-15 *Starfish*.

Soviet SSBN Patrol Areas

Hotel-class with SS-N-4 *Sark*
Yankee-class with SS-N-6 *Serb*
Delta-class with SS-N-8 *Sawfly*
Typhoon-class with SS-N-20 *Sturgeon*

In the 1970s, something odd happened in the submarine Cold War – a Soviet submarine in a different class. In itself this wasn't peculiar, the Soviets introduced new classes of submarines on a fairly frequent basis, but this one was different because it was particularly quiet, resulting in the NATO navies losing their advantage in the submarine cold war. The *Victor*-class, Project 671, was a new generation of Soviet attack submarine that entered service in the late 1960s as the *Victor* I. The second generation, *Victor* II, was under development when work was stopped. The next generation, *Victor* III, while superficially similar to the previous *Victors*, had much-reduced noise signatures and improved sensors such as a towed-array sonar. This was the submarine that had the intelligence officers of the NATO navies somewhat intrigued.

The answer lay in two developments: increased R+D on hull design in the Soviet submarine yards and, the source of the reduced sound signature breakthrough, the John Walker spy ring. Amongst many of the communications technology secrets passed over by Walker, a US Navy communications specialist, was the fact that Soviet submarines could be tracked via the SOSUS network though detection of their propeller and machinery noise. This prompted the Soviets to invest heavily in sound-proofing and soon led to the Americans discovering another aspect of the Walker spy scandal – the supply of Toshiba-Kongsberg computer-controlled milling machines to the Soviets, despite their export to the USSR being prohibited. The *Victor* III, on entering service in 1979

represented the finest example of the advances in submarine development made by the Soviets courtesy of the Walker spy ring and were known, unofficially, throughout the US Navy as the *Walker*-class.

The next generation of attack submarine was the *Akula*-class (Shark in English, an odd choice of NATO reporting name, *Akula* being a Russian word) and these, like the *Victor* IIIs, were much quieter than the earlier generations of Soviet boats, benefitting from new propellers produced on the high-technology milling machines. Developed as Project 971 as a follow-on to the *Victor*-class, the *Akulas*, particularly the *Akula* II, represented a quantum leap in Soviet submarine construction. *Akulas* were formidably armed with torpedoes, plus the

The *Typhoon*-Class was the boat that the Soviets viewed as impregnable. Designed to lurk under the polar ice cap before breaking through to launch its 20 SS-N-20 *Sturgeon* missiles. Six of these giants entered service and they are the quietest of the Cold War-era Russian Submarines. *US Navy*

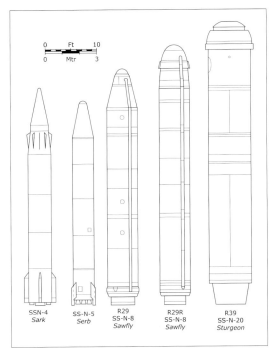

SSN-4 *Sark* — SS-N-5 *Serb* — R29 SS-N-8 *Sawfly* — R29R SS-N-8 *Sawfly* — R39 SS-N-20 *Sturgeon*

A Soviet Navy *Foxtrot*-Class submarine caught on the surface by a Coastal Command Shackleton MR.2. Diesel/electric submarines such as the *Foxtrots* need to surface to charge their batteries and as such could be detected by ASV radar in an aircraft. Using a snorkel reduced the chances of detection, but as radar developed, detection improved. Nuclear propulsion changed the game. *Author's collection*

Russian SLBMs grew in size and capability throughout the Cold War. The SS-N-4 *Sark* and SS-N-5 *Serb* were closely related to land-based missiles. The R29, one of the most effective SLBMs in the Russian inventory came in a number of variants, including the SS-N-20 *Sturgeon* carried by the *Typhoon*-Class.

A *Victor III* in the Atlantic 500nm (925km) off the coast of South Carolina. The boat was making slow progress, 2kt (4km/hr), suggesting it had suffered a mechanical breakdown. A submarine running on the surface isn't the most comfortable vessel in which to cross the Atlantic. *Author's collection*

The streamlining of the *Akula*-Class attack submarine is apparent in this view, as is the large housing on the stern for the towed-array sonar. The *Akula* replaced the early *Victor* I and IIs in the anti-ship role. The *Akula*-class benefited from the experience in sound-proofing the *Victors*, including the new propellers designed and made with Western equipment. *US Navy*

SS-N-15 *Starfish* and SS-N-16 *Stallion* anti-ship missiles which, when combined with the reduced noise signature, made the *Akula* a significant risk to carrier groups, NATO reinforcement convoys and, most significantly, US and Royal Navy SSBNs.

The Soviet Navy did not field the carrier groups that the Royal and US Navies operated but these were an ever-present and highly mobile threat to Soviet operations, particularly the US Navy's carrier groups that could roam the oceans at will. Until development of SLBMs and the Polaris submarines, it was the US Navy's carrier aircraft that carried the maritime nuclear deterrent. This made the carrier groups a prime target for Soviet submarine forces. High value assets such as carriers operated within their own ASW zone, complete with anti-submarine aircraft, helicopters and submarines, not to mention the destroyers and frigates that formed the battle group. The chances of a Soviet submarine penetrating this defensive perimeter were slim, so long-range submarine-launched anti-ship missiles were the weapon of choice against the carriers.

This need to attack such high-value surface units saw the development of the *Echo*-class, but it lacked the capability to target a fast-moving carrier group from outside the ASW 'bubble'. The first functional anti-ship missile submarine was Project 670, the *Charlie*-class, armed with long-range anti-ship missiles such as the SS-N-7 *Starbright*, which could be launched while submerged. The *Charlie* I carried eight missiles in two rows either side of the hull forward of the conning tower. The improved Project 670M *Charlie* II was longer, incorporating an improved fire control system for the SS-N-9 *Siren* and had a secondary hunter/killer role.

The 1970s saw improvements in guidance systems that allowed much longer ranges for anti-ship missiles; the Soviets developed the SS-N-19 *Shipwreck* missile that obtained its targeting data from satellites. *Shipwreck* was fitted to a new generation of missile boats, the Project 949 *Oscar*-class, commissioned in 1980 and capable of carrying an array of weapons for attacks on surface ships or submarines.

Although the *Charlie* and *Oscar* classes were capable of hunter/killer operations against other submarines, by the 1970s the *November* and *Victor* classes were becoming in need of replacement. For this the Soviets designed and built the innovative Project 705, *Alpha*-class, the fastest submarine built to date, capable of a submerged speed of 40kt (74km/hr). Such speed troubled NATO's navies, prompting The Admiralty to develop the Tigerfish torpedo and the US Navy the ADCAP torpedo. The *Alpha*-class was innovative in its use of titanium for its hull, reducing its magnetic signature, and was fitted with a lead-cooled reactor that was very powerful and compact, allowing a much smaller hull.

The Soviets placed great emphasis on the submarine throughout the Cold War and it was against these boats that the Coastal Command and later, 18 Group, Strike Command maritime patrol forces were ranged. The main concern, certainly from the 1980s, was detection, tracking and, if required, destruction of the Soviet attack and missile submarines tasked with attacking NATO submarine forces, convoys, carrier groups, and of course, NATO's SSBNs.

Blue Water Red Navy

Unlike the United States and Great Britain, the Soviet Union had no need to protect its trade routes with a large 'blue water navy' but did see the need to disrupt NATO reinforcement convoys in the Atlantic and counter the NATO submarine fleet, particularly once the *Polaris* SLBM had been deployed. Therefore the Soviets did not develop the carrier groups or even large surface ships that the NATO navies possessed, such as aircraft carriers, preferring to concentrate on submarines and guided-missile cruisers and destroyers. Soviet warships were armed with a variety of weapons to fit diverse roles, unlike the dedicated anti-aircraft or anti-submarine ships of the Western navies. The early 1950s saw the development of the *Sverdlov*-class cruisers that prompted much work in the UK on development of weapons to tackle these ships as it was thought that they were the beginnings of a move towards a 'blue water navy' by the Soviets. Nikita Khrushchev and his successor Leonid Brezhnev put paid to the Soviet Navy's aspirations as they saw large surface ships as very expensive and vulnerable, with the Soviet Navy being viewed by the politicians as a poor relation to the Red Army and Soviet Air Forces.

Despite this, the Soviets developed a few large ships, mainly for use in the anti-submarine warfare role, including the *Moskva* and *Kiev* classes of helicopter and aircraft carriers, although these were not on the scale of the American 'supercarriers'. They also developed modern cruisers, which like all Soviet surface units were very heavily armed, including some formidable multi-layered air defence systems. Examples of these are the Project 1144 *Kirov*-class nuclear-powered cruisers that appeared from 1980 and the gas turbine-powered Project 1164 *Slava*-class, which posed a great threat to NATO operations in the Atlantic.

Towards the end of the Cold War, the Soviet Navy had fielded its first carrier capable of operating conventional aircraft, the *Tbilisi*, later renamed *Admiral Kuznetsov*, (this became the name for the class, although only two were built). The *Kuznetsov* could operate the Sukhoi

This is the *Admiral Ushakov*, one of the *Sverdlov*-class that worried the NATO navies in the late 1950s. The Admiralty issued a number of guided weapons requirements specifically to attack the *Sverdlovs. Dave Forster Collection*

The *Kiev*-class cruiser *Baku* exemplified the Soviet Navy's doctrine of having all ships heavily armed for different roles. Two turrets for 100mm (4in) guns, launch bins for SS-N-12 *Sandbox* SSM and a number of launch silos for SA-N-9 *Gauntlet* SAMs can be seen. Also, ranged on deck are Ka-32 *Helix* helicopters.
US Navy

Su-33 *Flanker-D* and Mikoyan-Gurevich MiG-29K *Fulcrum-D* alongside the Kamov Ka-27 *Helix* series helicopters for ASW, plane guard/rescue and logistics. To complete the air group, training was provided by Sukhoi Su-25TG *Frogfoots* and the AEW role was to be filled by either the Yakovlev Yak-44E (no ASCC reporting name was applied) or the Antonov An-71 *Madcap* variant of the An-72 *Coaler* transport.

It was the arrival of the *Admiral Kuznetsov* and its battlegroup in the Moray Firth on that stormy morning in December 2011 that highlighted the lack of a credible maritime patrol capability in the UK armed forces. How did a maritime nation such as Britain come to lose such a vital asset, one that could have detected, tracked and monitored the *Kuznetsov* within half an hour of its arrival, rather than 'scrambling' HMS *York* from Portsmouth? Having sketched out the Soviet naval threat, *Nimrod's Genesis* aims to describe the development and ultimate demise of the Royal Air Force's maritime arm since 1945.

The variety of designs employed by the Soviet and Russian Navy is shown by this diagram. Only the main classes are shown, as some classes have a number of variants, such as the *Victor* class that came in at least three major versions. Of course, this view of a submarine was a rare event for a maritime patrol aircraft!

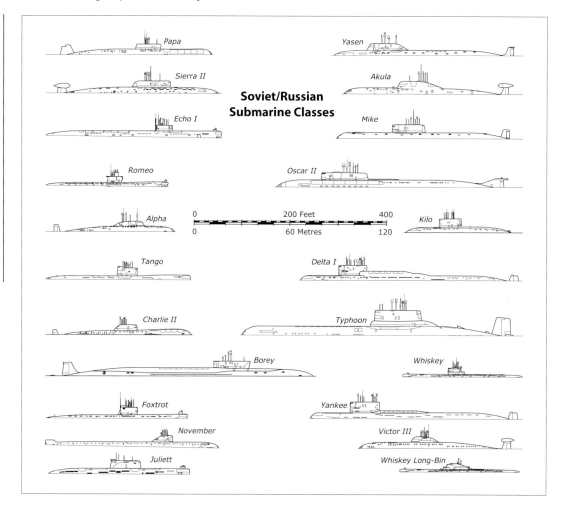

2 Hunter/Killer: Sensors and Weapons for Aircraft

'Of all the branches of men in the forces there is none which shows more devotion and faces grimmer perils than the submariners.' Sir Winston Churchill

'There is a touch of the pirate about every man who wears the dolphins badge.' Commander Jeff Tall RN

These quotes sum up the submarine and its crew as a weapon: hardy men with the means to destroy their targets while being a target themselves. Both world wars had shown that aircraft were a great hindrance to the operation of submarines. Submergence gave them the element of surprise and while described in 1901 by Admiral Sir Arthur Wilson as 'underhand, unfair and damned un-English', forty years on, the U-boats by 1943 had the full-force of Allied technical know-how ranged against them. Terms such as ULTRA, Huff-Duff, retro bomb, sonobuoy and MAD entered the lexicon of warfare and, with its roots in the Second World War; the postwar technological revolution pitted the submarine against aircraft yet again. Only this time the instant annihilation of entire cities was at stake.

Hunters – Finding the Submarine

The greatest advance in anti-submarine warfare in the Second World War was the cavity magnetron that led to centimetric radar. Maritime applications of this technology took a back seat to the development of bombing and navigation aids for Bomber Command, but once the H2S ground-mapping radar had been developed, application of the same techniques produced Air-to-Surface Vessel radars such as ASV.VI, ASV.VII and ASV.VIII. Postwar the Shackleton GR.1 (latterly MR.1) was fitted with ASV.XIII (later redesignated as ASV.13) in a Perspex radome under the nose but was prone to bird strikes and suffered from 'shadowing' – restricted field of view through airframe obscuration. The MR.2 and MR.3 Shackletons had their radar antennae moved to a retractable 'dustbin' under the rear fuselage. This had three positions: fully-retracted for take-off, transit and landing, semi-extended for search with minimum drag and fully-extended for making attacks. In search mode, under the right conditions, the ASV.13 could detect a *schnörkel* at a distance of 27nm (50km).

In March 1955 a new requirement, OR.3586, was issued for a maritime radar to replace the

Nimrod MR.2P XV254 of 120 Sqn RAF flying off the Caithness coast. The MR.2 was fitted with the state of the art in ASW equipment including the Searchwater radar and ASQ.901 acoustic suite. Weapons included the Stingray torpedo and Mk.57 NDB. As such the Nimrod MR.2 was suitably equipped to find and destroy Soviet SSBNs. *via Ron Henry*

Radar revolutionised anti-submarine warfare over the sea. Shackleton MR.3 Phase III WR979 (D) of 120 Sqn has its ASV.21 'dustbin' fully-extended in attack configuration with the weapons bay doors also open. Also of note on this MR.3 is the 'spark plug' for the Orange Harvest electronic support measures (ESM) used to detect submarines' radio communications. *Author's collection*

ASV.13 on Shackletons and the ASV.19 on Fairey Gannets. With the focus on maritime patrol in the late Fifties, firstly for the NATO MR type to NBMR.2 and then OR.350, a further requirement was issued in February 1963. This was ASR.827 and was to be fitted to the OR.357 aircraft and replace ASV.13 on the Shackleton MR.1, the ASV.21 for the Shackleton MR.3 and its ultimate replacement, the Nimrod MR.1. The ASV.21 radar was an X-band set that began life as the target acquisition radar for the Buccaneer S.1 and the Green Cheese missile. The ASR.827 was in its turn replaced by a radar to meet ASR.846 that eventually became called Searchwater. This benefitted from the computer revolution of the 1960s and 70s and on entering service in the 1980s gave the Nimrod MR.2 a sophisticated sensor that could be used to classify targets by their shape on the radar display.

One of the longest serving maritime radars also had its origins with Bomber Command's H2S series. This was the H2S Mk.9 that was to be modified for the maritime patrol role and fitted to the R.2/48 flying boat, with a capability on a par with the American AN/APS.20. Although this aircraft did not enter service (see Chapter Three) the radar was in use on the V-bombers as their primary bombing aid. The strategic reconnaissance HP Victor SR.2s of 543 Sqn used that radar for maritime reconnaissance until November 1973 when the Vulcan B.2MRR (Maritime Radar Reconnaissance) entered service in a sea surveillance role, providing wide area reconnaissance until 1982.

What ultimately did for radar, the primary sensor of the Battle of the Atlantic and the dominant method for finding submarines, was the postwar development of nuclear propulsion. The *Kreigsmarine's* improvement and deploy-

ment of the *schnörkel* at the end of the war had shown how difficult detection and tracking by radar of a small target at sea, particularly in rough conditions, could be. By removing the need for a submarine to take in air for its engines and crew through the development of nuclear propulsion and air scrubbers, the maritime patrol aircraft was more or less out of the submarine tracking game. However, submarines have other physical characteristics that can be used to detect, track and ultimately help destroy them from the air. This moves the story on from the radio spectrum to the geomagnetic, chemical and sonic spectrums.

The Earth's magnetic field allows navigation by use of a simple device, the compass, which dates back to the lodestones used by the Vikings in their voyages across the northern seas. By 1833 Carl Gauss had developed a magnetometer that could detect and measure the strength and direction of magnetic flux and so the science of geophysics was born. This should never be referred to as 'geofizz' around geologists; they get annoyed, it's a science not a soft drink. The magnetometer was mainly used for mapping the Earth's magnetic field and geophysical surveys of the Earth's crust soon revealed some interesting features. By creating maps of the magnetic field, areas of higher and lower magnetic flux within that magnetic field, were revealed. Called anomalies, these could be used by geophysicists to identify possible sub-surface ore bodies (whose magnetic flux is higher than the surrounding area) and exploited by miners. There was a major drawback to preparing a geomagnetic map and this author can affirm that geomagnetic surveying is a time-consuming process.

By the 1930s the more sensitive fluxgate magnetometer had been invented by Gulf Oil's

Victor Vacquier for oil exploration. By installing the fluxgate magnetometer in an aircraft and connecting it to a magnetograph to trace out the instrument's response, the magnetic field of the rocks below the aircraft could be recorded and analysed. While not as detailed as a ground survey, an aerial magnetic survey is a much quicker technique to cover large areas and was soon applied to finding areas of low magnetic flux that indicated the possible presence of oil deposits. The fluxgate magnetometer also allowed the angle of the magnetic field relative to the earth's surface to be measured and this, called the dip, could also be used to quantify the magnetic response. The magnetometer, fitted to an aircraft, allowed a large area to be surveyed for hidden objects that affected the Earth's magnetic field. It was only a matter of time before this technique was applied to find large steel objects under the sea – submarines.

The US Navy and the Imperial Japanese Navy pioneered the use of the fluxgate magnetometer, initially on ships, as far back as 1918 in the US Navy's case, but found the ship's hull interfered with the response. The equipment was soon fitted in aircraft as per mineral surveying and applied to detecting submarines. While the Royal Navy concentrated on ASV radar fitted to aircraft, with sonar and high-frequency direction-finding (HF/DF, known as 'Huff-Duff') on warships, the US Navy worked on what became known as a magnetic anomaly detector (MAD).

Consolidated Catalinas were fitted with MAD kit from 1944, the magnetometer being installed at the end of a long boom extending from the tail. The aircraft would patrol an area where submarines would be operating, possibly using intelligence gleaned from ULTRA or HF/DF. The MAD operator monitored the magnetograph and if a response was seen the aircraft would come about and line up for an attack on the submerged target.

The MAD equipment worked best at choke points such as the Straits of Gibraltar where the submerged submarines would be funnelled into a limited area of sea that lent itself to search by MAD-equipped aircraft. Such aircraft were also fitted with the retrobomb, a novel device tailored for accurate delivery at low level. The retrobomb was a 65 lb (29.5kg) weapon that carried 25 lb (11.3kg) of Torpex explosive and was fitted with a rocket motor that fired the bomb backwards, thus cancelling the forward motion of the aircraft and causing the bomb to drop straight down onto the target.

Postwar, MAD became standard kit on anti-submarine aircraft and helicopters, with the tail 'sting' being the defining feature of maritime patrol aircraft. The reason for the tail boom being that the response of the magnetometer was affected by the ferrous metals used in aircraft engines and structure. While most MR types carried their sensor in a tail extension, some types were intended to carry their sensors

Variations in the Earth's magnetic field had been used by geophysicists to find ore bodies for mining. The same principle was applied to finding submerged submarines. Although its range was limited it could be used in the final stages of attacks.

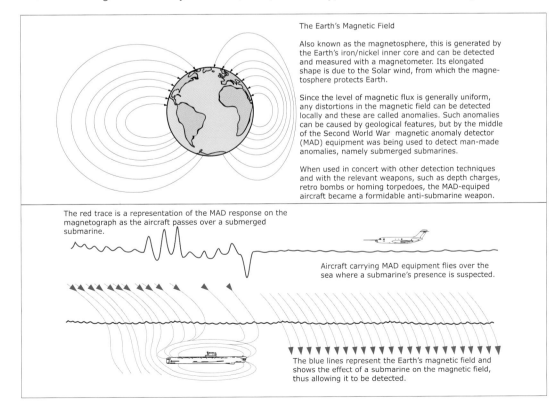

The Earth's Magnetic Field

Also known as the magnetosphere, this is generated by the Earth's iron/nickel inner core and can be detected and measured with a magnetometer. Its elongated shape is due to the Solar wind, from which the magnetosphere protects Earth.

Since the level of magnetic flux is generally uniform, any distortions in the magnetic field can be detected locally and these are called anomalies. Such anomalies can be caused by geological features, but by the middle of the Second World War magnetic anomaly detector (MAD) equipment was being used to detect man-made anomalies, namely submerged submarines.

When used in concert with other detection techniques and with the relevant weapons, such as depth charges, retro bombs or homing torpedoes, the MAD-equiped aircraft became a formidable anti-submarine weapon.

The red trace is a representation of the MAD response on the magnetograph as the aircraft passes over a submerged submarine.

Aircraft carrying MAD equipment flies over the sea where a submarine's presence is suspected.

The blue lines represent the Earth's magnetic field and shows the effect of a submarine on the magnetic field, thus allowing it to be detected.

in a long nose probe, retractable in the case of the BAC Ten-Eleven or on a probe projecting forward of a wingtip as in the Vickers VC10MR. One reason for this relocation was to prevent the tail boom striking the ground on rotation during take-off. Helicopters deployed their MAD 'bird' on the end of an umbilical, winched out to trail behind the helicopter, away from its magnetic influence.

As the use of magnetic anomaly detectors expanded in the two decades after the war, widespread magnetic surveys of the seabed were conducted to provide the NATO navies with a map of the expected magnetic responses in the Atlantic. The surveyors noticed a series of linear anomalies running roughly north/south. These 'stripes' showed areas of reversed magnetic polarity in the rocks of the seabed and indicated that the Earth's magnetic field had 'flipped' on a regular basis, with the last event around 780,000 years ago. Geologists advised that freshly erupted lava had taken up the contemporary magnetic field direction prior to solidifying. What the stripes showed was that the floor of the Atlantic was spreading from a centre, the Mid-Atlantic Ridge, pushing the continents of North America and Europe apart. From the need to find and destroy submarines arose the science of tectonics that explained how the Earth works and ultimately produced the conditions for life on Earth.

A new magnetic technique for finding submarines arrived on the scene in 1969 when the Massachusetts Institute of Technology (MIT) developed the Superconducting Quantum Interference Device (SQUID) for medical use in magnetic resonance imaging (MRI) scanners. The technique used very sensitive magnetometers and multiple magnetic sensors to produce a detailed image of the magnetic environment and was later applied to non-destructive testing of structures. As ever, the military saw a use for SQUID and it was considered a viable system that could be used from a moving platform to detect and provide bearings on a metallic target, such as a submarine. The development of SQUID sensors may even have prompted the development of the *Alpha*-class submarine and its titanium hull.

Autolycus – Sniffing Out Submarines

Rudolf Diesel, in developing his eponymous engine, aimed for the most efficient engine that was capable of burning the cheapest fuel available. Diesel was successful in this and his compression ignition engine soon found a place in marine propulsion culminating in the cathedral engines that have replaced the steam turbine engine in modern ships. Cathedral engines can burn the lowest quality fuel, but only the largest ships are powered by such engines, smaller ships use engines burning standard diesel fuel and this was the fuel of choice for submarines. Until the late 20th century and the increasingly widespread use of compression ignition engines in domestic vehicles, diesel engines were associated with dirty exhaust emissions. At some point during the Second World War attention turned to using this aspect of the engine against submarines. This was based on the fact that a submarine only ran on electric power when submerged but used its diesel engines for surface cruising and charging its batteries. So if the plume of exhaust emissions could be detected and used as a trail that could be followed to its source, the submarine could be attacked and destroyed.

Autolycus was a primitive ionization detector, a device that reacted to low concentrations of ions in the atmosphere and it could be credited as the first example of ion mobility spectrometry. This technique is used to identify ionised molecules within a gas sample, in this case the combustion products of diesel engines in the atmosphere. The basic technique could be described as a form of chromatography and involves measuring the time taken for an ion to travel through an electrically-charged tube to a detector. Larger combustion-product ions are slowed as they pass through the tube to the detector, which outputs a response on a recorder. The response was calibrated using known combustion products and the characteristic output of a diesel engine could be identified and quantified.

Initially fitted and tested on warships during the war, the Autolycus device was fitted to postwar maritime patrol aircraft such as Shackleton. While Autolycus could operate successfully, time and again the crews of maritime patrol aircraft found themselves following 'the scent' only to discover that the source was a deep-sea trawler. Whether the numerous small ships that complicated the maritime patrol task in peacetime would be operating in wartime was debatable, but such targets also affected Autolycus. As was pointed out, in a full-scale war, marine traffic would be much reduced and Autolycus may have been useful in the anti-submarine war. However, the fishing fleet was not the only problem that beset Autolycus. Humid

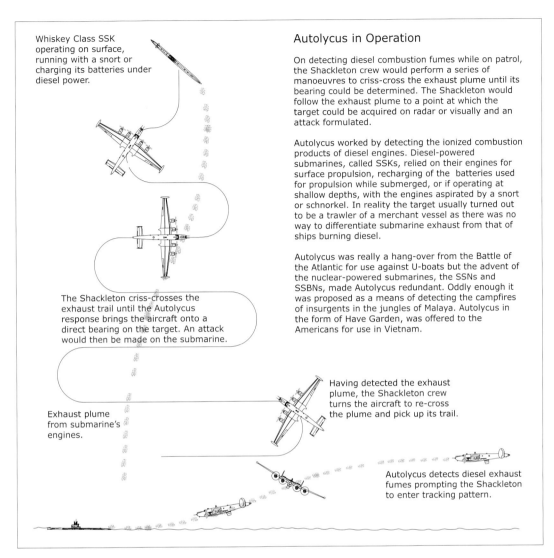

Whiskey Class SSK operating on surface, running with a snort or charging its batteries under diesel power.

The Shackleton criss-crosses the exhaust trail until the Autolycus response brings the aircraft onto a direct bearing on the target. An attack would then be made on the submarine.

Exhaust plume from submarine's engines.

Having detected the exhaust plume, the Shackleton crew turns the aircraft to re-cross the plume and pick up its trail.

Autolycus detects diesel exhaust fumes prompting the Shackleton to enter tracking pattern.

Autolycus in Operation

On detecting diesel combustion fumes while on patrol, the Shackleton crew would perform a series of manoeuvres to criss-cross the exhaust plume until its bearing could be determined. The Shackleton would follow the exhaust plume to a point at which the target could be acquired on radar or visually and an attack formulated.

Autolycus worked by detecting the ionized combustion products of diesel engines. Diesel-powered submarines, called SSKs, relied on their engines for surface propulsion, recharging of the batteries used for propulsion while submerged, or if operating at shallow depths, with the engines aspirated by a snort or schnorkel. In reality the target usually turned out to be a trawler of a merchant vessel as there was no way to differentiate submarine exhaust from that of ships burning diesel.

Autolycus was really a hang-over from the Battle of the Atlantic for use against U-boats but the advent of the nuclear-powered submarines, the SSNs and SSBNs, made Autolycus redundant. Oddly enough it was proposed as a means of detecting the campfires of insurgents in the jungles of Malaya. Autolycus in the form of Have Garden, was offered to the Americans for use in Vietnam.

Autolycus was basically a 'sniffer-out' of submarines, but ultimately sniffed out any diesel engine in the area. Later versions were more specific but the rise of the nuclear-powered submarine made Autolycus redundant.

conditions affected the Mk.II version, causing the detector baseline to wander and rain invariably rendered it unserviceable. Having operated similar equipment on drilling rigs for years, this author has sympathy for the frustrated Autolycus operator.

Autolycus Mk.III was much improved and did provide a degree of discrimination between submarines and other ships nor did it require 'drying out' after the aircraft had been flying through rain. It also had better resolution, with the Mk.III resolving signals that the Mk.II missed, which was '…a big advantage during homing.' The main problem encountered with Autolycus appears to have been electrical interference from other systems on the aircraft, particularly the Morse transmissions by the radio operator but this was fixed by the use of suppressors. While not a fool-proof method for detecting submarines, Autolycus was a useful tool in finding diesel-electric submarines. Unfortunately by the point in the 1960s where Autolycus was becoming perfected and useful,

nuclear-powered submarines were on the rise and, given their increasing threat in the Cold War, became the prime target for the ASW force. Not producing any tell-tale exhaust, these could not be detected by Autolycus and, along with its American AN/ASR.3 counterpart, its place in ASW became much reduced and therefore was not included in the sensor suite of the modern ASW aircraft. Despite their shortcomings in the ASW role, devices using ion mobility spectrometry are in everyday use around the world, particularly in the security industry where the technique is used in explosive detection systems in airports.

That was not the end of Autolycus as it, and a version developed by the Admiralty Research Laboratory (ARL) called Lochnicus, had a surprising extra-curricular use. As the war in Viet Nam escalated through the 1960s, US forces were having difficulty locating targets under the rainforest canopy. Having tried defoliation under Operation *Ranch Hand*, the USAF was interested in some means of detecting Viet

Nimrod MR.1 XV254 of 120 Sqn on exercise with a Royal Navy *Oberon*-class submarine. The *Oberons* were SSKs, diesel-electric powered, as were many of the Soviet submarines and the Shackleton and Nimrod MR.1 were adept at detecting these. Autolycus was much improved by the time the Nimrod arrived, but the SSN and SSBN made Autolycus obsolete. *Terry Panopalis Collection*

Cong and North Vietnamese Army laying-up areas. Optical methods, be they visible light or infrared, were defeated by the forest cover, so the USAF had relied on seismometers to detect the vibrations of vehicles and people on the move, and people sniffers that detected the ammonia compounds produced by humans. The latter were oversensitive and could be fooled by hanging urine-filled containers in the undergrowth. The seismic sensors only picked up movement, so to detect the laying-up areas some other method was required. One universal facet of soldiers is that as soon as they stop, they brew up. In the jungles of Viet Nam, the Viet Cong had various methods to minimise the visible smoke from their cooking fires, such as using charcoal-burning stoves. The USAF therefore became interested in using a detector on an airborne platform to detect the combustion products of vegetation – wood and charcoal fires – and commenced a project called Have Garden.

Have Garden

Have Garden was to compare the various methods available for detection of combustion products. The USAF conducted a series of trials in the Mojave Desert with the US Navy's AN/ASR.3, also known as Ash (not to be confused with the radar of the same name), but hoped that Autolycus and the improved Lochnicus might prove more reliable. A series of trials were flown at Eglin AFB in Florida in the summer of 1967 using a Lockheed C-121 Super Constellation fitted with a newly-developed carbon monoxide detector, Ash and Lochnicus. As the trials produced encouraging results, '…further trials over true tropical jungles…' were proposed. Later, in spring 1969, a Handley Page Hastings on loan from RAF Signals Command and fitted out by RRE Pershore with Lochnicus and Ash, was flown out to RAF Changi on Singapore Island for trials over the rainforest of the Malaysian state of Johore. Weather, ever the enemy of Autolycus, proved problematic, but once suitable trials area and weather conditions were available, the trials commenced. The traditional maritime search technique was found wanting in the detection of targets such as wood fires, so a new three-dimensional search pattern was developed. Accurate navigation was critical to establishing a good search pattern and location of the target. Ultimately, the Malaysian trials confirmed that targets could be detected under the forest canopy, but also demonstrated that contrary to popular belief, smoke did not accumulate in the canopy, but rose to altitudes where it could be detected by aircraft. A secondary goal of Have Garden was to investigate whether it was possible to distinguish wood smoke from that generated by burning oil. This was possible, as the smoke from oil fires contained sulphur compounds and as a result a Lochnicus was modified to discriminate between sulphur-bearing compounds and wood smoke only and this was called Valkyrie.

Another technique employed over the jungles of Viet Nam had its roots with ASW equipment, this time the magnetic anomaly detector. As the war progressed, the North Vietnamese forces and Viet Cong operating in South Vietnam needed logistics support and much of this was transported along the Ho Chi Minh trail. Famously portrayed as being carried on peoples' backs or on bicycles, much of the heavier equipment was carried on trucks. As noted above, the jungle canopy concealed much of the activity on the ground and some means of targeting the transport on the 'Trail' was required. A sensitive fluxgate magnetometer was used to detect the localised change in magnetic field caused by current flowing through a vehicle ignition coil. Called Pave Mace while under development, this became the AN/ASD.5 Black Crow when installed in the Lockheed AC-130A Spectre gunship under the Pave Pronto programme. Black Crow wasn't used for targeting weapons, but more as an indicator of trucks operating in the area, allowing other targeting methods to be employed once the source of the Black Crow response was located.

Yellow Duckling

Before the advent of radar, Britain's defence scientific organisations had high hopes for infrared (IR) systems. Dr R V Jones, prior to becoming Assistant Director of Intelligence (Science) at the Air Ministry, had worked on such systems for the Royal Aircraft Establishment, but with radar becoming the keystone of British air defences, infrared took a backseat, relegated to such roles as night-driving equipment for army vehicles. Not so in Germany, where IR was given pride of place in the Reich's technology activities and when Allied technical intelligence teams entered Germany at the end of the war they were impressed by the extent of German IR developments. These included *Zielgerät* 1229 *Vampir* night-sights for rifles such as the *Sturmgewehr* StG44, the *Sperber* night-fighting equipment for armoured fighting vehicles and *Spanner Anlage* tracking systems for Luftwaffe night-fighters.

In the search for and subsequent examination of German IR technology, Air Commodore Roderick Chisholm, a specialist in nightfighter operations and attached to 100 Group, the bomber support unit, was examining a Junkers Ju 88G and became intrigued by a piece of kit called *Kielgerät*. This was an IR tracking device

that allowed the *Nachtjägers* to stalk, undetected, RAF bombers using the heat signature from their exhausts from up to 3.5nm (6.4km) away. The *Kielgerät* soon found itself being examined at the RAE Farnborough and TRE at Defford, where its key component, a cooled lead sulphide (PbS) detector, was revealed. This 'lead salt' detector was much more sensitive that anything the British IR researchers had been using to date and soon found its way into British research labs. The PbS detector was quickly replaced with a lead telluride (PbTe) cell cooled by liquid nitrogen. This was incorporated with a lock-follow system called Blue Lagoon that allowed the sensor to track a target and these developments made possible the IR-guided missile, but there were a couple of further applications for the technology.

Earlier British attempts to produce an IR detector for ASW use had been prompted by the German development of the *schnörkel*, whose radar signature was so low that it could only be detected by ASV sets under ideal conditions at much shorter ranges than before. The Admiralty hoped that an IR system would pick up the heat from a submarine's exhaust being expelled from the *schnörkel*. Unfortunately, until the discovery of the lead salt detectors in the German *Kielgerät*, the Admiralty's efforts came to nought.

While much of this technology was applied to British weaponry, most notably the Trafficator and Bay Window equipment that led to the Green Thistle and Violet Banner seekers for the de Havilland Blue Jay and Red Top air-to-air missiles, it also had an application in ASW. Another problem that was exercising the scientists at the Royal Aircraft Establishment and Royal Radar and Signals Establishment was the navigation and guidance of long-range aircraft and missiles such as the Blue Moon flying bomb. The

A USAF Pave Pronto AC-130A Spectre gunship fitted with the AN/ASD-5 Black Crow vehicle magneto detector. This was an interesting offshoot from the magnetic anomaly detector used for ASW. The similarity between finding submarines under the sea and trucks beneath the jungle canopy also led to the Have Garden project based on Autolycus detectors.

The Yellow Duckling IR wake detection system used infrared technology derived from heat-seeking missiles such as the Violet Banner seeker used on Red Top, shown here. Yellow Duckling scanned normal to the flightpath to detect variations in the sea temperature caused by the submarine's passage mixing warm surface water with the colder water below.

lock-follow system mentioned above was also used on a pair of astro-navigation systems called Orange Tartan and Blue Sapphire being developed for Blue Moon. These used the Blue Lagoon equipment to lock onto and track specific stars, and having measured the elevation and bearings of those stars, allowed the astro-navigation system to calculate its position. By mounting such equipment on the underside of an aircraft to look down rather than up and converting the lock-follow equipment to scan transversely across the ground track, the PbTe detector could produce a heat map of the landscape below the aircraft.

When the Admiralty became aware of this work, they still had hopes for using IR techniques to detect submarines, and so became interested in a Radar Research Establishment project called Yellow Duckling. The thinking behind Yellow Duckling was that a submarine moving on the surface, at periscope or schnörkel depth would mix the warmer, surface water with colder, deeper water, thus leaving a trail of cooler water in its wake. If that wake could be detected and followed, the submarine could be found – and destroyed. This became more significant with the advent of the nuclear-powered submarine as, being true submersibles, they did not need to surface to recharge batteries and thus could not be detected as readily by radar. Nor could their exhaust plumes be detected with Autolycus exhaust sniffers, as they emitted no exhaust, but nuclear submarines did disturb the thermal layers of the water as they passed though the sea.

The original Yellow Duckling utilised the PbTe detector derived from those developed for Blue Lagoon and its derivatives. As the anti-submarine project developed it was tested in 1953 aboard Handley Page Hastings WD484 in the waters around Britain. After the loss of WD484,

updated equipment was installed in Hastings TG514 and flown against HMS Sea Devil, the last of the wartime S-Class submarines in service, in the Mediterranean around Malta. These tests showed that the PbTe detector, while capable of detecting submarines on the surface, wasn't sensitive enough to discern the wakes left by a submerged submarine, a submarine at periscope depth or one using a schnörkel. To detect these key temperature changes, a more sensitive detector for longer wavelengths would be needed. This much larger (15mm x 15mm rather than 6mm x 6mm for the PbTe) copper-doped germanium unit had a minimum detectable temperature difference of 1/2000 of a degree Celsius. Mounted in an assembly comprising a 24in (61cm) diameter concave mirror with a focal length of 12in (30cm), the scanner rotated at 150rpm at an angle 30° to the vertical. This sensor, when combined with improved electronic filtering systems and cooling to the temperature of liquid hydrogen, was applied to the original Yellow Duckling apparatus and tested in the Mediterranean during 1956. Results were patchy, if not disappointing, with no detection of submerged submarines with a keel depth greater than 100ft (30m). Further disappointment came when out of twenty six tests against schnörkelling submarines; the wake was detected only five times, even when the track of the submarine was known.

The germanium detector was shown to be only as sensitive as the PbTe detector, but did prove easier to operate and maintain. Given the 20% success rate in trials and the fact that it worked best over warm waters, in clear weather and at night, conditions not very common in the North Atlantic where it would be mostly used, Yellow Duckling was terminated. The IR wake detection work was resurrected in the early 1960s in an Admiralty study called Clinker, which was specified as a part of the sensor suite for the OR.350 type. Canada became interested in Clinker for their Canadair CL-28 Argus and many of the proposals to meet OR.357, the Shackleton replacement, include an unspecified wake detection system that may be Clinker.

As a postscript, the methods and equipment developed in the Yellow Duckling project were applied on active service in the late 1950s for land reconnaissance against EOKA forces on Cyprus. Yellow Duckling went on the form the basis of the IR Linescan systems that became a critical tool in aerial reconnaissance in the 1960s and 70s.

Exotica

The Royal Aircraft Establishment in 1962 sponsored a series of extra-mural studies based on their own proposals for new submarine detection systems. These studies were to '…examine the potentialities and feasibility of all possible scientific means of detection, classification and localisation applicable to this need…' This was not to be blue-sky research, but based on existing work that was showing promise and one such technique was described as a 'laser radar'. This combined a pulsed blue-green laser and a photosensitive receiver that scanned the sea below the aircraft. The basic premise was that the laser light would bounce off the sea surface, but being blue-green, would also penetrate the water and reflect off a submerged submarine. The system would differentiate between the surface and submarine return by time difference.

One sweep would cover around 2,000ft² (186m²) and a fixed-wing aircraft would need 10 sweeps per second. The water itself would cause returns from scattering in the water column, but an obstruction such as a submarine would cause a shadow. If the system was set to scan a specific depth, then a submarine would be seen as a shadow on the return display. It was known that a layer of plankton formed at or near the thermocline (a depth where water temperature changes) and this would be readily detectable, with any disruption by a submarine moving in the vicinity showing up as a disturbed layer. Unfortunately, clumps of plankton might have a similar signature to a submarine and, despite being blue-green, the light

would be attenuated with depth and simply increasing the power would have little effect. Despite this, the RAE took the view that it was a feasible system that could detect a submarine in 600ft (193m) of clear water and 100ft (30m) in 'turbid water'. It appeared that this had the potential to cover large areas very quickly, with sweep widths of several miles possible from 10,000ft (3,048m) over clear ocean water.

As a result, ASR.3557 (the Admiralty's AW.111) was issued to cover development of an airborne laser radar under the name ORADS (optical ranging and detection system). It was hoped that ORADS would provide simultaneous detection and localisation in three dimensions, become the primary sensor for ASW operations and increase the effectiveness of sonobuoy arrays. This eventually became a joint Naval/Air Staff requirement N/ASR.807, but laser radar had another application in ASW. By using infrared light, the laser radar could detect artefacts of the movement of a submarine through water. The first was the Bernouli Hump, a bulge on the sea's surface produced by water displaced ahead of the submarine as it moved. The other artefacts, Kelvin Waves, are left behind any vessel moving through the water and these can also be detected by infrared laser radar. Research into blue/green and infrared laser systems continues, with the US Navy deploying the AN/AES-1 Airborne Laser Mine Detection System (ALMDS) on MH-60S Knight Hawk helicopters. While used for sea mine detection, a variant with a more powerful laser could form the basis of a new field of submarine detection and tracking, par-

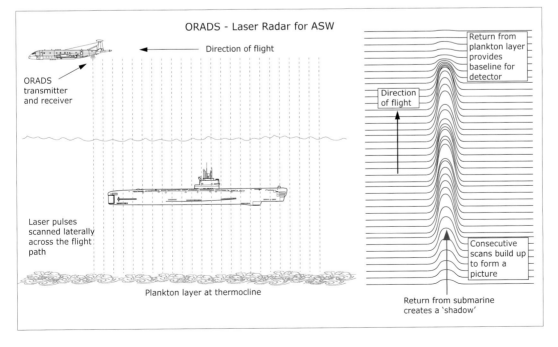

ORADS - Laser Radar for ASW

Direction of flight

ORADS transmitter and receiver

Laser pulses scanned laterally across the flight path

Plankton layer at thermocline

Return from plankton layer provides baseline for detector

Direction of flight

Consecutive scans build up to form a picture

Return from submarine creates a 'shadow'

Laser radar (Lidar, named after the physicist Daniel Lidar) was to scan the sea below the aircraft and detect reflections from a plankton layer at the thermocline. This would provide a background signal and any object above that layer would be discernable as a 'shadow' on the display.

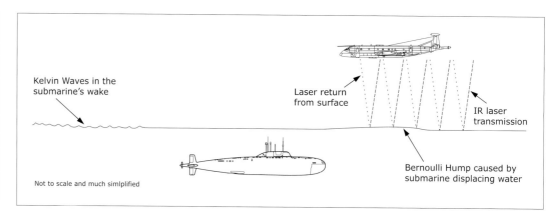

Kelvin Waves in the submarine's wake

Laser return from surface

IR laser transmission

Bernoulli Hump caused by submarine displacing water

Not to scale and much simlplified

Infrared could be used for laser radar and if applied like a rangefinder, can detect small variations in the sea surface such as the Bernoulli Hump ahead of a submerged submarine or the Kelvin Waves in its wake.

ticularly against the ultra-quiet conventional submarines that are being sold to navies around the world for use in littoral waters.

Jezebel, Barra and Woodford's Reusable

Since the 1920s, sound has been used to find, track, classify and direct attacks against submarines and the Royal Navy called it ASDIC until after World War Two when the US term sonar (SOund Navigation And Ranging) came into use with British forces. ASDIC was what is now called active sonar, emitting a pulse of sound and receiving any reflections of that sound from a target. On the other hand, a passive sonar uses an underwater microphone, a hydrophone, to listen for the sound of a submarine operating in the ocean. Passive sonars have evolved since the days of a crewman listening on headphones for the tell-tale clangs and hums of a machine. Modern sonar systems have developed into sophisticated sound detection suites using some of the most powerful computers ever put to sea, applying signal processing techniques to identify patterns within the mass of sound waves collected by the hydrophones. Despite this quantum leap in technology, a crew member still monitors the system though their headphones.

As ever, whenever a new technology is developed new problems appear mainly because the new systems reveal them. One such problem was the thermocline, a layer in the ocean from 300-500ft (91-152m) where a rapid decrease in water temperature creates a layer that reflects and refracts sound waves. This led to surveys, as with the earlier magnetic data acquisition, to map the thermal characteristics of the oceans, resulting in the establishment of oceanographic labs, civil and military, which played a vital role developing postwar techniques in ASW. Other areas that were investigated due to their effects

of sonar were oceanic salinity and density variations, while other surveys examined currents and seabed topography. Only revealed with the end of the Cold War, the extent of the various navies' efforts in the oceanographic field came as a surprise to the civilian scientists involved in oceanography and the declassified military research has since added a great deal to our knowledge of the oceans.

Ships and submarines hunt each other using sonar and being in the water; appear to have their hydrophones ideally placed to locate each other, whereas an aircraft most certainly does not. Aircraft have two methods of using sonar to detect and track submarines: sonobuoys and dipping sonar. Sonobuoys can be deployed by fixed- and rotary-wing aircraft, while since the demise of the flying boat, dipping sonars are used principally by helicopters.

Dating back to its first trials in 1951, dipping sonar is a sonar array lowered into the water on a winch from a hovering helicopter and used to detect and triangulate the position of submarines. Having dipped the sonar and taken a bearing on a potential target, the array is retrieved and the helicopter flies to another location and repeats the process. After a few repetitions the sonar plotter can use the bearings to triangulate the target's position. This is a much-simplified description, with the operation being much more involved. Saunders-Roe proposed fitting its P.162 flying boat with a form of dipping sonar that could be deployed through a hatch in the hull after the boat had alighted on the water. This of course assumed that the sea state was moderate enough to allow the aircraft to touch down. The later Saro P.208 flying boat was to be fitted with hydroskis on the hull and under the stabilising floats to allow it to skim across the surface at high speeds, towing a sonar array.

Sonobuoys on the other hand are small cylindrical stores dropped in a pattern by MR aircraft. On entering the water they deploy a hydrophone array from their base and an antenna from the top. The hydrophone array can be set to open at different depths to optimise its performance by placing it above or below the various layers in the sea such as the thermocline. In addition to hydrophones, sonobuoys are also equipped with sensors to provide temperature, salinity and density data. The sonobuoy operates principally as a sonar sensor and transmits its sonar signals back to the aircraft for processing. Battery operated, they have a limited life, up to eight hours, and their final act is to flood their flotation chamber and sink to the bottom, depriving an enemy of very useful intelligence material.

As noted above, sonobuoys are dropped in a pattern that will allow any contacts to be triangulated. The main types of sonobuoys are: passive, non-directional, passive directional, active non-directional and active directional. Active sonobuoys emit sound pulses and use the reflected sound to determine range to a target, a single directional sonobuoy can pinpoint a target, while three active, non-directional sonobuoys can provide a fairly precise location for a submarine. Passive sonobuoys listen to the sound of the sea and transmit that back to the aircraft signal processing system. Passive, non-directional sonobuoys are the most basic of all but can also be used to triangulate submarines, using the sound 'volume' to compute the distance and bearing of the sound's source from the sonobuoy. Three passive buoys can provide a good location for the target. Good enough for a nuclear depth bomb if necessary. A single directional sonobuoy can give an idea of what bearing the target is on, while a modern high-technology active directional such as CAMBS can provide range, bearing and relative speed.

Known as High Tea, sonobuoys were first used in 1942 and were, despite being 6ft (1.8m) long and 2ft (0.6m) in diameter, wonderfully compact units given the electronics of the time. They were manhandled out of a door or hatch, but by the time the R.5/46 and R.2/48 specifications were drawn up, dedicated launch systems were being specified, with Saro being particularly praised for their mechanical system on the P.104 flying boat. Generally the British preferred a pair of launch chutes within the cabin, into which sonobuoys were placed by the crew before being dropped. The Americans on the other hand carried their sonobuoys in an array of chutes on the underside of the aircraft, preloaded with sonobuoys on the ground, although the Vickers Vanguard MR to OR.350 had such a system for twelve Size C sonobuoys plus up to 120 Size A internally.

One of the principal aspects of the various Operational Requirements issued by the Air Staff for maritime patrol aircraft was their sonobuoy payload. Once ASV radar had been robbed of its effectiveness in ASW operations by the nuclear submarine, sonar became the primary sensor in submarine hunting. There was little point of a MR aircraft that could patrol for eight hours 1,000nm (1,852km) from base if it ran out of sonobuoys and had to break off a contact to return for more.

High Tea was the first sonobuoy to be used in action, essentially a hydrophone on a cable with a radio transmitter to send the sound signal back to an operator in the aircraft. The post-war electronic revolution changed that, with the science of signal processing, more associated with radar, bringing benefits and a new approach to sonar. Improved computers on aircraft allowed the systems to sift through the oceanic noise to find the 'clanks and hums' of the submarines lurking below.

The main sonobuoy in use by the RAF was the Jezebel, developed by Ultra Electronics as a

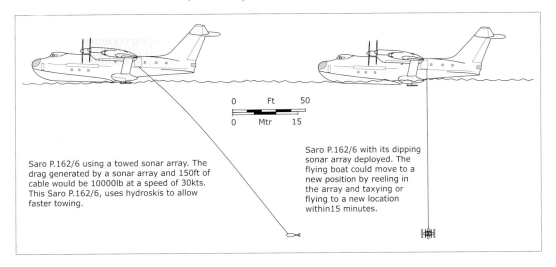

Saro P.162/6 using a towed sonar array. The drag generated by a sonar array and 150ft of cable would be 10000lb at a speed of 30kts. This Saro P.162/6, uses hydroskis to allow faster towing.

Saro P.162/6 with its dipping sonar array deployed. The flying boat could move to a new position by reeling in the array and taxying or flying to a new location within 15 minutes.

0 Ft 50
0 Mtr 15

Dipping sonar, now a standard sensor on ASW helicopters was originally intended for use by flying boats. The Saro P.162/6 could also draw a sonar fish through the water as it motored on the surface. Deploying hydroskis from the floats allowed much faster towing speeds

This highly simplified diagram shows how a non-directional sonobuoy array operates. Generally a pattern of four sonobuoys will be dropped in a square pattern, with a fifth in the middle.

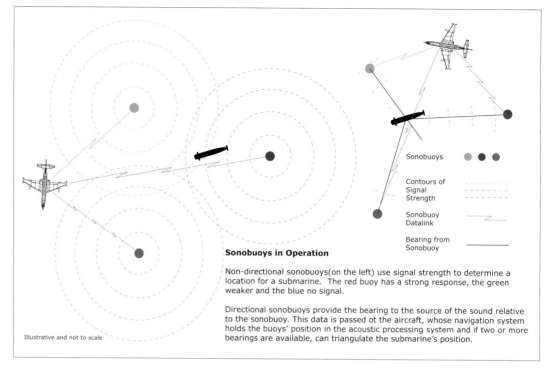

Sonobuoys	⬤ ⬤ ⬤
Contours of Signal Strength	
Sonobuoy Datalink	
Bearing from Sonobuoy	

Sonobuoys in Operation

Non-directional sonobuoys(on the left) use signal strength to determine a location for a submarine. The red buoy has a strong response, the green weaker and the blue no signal.

Directional sonobuoys provide the bearing to the source of the sound relative to the sonobuoy. This data is passed ot the aircraft, whose navigation system holds the buoys' position in the acoustic processing system and if two or more bearings are available, can triangulate the submarine's position.

Illustrative and not to scale

basic passive non-directional sonobuoy to meet OR.3458. Jezebel was to be part of the sensor suite of the OR.357 type, used with a signal processing system called APOJI (Automatic Processing of Jezebel Information). A later development, the Size F Mini-Jezebel was one third of the size of the original allowing three times the payload or a smaller aircraft to carry a similar sonobuoy capability as a larger type. In the active role, the RAF adopted an Australian model called *Barra*, which wasn't named after the Hebridean Island, but is an indigenous Aus-

tralian word for 'listening'. Its development began as a project called *Nangana*, which was an Australian project examining miniaturised electronics. This evolved into the SSQ.801 developed by Amalgamated Wireless from 1969, but only got under way in earnest in 1974, with the British SSQ.981 being built by Ultra Electronics. *Barra* is a marvel of engineering and packaging, dropped from a Nimrod; it plopped into the drink and lowered its sonar array on a cable by a winch that could place the array at depths from 80ft (25m) to 440ft

Sonobuoy Sizes and Numbers by Operational Requirement

Sonobuoy Type	Dimensions
High Tea	24in x 72in (61cm x 183cm)
Size A	5in x 36in (13cm x 91cm) a Western standard size: Jezebel, *Barra*
Size B	6in x 60in (15cm x 152cm) SSQ.75 ERAPS
Size C	9in x 60in (23cm x 152cm) (UK only)
Size F	5in x 12in (13cm x 30cm) Mini-Jezebel
Size G	4.75in x 16.5in (12cm x 42cm) SSQ.955

Requirement	Number of Sonobuoys
R.5/46	Eight *High Tea*-sized sonobuoys in weapons bay
R.2/48	25 sonobuoys in weapons bay or a reloadable dispenser with six at readiness
Medium MR	12 non-directional, 14 directional
OR.350	120 Size A, 24 Size B, 12 Size C, in cabin, loaded manually into two chutes
OR.357	100 Size A, 30 Size B, 20 Size C, in cabin, loaded manually into two chutes
OR.381	62 Size A, 10 Size B, 15, preferably 35, Size C, loaded manually into three chutes with the Size C in the weapons bay.

(135m). Once the array reached the required depth, defined by information from the thermal and density sensors, the array unfolded its five telescopic arms that carry five hydrophones, and extended the arms out to a distance of 9ft (2.7m). *Barra* uses beam steering to provide its directional capability and its electronics are completely digital, with a digital datalink to the aircraft and the Marconi Avionics ASQ.901 acoustic processor suite and ASWEPS (ASW Environmental Prediction System). The ASQ.901 system was installed in the Nimrod MR.2, upgraded from the Tandem system fitted on the MR.1. Another proposed method was for an array of *Barra* sonobuoys to be operated in conjunction with an active sonar emitter called Rasputin. Its signals would reflect off the target and allow the ASQ.901 to triangulate the target with greater accuracy.

Barra has been continually upgraded, but by the Eighties computer power and the signal processing capabilities on the aircraft meant that a new sonobuoy with much more flexible capabilities could be developed. This was the Command-Activated Multi-Beam Sonobuoy (CAMBS) that was used in concert with *Barra*. CAMBS is an active sonobuoy that can produce bearing, range and target velocity, through the use of Doppler analysis in the signal processor. The sonobuoy can be controlled by the aircraft to adjust the sound pulse characteristics as conditions change and provides the parent aircraft with a formidable sonar capability.

The RAF also used American sonobuoys that used DIFAR (DIrectional Frequency Analysis And Recording – two beams normal to each other, allowing direction to be ascertained) and Julie active sonobuoys, particularly on the US-supplied Neptune MR.1. Julie was used in conjunction with Jezebel, with early versions firing a small explosive charge as a sound source with the returns picked up by the Jezebel sonobuoys. Sonobuoys were expendable and as technology developed, their cost increased in proportion to their capability. For a cash-strapped RAF in the 1970s, this was a major expense and so the search for an alternative began.

Woodford's Reusable

Imagine, for a moment, the response of the financial controller at the MoD in the mid Seventies when the bill for sonobuoys landed on your desk: 'How much? And they sink to the bottom of the sea after eight hours?' In 1980 a *Barra* sonobuoy cost something in the region of £1,550 (approx. $3,300) and equipping a single

Nimrod with a 62-round payload could cost around £96,000, all of which could be lying on the seabed by the end of the day. Hawker Siddeley at Woodford had in 1976 a potential solution to this costly operation in the form of the HS.828 flying sonobuoy. This might sound familiar and was in essence an unmanned version of Saro's P.162 and P.208 flying boat with dipping sonar, which through the miracle of miniaturised electronics could be made small enough for a number to be carried by a maritime patrol aircraft.

Woodford's concept was for a maritime patrol aircraft to carry a pair of HS.828 out to the patrol area on underwing pylons and then act as a mothership providing the data processing and refuelling for the duration of the mission, before retrieving the sonobuoy and returning to base. Hawker Siddeley at Woodford estimated that over a 10 year period, and taking development to meet an evolving threat into account, the cost of a fleet of reusable sonobuoys would be 12% that of *Barra*. Being able to move to a more advantageous position, effectively tracking the target and maintaining constant contact, a pair of HS.828s could pro-

The high cost and increasing complexity of sonobuoys prompted Hawker Siddeley to design the re-usable HS.828 around the innards of the Jezebel sonobuoy. A pair of HS.828 were to be carried by maritime patrol aircraft, released to fly to an area where a submarine was suspected, land on the sea and deploy their sonar arrays. They could then take off and 'hop' alternately to their next location, refuelling in flight from the aircraft if required.

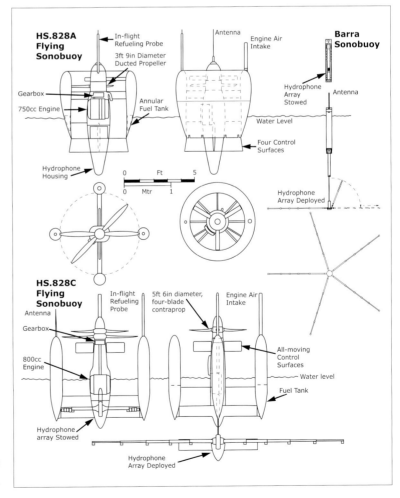

vide better cover than the continual deployment of expendable *Barra*. Typically five *Barra* were deployed in a line to gain the initial contact with a nuclear submarine travelling at 22kt (41km/hr), followed by three *Barra* for every 44nm (81km) the submarine travelled. Jezebel needed 18 sonobuoys for initial detection followed by 13 buoys for every 77nm (143km) progress. The HS.828 utilised two sonobuoys for the entire detection and tracking operation, each working a line 5nm (9km) parallel to the submarine track. The two HS.828 sonobuoys were to perform alternate hops of 35nm (65km) to keep the submarine in contact, with up to nine hops being made before the HS.828 required refuelling. On completion of the task, or destruction of the target, the HS.828 sonobuoys would be recovered for later use.

Unfortunately, one major problem that pretty much scuppered the HS.828 was sea state. The sonobuoys could operate in sea states up to World Meteorological Organization State 5, which is described as 'rough', with wave heights of 8ft to 13ft (2.5m to 4m), which is not an unusual condition in the north east Atlantic and may have been what saw the concept dropped. The engine drew its air through an extended intake that projected 4ft (1.2m) above the vehicle's body; however in rough seas the HS.828 might roll enough to swamp the intake. Take-off and landing would have posed a problem if attempted at the wrong point in the wave cycle.

The *Barra* sonar equipment was to be fitted in a remotely piloted vehicle that could alight and take-off and act like a dipping sonar for the MR aircraft. Four versions of the HS.828, A to D, were proposed, with AUWs ranging from 445lb (20kg) for the D model to 575lb (261kg) for the B. A small reciprocating engine with a capacity of around 800cc drove propellers that provided lift and propulsion, either tandem for the D or in the case of the A and B, a ducted fan within an annular wing or C using four-bladed contraprops. The HS.828 was to perform short hops, land on the sea in conditions up to Sea State 5, and deploy its hydrophone array to listen for targets. On detecting a submarine, the HS.828 could retrieve its array, take-off and move to a better position to track the target. The HS.828 was to carry out nine such hops before making a rendezvous with the MR aircraft to refuel using its fixed refuelling probe, before returning to the fray.

HSA Woodford proposed combining the HS.828 with the maritime patrol aircraft derived from the Large Aircraft Replacement Concept (LARC) that Woodford intended as the basis of a complete re-equipment of the RAF's large fleet of support aircraft such as AEW, tanker and maritime patrol. The Multi-Role Support Aircraft (MRSA) that was to combine these roles in a single airframe would no doubt have been a suitable platform for the HS.828. The MR version was to carry a HS.828 on pylons under the wing between the inboard engines and the fuselage. Refuelling pods, no doubt Flight Refuelling's Mk.20, but how these would handle the much lower volumes of fuel, typically 70lb (32kg), involved with the HS.828 is not documented, but the HS.828 may have required a new small volume/low flow rate pod.

Killers – Destroying the Submarines

'The first use of the submarine weapon will always, therefore, be an important step in escalation, to be decided by government, and so probably will any decision to prosecute submarine contacts to destruction.' Rear Admiral Richard Hill, 1984

By 1945 all the weapon types that would be used in the forthcoming Cold War were available to Coastal Command either in service or under development. Coastal Command's weapon of choice against surfaced submarines was the 3in (7.6cm) rocket projectile, with a 60lb (27kg) high explosive or 25lb (11kg) armour piercing warhead but generally it was the Mk.11 depth charge that was dropped on a 'crash-diving' U-boat. The late Fifties saw that last great technological development of the Second World War: the atomic bomb. When applied to anti-submarine warfare it made depth charges supremely lethal and quite possibly the weapon whose use would trigger the tripwire of nuclear warfare across the globe.

With the Shackleton coming into service in the early Fifties, the Air Staff looked to optimising its weaponry for a third Battle of the Atlantic and in 1951 listed the weaponry that was either available or destined for service with Coastal Command in the mid-Fifties. The dedicated anti-submarine weapons were the homing torpedoes; Dealer-B and heavier Pentane, and the Mk.11 depth charge. Mines were the weapon that the Admiralty feared the Soviets would spread around the UK, but the RAF also had mines at its disposal in the shape of the 1,000lb (454kg) and 2,000lb (907kg) ground mines and the 800lb (363kg) buoyant mine. Coastal

Command could call on a range of freefall bombs ranging from the 500 lb (227kg) general purpose bomb to the 22,000 lb (9,979kg) Grand Slam. The 12,000 lb (5,443kg) Tallboy had been Bomber Command's preferred weapon against capital ships, but for a number of reasons these were never adopted by Coastal Command. Firstly the Tallboy and Grand Slam required specially modified aircraft, secondly they were being superseded by guided bombs such as Blue Boar and Yellow Sand and thirdly, by 1954 they were in very short supply. Interestingly, although initially earmarked for the V-Force, they were soon deleted from their armoury. The bombs would require new high-speed/high-altitude tail units as the delivery speed was much higher than the Lancaster. Added to that, the Air Staff's analysis of wartime use showed that 52 Tallboys would be needed to sink a cruiser and since Bomber Command would be using its aircraft to carry 'special weapons' against the Soviets, there would not be the aircraft available for Tallboy attacks on warships. Not that Coastal Command would be fielding 'Special Weapons', as in 1954 these were reserved for use by Bomber Command in the deterrent role. Of course, this would change as the stakes were raised with the appearance of the ballistic missile submarine at which point Coastal Command adopted nuclear depth bombs.

Rockets were a firm favourite with Coastal Command, both on the reconnaissance squadrons and with the Strike Wings. By 1951 the Strike Wings were gone, the role now performed by the FAA, but the reconnaissance aircraft still had a requirement for rockets such as the 3in (7.6cm) RP and what was described as the 'Improved Aircraft Rocket' to OR.1073. A decade or so later, rockets had been displaced from the stores pylons of the reconnaissance aircraft by the guided missile, initially the SS.12, then Martel, and ultimately the Harpoon. As in the Great War and World War Two, the aircraft versus submarine struggle was a game of move and counter-move, with weapons playing the decisive role. In short, having found the enemy submarine, how is it destroyed?

Depth charges, basically an explosive device with a pressure-sensitive (hydrostatic) fuze, have been used against submarines since The Great War and were generally effective, but only if used in numbers and in a pattern. The aim was to put at least one charge close enough to the submarine's hull to cause sufficient damage to either sink the vessel or force it to surface where it could be dealt with by gunfire or rockets.

Although water is a very good medium for the transmission of pressure waves, submarines are built to withstand the pressure of deep dives. Therefore they have an outer hull and an inner pressure hull and to be effective, a depth charge needed to detonate very close to – within a few feet – of the submarine's outer hull to damage the pressure hull within. In the case of the Royal Navy's Mk.VII depth charge, this carried a 290 lb (131kg) explosive charge, but needed to be within 20ft (6.1m) of a submarine to damage the pressure hull.

Obviously a bigger explosive charge has a better chance of causing damage, but a large capacity depth charge, such as the British Mk.X, could also damage the warship that had just dropped it. As carrier-borne aircraft became involved in ASW operations, they were seen as ideal for delivering depth charges; however the larger weapons could not be carried by carrier-borne aircraft, particularly helicopters. The RAF began the Second World War with its Avro Ansons carrying the 100 lb (45.4kg) air-delivered depth bomb, but this was soon replaced by a modified naval Mk.VII fitted with an aerodynamic nose and tail fins. These saw success from 1942 onwards and the air-delivered depth bomb, the BAE Systems Mk.11, continues in service to this day, with the Fleet Air Arm (FAA) rather than the RAF. The Mk.11 dates back to the early 1950s, having appeared on the Coastal Command list of anti-submarine weapons drawn up in 1951. By its nature, the depth charge's success very much hinged on chance, as the target would be manoeuvring and the weapon had to be close to, if not in contact with, the target in the case of the retrobomb. In fact a 100 lb (45.4kg) explosive charge has an underwater lethal radius against a submarine of less than 12ft (3.6m) and increasing the size of the charge does not produce a proportional increase in destructive power, so ideally the weapon should hit, or detonate extremely close to, the submarine.

The answer to this proximity problem came in two forms; either more precise delivery of the

The Mk.30 homing torpedo, also known as Dealer-B, was in widespread use in NATO air forces and navies until replaced by the Mk.44 and Stingray in the Eighties. The pack for the retarder parachute can be seen aft of the propeller and control fins.

weapons, which was how the MAD and retrobomb system worked, or increase the explosive capacity by use of nuclear warheads to such an extent that the lethal radius became a hazard to friendly vessels in the area. Therefore delivery of nuclear depth charges, always referred to as nuclear depth bombs, directly from warships was frowned upon, so development of the nuclear depth bomb went hand-in-hand with the embarked ASW helicopter.

Nuclear warheads are wonders of miniaturization, with the original Gadget tested at the Trinity Site in July 1945 being 6ft (1.8m) in diameter, shrinking within five years to be less than half that diameter and capable of being carried by the larger carrier-borne aircraft. After a further five years the weapons were small enough to be launched on a rocket-powered delivery vehicle such as ASROC or Ikara (also known as Blue Duck) and carried by the smallest of helicopters such as the Westland Wasp. In the era of the Trip Wire, the Cold War was expected to rapidly turn hot and as Rear Admiral Hill pointed out, anti-submarine operations would be likely to see the first use.

The RAF's Shackleton force relied initially on US nuclear depth bombs supplied under Project N. This involved the use of American nuclear depth bombs by the RAF under a dual-key system whereby the weapons were kept under US control until released for use in wartime. The Project N weapons were stored at Special Ammunition Sites (SAS) at RAF St Mawgan in Cornwall and RAF Machrihanish at the Mull of Kintyre from where US-supplied Mk.101 Lulu and B57 nuclear depth bombs were to be col-

lected at times of increasing tension by British and Dutch maritime aircraft. The Air Staff in 1967 took a dim view of this concentration of its maritime aircraft in two places and suggested that two other bases, RAF Ballykelly in Northern Ireland and RAF Kinloss in Morayshire be added. The US government disagreed and the two-base system was retained, much to the annoyance of the RAF.

The Project N weapons comprised the 11KT Mk.101 Lulu (Bomb, AS, 1,200lb, MC to the RAF) and its replacement, the 15KT Mk.57 free-fall bomb. The Mk.57 (also known as the B57 post 1968) was a flexible weapon intended for a variety of roles, including anti-submarine, with delivery methods ranging from toss-bombing to laydown. As a nuclear depth bomb on the RAF's Nimrods the Mk.57 had selectable yield up to 10KT and hydrostatic fuzing. A home-grown nuclear depth bomb was the WE.177, developed for the anti-submarine role as the WE.177A that used a boosted fission warhead with selectable yield of 0.5KT or 10KT. WE.177A weighed in at 600lb (272kg) and was thus named Bomb, Aircraft, HE, 600lb, MC by the RAF, HE being a red herring (unless it referred to the explosive lenses within) and MC defining a kiloton-range weapon, its maximum yield being 10KT.

Research by Brian Burnell at the National Archives at Kew has turned up some interesting aspects of the use of WE.177A in the anti-submarine role, particularly on the matter of how the selectable yield was used. The 0.5KT yield was selected if the target was in coastal waters shallower than 40m (130ft) or if friendly vessels

A Westland Wessex HAS.3 is seen here with a WE.177A trials round. Although intended for the OR.357/381 aircraft, the ultimate result, Nimrod, never carried the WE.177A nuclear depth bomb. *via Michael Fazackerley*

were in the area. The full 10KT yield was selected if the target was in deeper waters. In the depth bomb role the WE.177A body was configured to flood and sink at 20ft/sec (6m/sec), with the warhead capsule pressurised to remain watertight. A single WE.177A could be carried by the Westland Wasp HAS.1, slung under the cabin. The doors were always removed, not due to weight, but to reduce the effect of any overpressure on the aircraft when the weapon detonated. The delivery procedure was to arm the WE.177A, release it and turn the helicopter into the wind and accelerate away at maximum permissible speed (V_{NE}).

While the Wasp operated from destroyers and frigates, the larger Westland Wessex HAS.3 operated from carriers and helicopter cruisers and carried a single WE.177A. The WE.177A was in the suite of weapons for the OR.350 and OR.357 maritime patrol aircraft, Atlantic and HS.800 respectively, and earmarked for the HS.801 Nimrod to meet the interim OR.381. The WE.177A weapon was to be deployed at low altitudes, typically around 1,000ft (305m) and at speeds as low at 150kt (278km/hr). The BAe Sea Harrier could also carry the WE.177A, on the port inboard pylon. In the event, the WE.177A was not used by the HS.801 Nimrod, the American Mk.57 being the preferred option, and no British nuclear weapons were used for the Nimrod as those WE.177s that were allocated were issued to SEPECAT Jaguar units.

Underwater Hittiles

Coastal Command's weapons were found wanting against the new generation of submarines that could stay submerged longer and, through use of the *schnörkel*, came up to surface less frequently. Then, of course, with the deployment of the USS *Nautilus* in September 1954 the prospect of a Soviet nuclear-powered submarine would herald a completely different challenge to the RAF's MR and anti-submarine force. The only perceivable solution to the nuclear submarine was the homing torpedo working in conjunction with sonobuoys to locate the vessel.

May 1943 had seen first blood to what was effectively the first guided weapon for use against submarines. Delivered by a US Navy Catalina in an attack on a U-boat, Fido was a cover name for an acoustic homing torpedo that had been developed by Harvard University's Underwater Acoustics Laboratory. In fact even that cover name was deemed insufficient

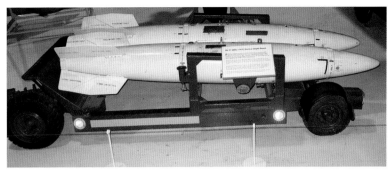

for what was a truly-secret weapon and the nondescript term Mk.24 Mine was used for the device. A further deception was that the weapons could not be dropped while U-boat crews were on deck and could see what was being delivered. Fido used four hydrophones to home in on the target and was three times more effective than conventional depth charges.

Postwar, the acoustic homing torpedo, like all military technology, benefitted from advances in science and engineering, particularly miniaturisation of components and improvements in hydrophone sensitivity. Two projects were of interest to the RAF: Dealer, a joint project with the Navy for an 18in (46cm) homing torpedo for helicopters and short-range types such as the Shorts Seamew and Shorts Sturgeon, and Pentane which was a 21in (53cm) diameter weapon to be carried by larger aircraft such as the R.2/48 flying boat and the Shackleton. Pentane was also earmarked as a weapon for the medium-range MR types such as the Varsity MR and Airspeed AS.69 while the short-range Seamew could only carry the 17ft (5.2m) Pentane if the radar apparatus in the forward portion of the weapons bay was removed.

These developments led to the Mk.30, also known as Dealer. This had the usual diameter for British air-launched torpedoes of 18in (46cm) but was only 8ft (2.4m) long, lending it a somewhat squat appearance. However, Dealer failed to prosper, one of those projects that fell foul of the immediate postwar reduction in defence spending to support the Blue Danube atomic bomb project. This decision was reversed and a second version, Dealer B, was developed from 1954 to meet OR.1194 (Admiralty Weapon requirement AW.60) becoming the Mk.30 Mod 0 which was very successful, with 1200 examples built and serving until 1970. A further development with improvements to its components was the Mk.30 Mod 1, but this was cancelled in 1955 when the Royal Navy opted for the American Mk.43, although only 50 were acquired.

A pair of Mk.57 nuclear depth bombs on their handling trolley. The American Mk.57 replaced the Mk.101 Lulu in RAF service on Nimrods. American weapons were held in storage in the UK for use by RAF Nimrods.

The Z-Series Torpedoes

Despite having used the 18in (46cm) torpedo for aircraft since the Great War, by the end of the Second World War the RAF and FAA were looking at arming their maritime reconnaissance aircraft with a larger torpedo. From 1945 a new range of torpedoes, the Z-Series, was under development and these revelled in names like Zombi, Zonal, Zeta and Zannet. The Z-series was described in a note from the Joint Chiefs of Staff Committee dated June 1946 that contained a 'Consolidated List of Staff Requirements in the Guided Missile Field'. This document provided brief details of the earliest versions on the Z-Series and other anti-ship and anti-submarine weapons. Zombi was a heavy torpedo, around 4,000 lb (1,814kg), for use by 'future submarines' against surface vessels and other submarines. It was to be capable of high speeds, 60kt (111km/hr) over a range of 10,000 yards (9,144m) or double that range at a speed of 40kt (74km/hr) while homing in on a target. A related torpedo, called Fancy was to be used by the current RN submarine force and was to be a faster version of the conventional 21in (53cm) torpedoes in service.

Zonal could not really be described as a torpedo, but was, in the words of the draft requirement 'A ship-launched aerodynamically supported anti-ship weapon which enters the water outside effective AA close-range fire and homes on the target.' Its flight range was to be 50,000 yards (45,720m) while the 'effective range' of ships' AA defences was deemed to be 5,000 yards (4,572m). A minimum range was also stated, dispensing with the airborne phase and entering the water on launch to home in on a target 1,000 yards (914m) away. All this in a weapon that was to weigh no more than 2,200 lb (998kg) and fly at 450kt (833km/hr) while homing in on a target as it flew and also once it had entered the water. Not that it was to use a ballistic trajectory either, it was to fly at 200ft (61m) for the first 5nm (9km) to clear any friendly ships in the launch area, then as low as possible to the target area. To attack capital ships, at least four should be fired within eight seconds. All-in-all a somewhat complex item to develop, especially in the late 1940s.

As an air-launched homing torpedo, Zeta was to be '...released from an aircraft in the vicinity of a submarine which has been sighted or located by any form of air or ship-borne detecting gear.' Weighing in at 2,000 lb (907kg), Zeta was to home in on and destroy any submarine within 2,000 yards (1,829m) of its point of entry into the water by searching in both planes. The explosive charge was to be '...sufficient to rupture the pressure hull of a submarine designed to dive to 1,500ft (457m) in whatever position relative to the target the charge is made to explode.'

Zeta could only be described as a strange beast, 14ft (4.3m) long and with an elliptical cross-section 30in (76.2cm) deep and a beam of 22.5in (57cm), with the shape intended to reduce the weapon's roll characteristics. In plan view, Zeta exhibited an ichthyoid shape, with a pair of ellipsoid-shaped flip-out wings 59in (150cm) aft of the nose at the thickest point of the shape. A pair of small canard control planes to control pitch and roll in flight and underwater projected from the forward body and these were considered a better bet than rear-mounted controls. The dual-mode propellers were to be used in both airborne flight and under the water, with the propellers rotating within a shroud ring. Aft of the shroud ring was a vertical support that doubled as a stabilizer fin fitted with rudders.

One interesting aspect of Zeta was that its developers hoped to use it to '...solve many of the problems necessary for the production of the ship-launched homing anti-submarine weapons'. Pending development of Zeta, 'An interim anti-submarine homing weapon (Code-named Bidder) is partially developed'. Zeta would form the basis of a further development in the British homing torpedo programme, a weapon called Pentane.

Air-launched torpedoes were invariably 18in (46cm) or less in diameter. Torpedoes were no longer carried by strike aircraft but moved to large maritime patrol aircraft. Shown here are artist's impressions of Pentane and Zeta, as drawn on various design studies for the R.2/48 flying boat. Note that these show no propulsion systems. The smaller Stingray replaced the Mk.44, which had previously replaced the Mk.30.

Marconi Stingray

Mk.30/Dealer B

Mk.44

Pentane as shown on Supermarine drawing

Pentane as shown on Saro drawing

Zeta as shown on Shorts drawing

0   Ft   10
0   Mtr   3

Pentane was the codename for the Vickers/Whitehead Mk.21 air-launched anti-submarine homing torpedo to meet OR.1058 and AW.59, issued in October 1954. An interesting weapon, Pentane is identified in a number of maritime aircraft design studies as the principal weapon for use against Soviet submarines and to be carried by aircraft as diverse as the Shorts Sturgeon, the Fairey Gannet, Avro Shackleton and all of the R.2/48 flying boats. Its size posed a problem for maritime patrol aircraft, especially types designed at the end of the war, such as the Shackleton and Varsity MR as these could not carry the preferred mix of torpedoes and depth charges due to space constraints in the weapons bay – it was either Pentane or depth charges, not both.

Although the documents state that Pentane was derived from Zeta, it shared none of Zeta's odd characteristics, unless of course the cylindrical shapes shown in the various drawings included in the tender brochures are merely there to preserve security. Or, as is more likely, the highly complex Zeta was scrapped and the systems developed for it applied to the more conventional Pentane. Weighing in at 1,950 lb (885kg), Pentane was 17ft (5.18m) long with a maximum diameter of 21in (53cm) the tail assembly adding a further 3ft (0.9m). Oddly enough, Pentane required a certain amount of forward speed on entering the water to start its propeller, something that posed a problem when it was proposed as a weapon for the Bristol Type 191 helicopter to meet NA.43 and the Type 192 to meet OR.325 (Specification HR.146).

While not for use against submarines *per se*, Zoster was described and 'An air-launched aerodynamically supported anti-ship weapon. Using the Zeta weapon for its terminal stage, Zoster was to be launched at up to 10nm (19km) from its target and fly at low level on a flat trajectory before entering the water at around 3nm (5.5km). The weapon was to use active radar homing in flight, before changing to acoustic homing in water.

Another variation on the air-launched anti-ship weapon was Vickers Armstrong's Bootleg which was intended to meet OR.1060 for toss delivery at 400kt (740kph). The weapon entered the water and was propelled under the surface at 70kt (130kph) by a rocket motor integrated with the body. Bootleg was cancelled in 1947 after trials had shown that, unsurprisingly, it could not be used against land targets and was too heavy for the contemporary strike aircraft. Bootleg was 12ft 6in (3.8m) long, with a diameter of 18in (46cm). As noted above it was found to be too heavy at 1,730 lb (785kg), of which 575 lb (261kg) was the warhead. Bootleg was stripped of its rocket motor and fitted with a conventional propulsion system to be reincarnated as the BA.920 torpedo, also known as Bootleg.

Air Vice-Marshal Geoffrey W. Tuttle was by February 1952 discussing the future of the dedicated torpedo bomber which was now considered obsolete. Nor would the 18in (46cm) Mk.XVII torpedo be replaced by a conventional torpedo and it, like the torpedo-carrying aircraft itself, was extinct. The Air Staff had decided that anti-shipping strikes would be conducted by '…light bomber and fighter aircraft … employing their standard armament i.e. bombs, rockets and cannon.' However, the FAA still had a need for an aircraft that could carry the conventional torpedo. So while the FAA continued to consider the Mk.XVII torpedo adequate, the RAF did require '…unconventional torpedoes (Dealer B and Pentane) for use as anti-submarine weapons…' So, as early as 1952 the RAF were looking at the maritime patrol aircraft as a dedicated anti-submarine platform, rather than one for use against surface shipping, which would henceforth be the domain of fast jets.

Dealer B was in February 1952 seen as the most important weapon, mainly because it would be available sooner than Pentane, which showed how much prominence Coastal Command placed on the homing torpedo. This was mainly due to the development of the *schnörkel* and its use by the Soviets on their *Whiskey* and *Zulu* class submarines that drew much from the German Mk.XXI submarines of last year of the war. These submarines could not be detected by ASV radar and 'The only hope of doing so is by a sonobuoy pattern which gives the fix so inaccurately that a directly aimed weapon is of no use.'

Another aspect under consideration by the Committee was the need to produce the torpedoes as cheaply as possible, with the adjective 'expendable' being used, particularly with reference to electronics. An odd expression, given a torpedo's one-way trip, but in this context 'expendable' might just be a euphemism for 'as cheap as possible'. This is in keeping with the climate of austerity in Britain, with the RAF also considering an expendable bomber as a cheaper alternative to the V-bombers. ACAS(OR) Air Vice-Marshal Tuttle emphasised this austere approach to equipment design, stating '…great stress is now being laid upon the need for having expendable instruments and electronics for guided weapons and the expendable bomber.'

As noted above, the air-launched torpedo was becoming consigned to carriage in the flying boats and piston-engined MR types such as the Shackleton and Neptune, mainly because delivering torpedoes from faster jet-powered aircraft led to increased failures. It had been observed that a major cause of these failures was damage to the torpedo as it entered the water and structural damage from shock as the retarding parachute opened. The Torpedo Experimental Establishment (TEE) at Greenock had conducted tests and found that to minimise damage, particularly to the electronics, the torpedo should enter the water at less than 60ft/sec (18m/sec). Captain G O Symonds RN, Superintendent of the TEE, in September 1957 outlined a possible solution to this failure on water entry problem. The most obvious solution was to use a parachute, but having tried this, soon discovered that the shock loads on the torpedo when the parachute opened required the tail structure and the propeller shaft and bearings to be much stronger. This beefed-up structure added weight and was only required for the short period – milli-seconds – as the parachute deployed. The TEE proposed fitting a solid rocket motor and using that as a retro rocket. The motor would be triggered by '…a pilot parachute and attached by a hawser to the propeller boss of the weapon and thrusting in the opposite direction to the line of flight.'

In the case of the 18in (46cm) Mk.30 torpedo, TEE calculated that a rocket motor producing 1,350lbf (6kN) for 4 seconds would retard the torpedo to below 60ft/sec (18m/sec) and '…allow it to "plop" gently into the water without damage.' Further benefits of this deployment method included that the protective nose cap could be dispensed with, as could the heavier structure, producing '…a much lighter design than to date'. The rocket motor would need four nozzles, angled to miss the towing hawser, with the motor built into the existing parachute pack.

As time progressed a need arose for a small, air-delivered active-homing torpedo to be used by helicopters (which were replacing the Fairey Gannet ASW type in the FAA) as an alternative to Dealer B. Development of the Mk.31 torpedo began in 1955 and continued up until its cancellation in 1971. Prompted by this long development time and with a failure to deliver a satisfactory lightweight torpedo, the RN and RAF turned to the USA for its lightweight torpedo needs, purchasing the Mk.43 and Mk.44 torpedoes as interim weapons pending availability of the Mk.31. The 1971 cancellation of the Mk.31 had been on the cards since the mid-1960s when the Air Staff and Admiralty saw the need for a new British-developed airborne torpedo to reduce dependence on American weapons. Prior to this the Air Staff in 1959 had become frustrated with the Mk.31 project and particularly the Admiralty's emphasis on its deployment on helicopters, which necessitated a lighter weapon than the RAF required. As a result the Air Staff drew up OR.1163 for an improved airborne torpedo using active and passive homing to be carried by the new generation of ASW aircraft to meet ASR.357.

By 1961 OR.1163 had been re-written as OR.1186 and, with the Admiralty becoming equally frustrated with the Mk.31 saga, all this led ultimately to the issue in 1964 of NASR.7511 and the commencement of Marconi Project 7511. This weapon would eventually enter service as the Stingray in 1983 and had very interesting characteristics that were to be incorporated in the design. One of the most interesting of these was the use of polyethylene glycol as a drag reduction system. By injecting polyethylene glycol at the torpedo's nose, a reduction in drag in the order of 25% could be obtained, providing increased range and speed. However, the glycol had to be carried in a tank behind the warhead and the added weight and complication of the tank and injection system made this less attractive than the simple expedient of installing a larger, higher capacity battery. By 1976 the original concept of the Stingray, particularly the effectiveness of the

The speed of a new generation of maritime patrol aircraft prompted concerns about the effect of impact with the sea surface on a homing torpedo's nose cap. A retarding system was investigated, using a quartet of rocket motors to brake the torpedo and slow its impact speed. Interestingly, torpedoes delivered by helicopter had a different problem – not enough forward speed to start the motor.

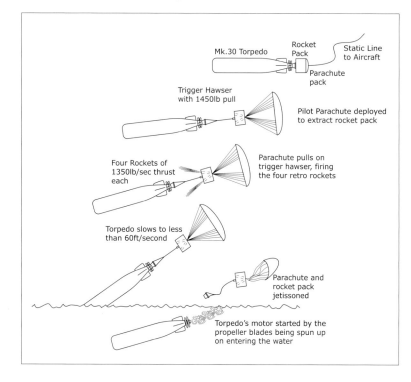

Mk.30 Torpedo
Rocket Pack
Static Line to Aircraft
Parachute pack

Trigger Hawser with 1450lb pull
Pilot Parachute deployed to extract rocket pack

Four Rockets of 1350lb/sec thrust each
Parachute pulls on trigger hawser, firing the four retro rockets

Torpedo slows to less than 60ft/second

Parachute and rocket pack jetissoned

Torpedo's motor started by the propeller blades being spun up on entering the water

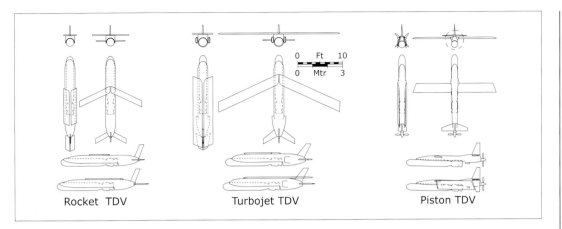

Rocket TDV   Turbojet TDV   Piston TDV

The various configurations of the Torpedo Delivery Vehicle (TDV). These were carried with wings folded in the aircraft weapons bay or on pylons. Three versions were considered: rocket, propeller and turbofan, all carrying a single Stingray torpedo. The rocket-powered version had the fastest reaction time, while the propeller version had a decent loiter/search period.

warhead against the new generation of fast deep-diving Soviet submarines, was found wanting and the RAF and Admiralty purchased the American Mk.46 as an interim weapon. Meantime Marconi had gone back to the drawing board and redesigned Stingray with solid-state electronics, to reduce failures due to shock when the weapon entered the water, and fitting a new shaped-charge warhead weighing 100 lb (45kg).

Stingray weighs in at 590 lb (268kg) and is 8ft 6in (2.6m) long with a diameter of 12.5in (38cm). It is propelled by a seawater battery-powered pump-jet, essentially a propeller operating within an annular shroud which reduces the torpedo's acoustic signature. The pump-jet powers Stingray to 45kt (83km/hr) for a range of up to 6nm (11km) and makes the torpedo particularly agile against targets to a depth of 2,625ft (800m). Stingray Mod 0 was the model initially deployed on Nimrod, whose ASQ.901 signal processor systems provided the initial targeting, as well as the ASWEPS data, for Stingray, which upon entering the water, carries out an autonomous active sonar search and

homing before detonating on the target. Stingray Mod 1 was for use against smaller submarines in shallow water.

Stingray was to be the weapon carried by BAe Woodford's Torpedo Delivery Vehicle (TDV) that was examined as part of the Large Aircraft Replacement Concept (LARC) aimed at replacing all the RAF's large support aircraft with a single type. LARC studies included Elint, air-to-air refuelling tanker and AEW aircraft, but the main role envisioned for LARC was maritime patrol. As described above, LARC was to carry a pair of HS.828 flying sonobuoys, but it was also to carry a number of TDVs on underwing pylons or in the weapons bay, as shown on one LARC drawing. The TDV may have its origins in the 1962 discussions between the RCAF and RAF while formulating a joint Staff target for a Shackleton and Argus replacement, which emerged as OR.350. The Canadians were concerned that submarines might soon attain some form of anti-aircraft capability reminiscent of the heavily-armed *U-flak* boats fielded by the *Kriegsmarine* in the Bay of Biscay in June 1943. The RCAF considered the potential threat to be a SAM that

Some Soviet submarines were thought to be armed with SAMs and the TDV would allow the aircraft to stand-off out of SAM range. Although the threat was real, any submarine taking a shot at a ASW aircraft would need to make sure of a hit as the attack would give away its position.

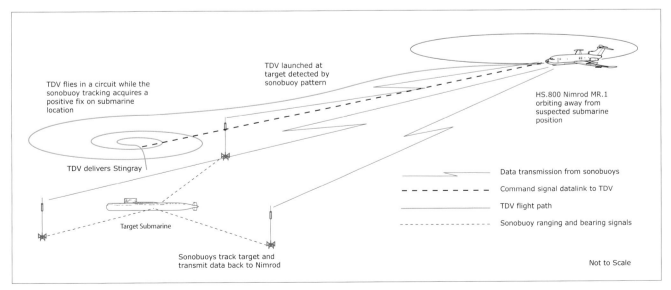

TDV launched at target detected by sonobuoy pattern

TDV flies in a circuit while the sonobuoy tracking acquires a positive fix on submarine location

HS.800 Nimrod MR.1 orbiting away from suspected submarine position

TDV delivers Stingray

Target Submarine

Sonobuoys track target and transmit data back to Nimrod

———〈〈〈 Data transmission from sonobuoys

– – – – Command signal datalink to TDV

——— TDV flight path

- - - - - Sonobuoy ranging and bearing signals

Not to Scale

Hawker Siddeley's torpedo delivery vehicle (TDV) allowed a MAD attack to be prosecuted much faster. The TDV could make a reciprocal turn and deliver a torpedo much quicker than the launch aircraft. In this diagram the colours represent a stage in the process and so when the TDV is delivering its torpedo the aircraft is only halfway through its turn.

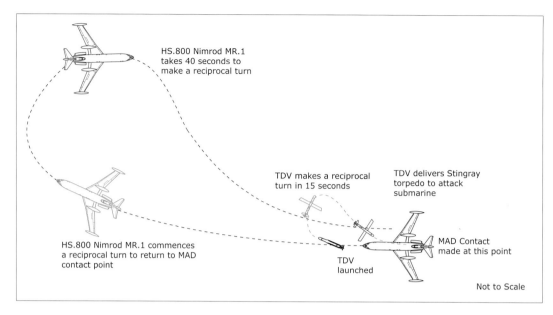

could be used while submerged, making the ASW aircraft vulnerable, and so proposed a stand-off weapon. Vickers' SLAM (Submarine Launched Air-flight Missile) was one such anti-aircraft systems that was developed, but SLAM was not widely used. The Soviet Navy fitted some of its *Kilo*-class submarines with the SA-N-8 *Gremlin* or SA-N-10 *Gimlet* SAMs for air defence against maritime aircraft.

Three versions of the TDV were studied, powered by a Westlake 430 piston engine, Noel Penny Turbines NPT171 turbojet or Aerojet 15-KS-1000 rocket motor. These were small aircraft, typically 1,000 lb (454kg) AUW, with folding wings and tail surfaces. With a length of 13ft 6in (4.1m) and wingspans ranging from 17ft 6in (5.3m) to 10ft 6in (3.2m), the compact TDV could be carried by most maritime aircraft. The concept was for the TDV be launched when a submarine was detected on MAD and, because it was more manoeuvrable than the large ASW aircraft, could home in on the submarine's position much faster than the MR aircraft. This time to return to a reciprocal heading was specified in OR.357 as being a maximum of 40 seconds to perform a 180° turn, whereas the TDV could turn on a reciprocal heading in 15 seconds. The TDV could deliver the Stingray under the direction of the parent aircraft after a MAD contact or '…could act directly in conjunction with the (sono)buoys as and when required. The aircraft could maintain a safe distance from any enemy surface units or leave the TDV to attack while the aircraft continued its search. The piston-powered TDV had an estimated range of 150nm (278km) or an endurance of one hour at a speed of 150kt (278km/hr). Unlike the HS.828 flying sonobuoys, the TDV wasn't recoverable.

Air-launched Anti-ship Weapons

The British had a fascination with the 3in (76mm) rocket projectile (RP, which may also stand for rotated projectile) throughout the Second World War, possibly to the detriment of other weapons, but it was available and provided a weapon to attack armoured vehicles and other ground targets, while a further variant was used as an anti-aircraft weapon in the early war years. Against shipping and U-boats, the 60 lb (27kg) high explosive/semi-armour piercing warhead was replaced by a 25 lb (11kg) solid warhead to punch through hulls. The 3in RP soon found itself fitted to aircraft as diverse as the Fairey Swordfish and Consolidated Liberator for use against submarines, particularly those that were fitted with extra flak guns to fight it out with aircraft on being caught on the surface by RAF aircraft fitted with Leigh Lights and radar.

Its use against shipping was made most famous by the Banff Wing, with 18 Group, Coastal Command flying DH Mosquitoes and Bristol Beaufighters from north east Scotland against Axis shipping in the North Sea and off Scandinavian coastlines. The use of the 3in RP revealed many a useful quirk in its fight characteristic, but none odder than its behaviour underwater. Coastal Command discovered that after entering the water, the rocket had a tendency to curve towards the surface. While this was very useful in attacking ships fitted with belt armour, it also had the effect of enlarging the apparent area of the target and thus increasing the chance of a hit.

The success of the 3in RP prompted the development of a larger anti-ship unguided rocket, Uncle Tom, which appeared too late for service in the war. Designed to meet OR.1009, Uncle Tom was a 9ft (2.74m) long rocket projectile with a diameter of 11.5in (29.8cm) and propelled by a sextet of 3in (7.6cm) rocket motors. Postwar, a further development in unguided rockets produced Red Angel. This was longer and replaced Uncle Tom's four large fins with six flip-out vanes but Red Angel's main difference was its warhead: an armour-piercing type shaped to penetrate the deck armour of warships such as the *Sverdlov*-class cruisers. It was thought that a salvo of six Red Angels could cripple a *Sverdlov* cruiser. Red Angel was 10ft 9in (3.28m) long, with a diameter of 11.25in (28.58cm) and weighing in at 1,055 lb (478kg), 88 lb (40kg) of which was explosive. Red Angel was tested on Westland Wyverns and was earmarked for the Blackburn NA.39 Buccaneer, with four rounds mounted on the rotary bomb door, providing a formidable barrage from a single aircraft.

Red Angel, like Uncle Tom and the 3in RP, suffered from short range and the attacking aircraft had to enter the ship's air defence zone and thus was vulnerable to defensive fire, something that would require nerves of steel in a pilot barrelling-in to launch point. To address both of these, the Air Staff and Admiralty drew up an outline requirement, possibly leading to

OR.1057 for a longer-ranged weapon called Nozzle: 'An air-launched flat trajectory weapon which can be carried by both Naval and RAF aircraft, can be released at long range, does the majority of its travel in air at a supersonic speed, and enters the water close to the target to secure underwater non-contact detonation.' Nozzle's main targets were surface ships, with a possibility of use against ground targets.

On the matter of range, the outline requirement stated 'Range of release is governed by the acceptable casualty rate to the launching aircraft, which in no circumstances must be required to approach within 10,000 yards [9,144m] of the target.' There was also an option to increase that range '…to ensure immunity of the aircraft from the maximum range of medium range GAP.', GAP being a guided anti-aircraft projectile, the mid-Forties British term for what is now known as a SAM. Nozzle was to weigh no more than 2,000 lb (907kg) and possess supersonic performance in its flight phase. Guidance was to be self-contained and automatic, with no input from the launch aircraft once 'off the rail' with entry into the water controlled automatically as part of the air-to-ship homing system. The warhead was to be of a non-contact type to save weight, but capable of impact fusing to allow attack on ships in shallow waters. As described here, Nozzle was an incredibly complex weapon for the late 1940s and this version was soon super-

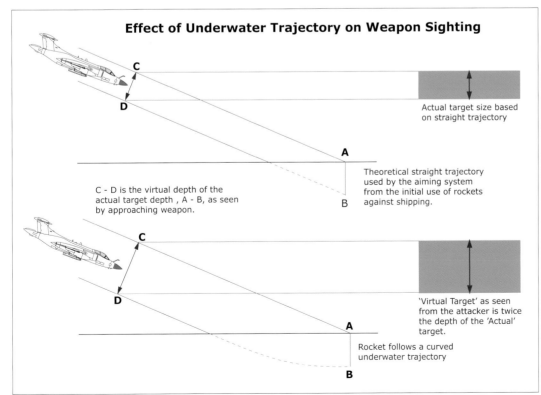

Effect of Underwater Trajectory on Weapon Sighting

C
D

Actual target size based on straight trajectory

A

Theoretical straight trajectory used by the aiming system from the initial use of rockets against shipping.

B

C - D is the virtual depth of the actual target depth , A - B, as seen by approaching weapon.

C
D

'Virtual Target' as seen from the attacker is twice the depth of the 'Actual' target.

A

Rocket follows a curved underwater trajectory

B

One surprising discovery from the anti-shipping strikes in World War Two was the 'virtual target'. Coastal Command discovered that its 3in (7.6cm) rockets followed a curved trajectory after entering the water. This effectively increased the size of the target if an impact below a vessel's waterline was required.

seded by what was more or less a primitive guided bomb that used a 'bonker' rocket for final correction of its trajectory. Whether this was done automatically or remotely by the launching aircraft is not detailed.

Green Cheese

The high hopes that the Air Staff had for guided weapons in the late Forties were dashed by the development of Green Cheese. A radar-guided, rocket-powered anti-ship missile, Green Cheese was intended to be the main weapon of the Fleet Air Arm's M.148 strike aircraft, met by the Blackburn NA39 Buccaneer.

Originally known as Fairey Project 7, Green Cheese was developed for use by RAF Vickers Valiants and Avro Shackletons to meet OR.1123, issued in July 1953, and to the January 1954 requirement AW.319 for Royal Navy Gannets but ultimately on the Navy's forthcoming Buccaneers. The target for Green Cheese was the new Sverdlov-class cruisers entering service with the Soviet Navy and as noted above, viewed as a significant threat by Western intelligence services. Right from the off, launch conditions for Green Cheese from these aircraft in very different flight regimes, cast doubt on a joint requirement.

Green Cheese began life as a glide bomb, like the TV-guided Vickers Blue Boar, and comprised a shortened Blue Boar casing fitted with the X-band active radar seeker from a Vickers Type 888 Red Dean air-to-air missile. Blue Boar had suffered from high drag and so to give a better glide angle for Green Cheese, a number of drag reduction modifications were made. The wings were cut back and rear fuselage tapered, but the latter would reduce the volume available for the rather bulky electronics. Fairey soon gave up on the glide bomb idea and after examining different rocket motor types, fitted a cut-down

Smokey Joe boost/sustainer motor, based on that used on the English Electric Red Shoes SAM, to provide the correct 30° glide angle. A new rocket motor designed by the Summerfield Research Station, the Corgi, was to be used for the ultimate version of Green Cheese.

Two versions of Green Cheese were envisaged, one for the RAF and one for the RN. The RAF variant was to meet OR.1123 with a fixed wing version to make 'dry hits' on the target's superstructure. The Vickers Valiant was to carry up to four of the fixed-wing weapons externally on underwing pylons or in bomb cocoons, with initial targeting carried out with the Valiant's Navigation and Bombing System/H2S Mk.9A. More realistically a pair of Green Cheese could be carried internally '…on a suitable lowering mechanism'…'but with very little to spare in length' according to Fairey's Chief Project Engineer, G T Dobson. However, as noted above, the RAF's interest in using maritime patrol aircraft against surface ships had been waning since 1952, the preference being to use faster, smaller strike aircraft for such operations, so the RAF's interest in Green Cheese soon lapsed. This may have been prompted by a June 1954 RAE technical memo outlining the difficulties carrying Green Cheese on the different aircraft, summarising the problem 'Hence the Gannet and Valiant requirements are probably incompatible, and ideally lead to entirely different types of bomb.' The Shackleton had been ruled out as a carrier for Green Cheese as it flew too slowly and the missile would require strap-on boosters to achieve adequate stand-off range for the Shackleton.

The Navy version, to meet AW.319, was intended for 'wet hits' under the waterline and was to be fitted with flip out wings and control surfaces that allowed internal carriage in the weapons bays of the Gannet and Buccaneer. Internal carriage in the Gannet meant that some form of extending launch rail was required to

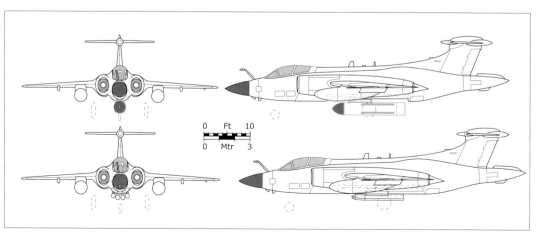

Sverdlov-busters. Shown here on a Buccaneer S.2, Green Cheese and Red Angel were designed to attack the Sverdlov-class cruisers. The Admiralty considered six hits by Red Angel to be sufficient to disable a Sverdlov.

lower the missile out of the bay to allow it to acquire the target. This added weight and complication to the entire system. The Buccaneer on the other hand had the weapon mounted on the rotating weapons bay door that allowed the seeker to 'see' the target. In the end the requirements merged to produce a weapon with flip-out wings for the Buccaneer .

Targeting of the weapon was to be controlled by the ASV-21 radar that was being developed for maritime patrol aircraft but was also to be fitted to the early Buccaneers. ASV.21 would provide initial data for the Green Cheese seeker but an early proposal to use the missile's seeker as a target acquisition radar was dropped, again due to the need to expose the weapon prior to launch. Green Cheese's seeker was being developed by GEC and EMI based on their X-band seeker for Red Dean but with an enlarged dish. This was an active homing system, allowing a launch and leave delivery. A semi-active seeker was proposed but this would require the target to be illuminated by the launch aircraft radar, thereby making the aircraft vulnerable to attack by the target ship's defences. The missile itself would be vulnerable to ship's defences (the requirement for the Orange Nell anti-missile system had been drawn up to counter weapons such as Green Cheese) so it was proposed to fit a 'jinker' to make the missile perform a random weave manoeuvre during its attack.

Green Cheese was to hit the water at an angle of 40° (after motor burn out the glide angle would have steepened) a distance of 150ft (45m) from the target ship. The impact was to shear off the radome leaving an angled surface that forced the missile upwards under the ship to impact under the waterline, a 'wet hit'. To ensure that the missile hit the water rather than the ship itself (the strongest reflector) the seeker was to be given 'squint'. However, it was pointed out that the angle of squint was dependent on the range to the target, so the squint would need to change as the missile approached the target. Having dismissed the squint, the Admiralty decided that the missile should dive at a predetermined distance from the target. Both suggestions were dismissed as adding complication to the homing system and a conventional anti-ship warhead was adopted.

Weighing in at 3,800 lb (1,724kg), Green Cheese was too heavy to be carried by the Fairey Gannet, mainly due to the wing deployment mechanism adding 800 lb (363kg) to the missile's launch weight. The seeker had to be exposed to acquire the target and since the Buccaneer and the Gannet had an internal bomb bay, the missile had to be lowered before acquiring the target. This was a particular problem with the Gannet as Green Cheese required a mechanism to lower the weapon, whereas the Buccaneer had a rotating bomb bay door. Another problem with the Gannet was that with a missile in the weapons bay, the doors could not close! This carriage problem was compounded by the need for Gannet to enter the area defended by the target's AA weapons prior to launch. The Admiralty graciously dropped the requirement for use by the Gannet and had also raised doubts about Green Cheese's ability to sink a major ship, due to its shallow angle of impact. Fairey had by this time also proposed a weapon called Sea-Skimmer, whose terminal phase included a vertical dive onto the target's deck.

Described as looking like a small aircraft, Fairey's sea-skimming anti-ship missile used a Roll-Royce RB.93 Soar expendable turbojet as its sustainer and was proposed for the same requirement as Green Cheese. Sea-Skimmer was not proceeded with due to guidance problems with the X-band seeker that was intended to be locked onto the target before launch, dive to sea level and fly just above wave-top height with the final approach being a bunt onto the deck of the target ship. This bunt attack was thought to have a more lethal effect than the shallow dive of Green Cheese.

The Rolls-Royce Soar sustainer turbojet required 30 seconds running up time prior to launch. This was a particular problem on aircraft such as Shackleton MR.2 or MR.3 and Gannet whose ventral radomes were aft of the weapons bay so the launch aircraft's Air-to-Surface Vessel (ASV) radome would need to be retracted, thus blinding the launch aircraft at a critical time. Aircraft such as the Bristol Type 175MR or Type 189 would have carried Sea-Skimmer on underwing pylons, and the unidentified weapon on the Bristol drawings of the Type 175MR shown in Chapter Five may be Sea-Skimmer.

The Sea-Skimmer used mid-course guidance with a terminal homing phase using radar. The Admiralty considered Sea-Skimmer to have a better chance of sinking a large ship than Green Cheese, due mainly to the angle of its terminal dive. The additional range, greater than 40nm (74km), allowed by the turbojet sustainer and mid-course guidance also appealed to the Admiralty, especially as it would keep the Gannet out of harm's way.

Green Cheese showed promise, but was essentially scuppered by cost and weight overruns, not to mention being overtaken by another Fairey project, the Sea-Skimmer. The writing was on the wall for Green Cheese by March 1955 and it was cancelled in 1956. The ultimate version, with improved capability, was called Cockburn Cheese and was intended to be in service from 1962. Cockburn Cheese was named after the PDSR(G) Dr Robert Cockburn. The replacement for Green Cheese was called Green Flash, which might allude to a nuclear warhead, but again this was cancelled in favour of toss-delivered Red Beard free-fall bomb. From this point, until the mid-1960s, the RAF's maritime patrol aircraft carried no weapons for use against surface ships apart from free fall bombs, their primary role being to direct air strikes by fast jets.

ASR.357, AS.12 and the Martel

By the late 1950s the miracle of miniaturisation, particularly of electrical components, had seen a reduction in the size of guided weapons, but it was only after the turn of the decade that the Air Staff returned to consider air-launched anti-ship weapons for its maritime patrol aircraft. In the USA, ASMs such as Bullpup were in service and the Fleet Air Arm had acquired this weapon to meet GDA.5, arming its Buccaneers and Scimitars. Meanwhile Vickers at Weybridge had Barnes Wallis working on the unpowered Momentum Bomb, which was also known as Apple Turnover, but again this was tailored to fast jets and was intended for low-level delivery by TSR.2. De Havilland Propellers was working on a rocket-propelled weapon called the RG10, intended for strike aircraft, but could also be launched from slower maritime patrol aircraft.

The initial operational requirement for a Shackleton replacement, OR.350, had no mention of an air-to-surface missile (ASM), the RAF sticking to its diktat that strike aircraft should be used against surface ships. Earlier studies involving fitting the Shackleton MR.3 with ASMs had come to nothing, with Air Vice-Marshal C H Hartley, ACAS(OR), commenting in a letter to the AOC, Coastal Command, that '…the arming of the Shackleton…with either Bullpup or AS.30, although desirable, is not considered to be practicable.' However on the suspension of OR.350 and the adoption of the more advanced OR.357, the guided weapon was back in favour for maritime patrol aircraft.

Concurrent with the discussions on OR.357 were a series of studies for a missile to arm tactical aircraft and these led to NASR.1173. These studies ultimately became the AJ.168 Martel and the Bullpup/AS.30, but many of the aircraft design studies for OR.357 show the aircraft carrying what appears to be the AS.30. This rocket-powered missile used MCLOS (Manual Command to Line Of Sight) guidance whereby the operator tracked the missile by using flares in its tail and used a joystick and datalink to control its flight. The missile weighed in at 1,150 lb (520kg), was 12ft 1in (3.2m) long and had a maximum range of 6nm (11km) and trials of the AS.30 in 1965 at the A&AEE using a Canberra B.15 were described as 'satisfactory'. The Martin AGM-12 Bullpup was another MCLOS missile, but it did enter service with the Fleet Air Arm. Larger than the AS.30, the Bullpup, weighing around 1,800 lb (110kg) and 13ft (4.1m) long, had a range of 10nm (19km). As it turned out, neither of these weapons were up to the RAF's expectations, so a new requirement for an ASM was drawn up – NASR.1168 – better known as Martel.

A Disappointing Weapon – Martel

Martel was an early Anglo-French project to meet OR.1168 that ultimately produced two variants, the TV-guided AJ.168 that the RAF was most interested in and the French-led anti-radiation version. The AJ.168 TV-Martel was a marriage of the Matra AS.37 anti-radiation missile and a Marconi TV guidance package with the airframe and motor being developed by de Havilland Propellers, which soon became Hawker Siddeley Dynamics in 1963. TV-Martel had a launch weight of 1,200 lb (545kg) was 13ft 6in (4.1m) long and had a diameter of 16in (40cm).

The RAF took both types, but that service's relationship with Martel was never a particularly happy one, principally due to their disappointment in the missile's capability. One memo dating from 1970 was particularly scathing of the Martel in both its guises, summarising the situation thus: 'Martel – this has been a disappointing weapon. Costs have risen sharply…and of course the TV missile is vulnerable to ground defences. However, there is no feasible alternative stand-off missile and we must proceed with our present commitment for 200. It is debateable whether we should place a further supplementary order, but we have several months to weigh this up.'

So what prompted this attitude towards Martel? When used on a strike aircraft, the missile was launched and the aircraft then had to perform a 180° turn while the operator attempted to initiate and maintain a datalink from a rear-

RAF Buccaneer S.2B XX895 is seen carrying the full suite of Martel equipment. Under the starboard wing is an AJ.168 TV-Martel on the outboard pylon and a TV datalink pod on the inboard pylon. Under the port wing is a TV-Martel inboard and an AS-37 anti-radiation Martel on the outboard pylon. The Air Staff were not impressed with either. *Author's collection via BAE Systems*

facing pod on one of the aircraft's pylons. This datalink pod took up a pylon, thereby reducing the weapons load-out if the TV-Martel was carried. Nimrods were wired to carry Martel on wing pylons, but it was never used operationally, although the wiring came in handy when the Nimrod MR.2P was fitted with AIM-9L Sidewinders during the Falklands Conflict. The intention was to carry the Martel on the wing pylons and once launched it was to be controlled by the Air Electronics Officer (AEO) in the cabin using a joystick while tracking the target via the video feed from the TV camera in the missile's nose. However, tests soon revealed that such a large store mounted on outboard wing pylons affected the aircraft's directional stability and the weapon was never used operationally.

In the 1970s the 'disappointing' Martel prompted a number of developments, one of which was the P3T, an air-breathing sea-skimming anti-ship missile that entered service as the BAe Dynamics Sea Eagle. The Sea Eagle was mainly carried by fast jets such as Buccaneer and Tornado, although the Indian Navy fitted it to their Ilyushin IL-38 (NATO reporting name *May*) and possibly the Tupolev Tu-142 *Bear* maritime patrol aircraft. While the Sea Eagle was never carried by Nimrods, they were armed with the McDonnell Douglas AGM-84 Harpoon in the post-Falklands re-equipment process. However, in the Avro Archive at Woodford there is a file that describes possible uses of the redundant Nimrod AEW.3 airframes. One drawing shows a Nimrod AEW.3 carrying six Sea Eagles in its weapons bay and a Searchwater radar in the enlarged nose radome. Another proposed carrier for the Sea Eagle was the Future Inter-

national Military Airlifter (FIMA), which carried 27 rounds on a carousel within the freight bay, although these may be the P4T cruise missile variant.

Another ASM associated with the Nimrod was the Nord AS.12, which was intended for use against smaller craft such as *Komar*-class fast patrol boats (FPB), Elint trawlers and surfaced submarines, rather as the 3in RP had been used in the past. While the Martel was the main guided weapon intended for ASR.357, the Air Staff were somewhat alarmed by its expense, particularly if used against low threat or small targets such as FPBs and so the AS.12 was selected. The AS.12 was a scaled-up SS.10/11 wire-guided anti-tank missile modified for air-launch from helicopters, such as the Westland Wasp and Wessex in Royal Navy service, most famously in an attack on the Argentinean Navy submarine *Santa Fe* in 1982. The Nimrod MR.1 could carry four AS.12, in pairs, on the wing pylons and these could be fired singly, guided visually by the co-pilot using a small joystick and tracking the missile via a pair of flares in the tail. The co-pilot's field of view was the limiting factor in the launch envelope, with an altitude of

The Nord AS.12 was one weapon intended for the Nimrod, and is seen here on the twin mounting on the wing pylon of the prototype Nimrod. The AS.12 could only be used against undefended targets as it lacked the range to keep the aircraft outside a warship's anti-aircraft defence. *Avro Heritage*

A Lockheed P-3C Orion carries a quartet of AGM-84 Harpoons on underwing pylons. The Nimrod carried up to three Harpoons, arranged linearly within its weapons bay. In the light of the Martel's poor performance, Harpoon conferred a much-needed over-the-horizon anti-ship capability on the Nimrod. *US Navy*

500ft (152m) and a range of 7,000 yards (6.4km) being recommended. Unfortunately these parameters brought the launch aircraft into the defended zone of Soviet warships armed with SA-N-3 *Goblet* and SA-N-4 *Gecko* SAM systems and AA guns, so the missile was restricted to use against undefended targets. However, having picked up a target on radar, a positive identification of a surface unit would require the Nimrod to approach close enough to become vulnerable. A 1972 report by the Central Tactics and Trials Organisation on the effectiveness of AS.12 recommended that a magnifying sight be installed for the co-pilot to allow the Nimrod to maintain a less vulnerable stand-off range.

The use against surfaced submarines was more or less rendered obsolete by nuclear propulsion, although the 1972 report noted that some Soviet submarines had to surface to launch their missiles. Surprisingly enough, the report recommended that the AS.12 be retained, but only for use against lightly-defended targets, although the report did state that the weapon was next to useless at night due to the flares. The cheaper SS.11 was used for training aircrew, but had different flight characteristics to the heavier AS.12.

The final ASM to enter service with the Nimrod fleet was the McDonnell Douglas AGM-84A Harpoon, acquired as an urgent operational requirement during the Falklands Conflict. Harpoon began as a sea-skimming weapon for use against surfaced submarines (those Soviet SSBNs sitting on the surface preparing to launch missiles must have made *very* tempting targets) by aircraft such as the Lockheed P-3C Orion. Entering service with the US Navy in 1979, the RAF acquired the weapon

for the Nimrod MR.2P in 1982. The air-launched AGM-84A Harpoon weighs 1,140 lb (520kg) and is 12ft (3.8m) long. The Harpoon is generally mounted on a pylon, but when used by the Nimrod MR.2P, the pylons were so far outboard that the Harpoon, like Martel, caused directional stability problems. As a result the weapon is loaded in the Nimrod's weapon bay, with up to three carried. The Nimrod MRA.4 with its re-engineered wings was fitted with a pair of pylons on each wing and could carry up to four Harpoons. With a range in excess of 50nm (93km) the Harpoon provided over-the-horizon capability with the terminal phase being guided by an active radar seeker.

Hunter/Killer

The original reason for having maritime patrol aircraft was protection of the nation's maritime trade, which during World War Two became critical to maintaining the ability to fight. Post-war, and particularly since the development of the ballistic missile submarine, the object was to prevent the annihilation of the nation state by nuclear weapons. The effort to develop the means to detect, track and subsequently destroy submarines became a primary task of the United Kingdom's research establishments. Technology played a major role in the Battle of the Atlantic but became ever-more important in the Cold War at sea. Thankfully the question of whether the modern anti-submarine aircraft would beat the nuclear submarine was never answered. Having outlined the systems and weapons to be carried by the RAF's maritime patrol aircraft, the aircraft themselves will be described. Like the weapons and systems, these had a protracted genesis.

3 Old Men and Airships: The Last Flying Boats

'The many firm believers in the flying boat are either very old men or those who have an affection for flying boats which no other type of aircraft has ever aroused.' Air Vice-Marshal Geoffrey Tuttle, Assistant Chief of the Air Staff (Operational Requirements), 28th May 1953

'I was recently drawn into an argument as to whether flying boats are as safe as their landplane counterparts; and to verify my argument that they are decidedly less safe, I have requested DFS to make a brief study of the question.' Air Marshal Geoffrey Tuttle, Deputy Chief of the Air Staff to Air Marshal Sir Edward Chilton C-in-C Coastal Command, June 1959.

There is possibly nothing more graceful than the sight of a flying boat 'on the step'. That is the point at which a flying boat makes the transition from a boat with wings, bobbing in the swell, to a less than sleek aircraft lumbering into the air. For those brief few seconds as it scythes through the wave tops, no matter how frightful it looks at a standstill, a flying boat looks superb. Whatever happened to the flying boat as a military aircraft? Once considered ideal for maritime patrol and

anti-submarine work, the boats had disappeared from the RAF's order of battle by 1959. What prompted the move from seaplane to landplane and the military flying boat to all but vanish from the skies?

Today, the argument quoted above would be settled by the protagonists whipping out smartphones and using the resources of a famous internet search engine. Two senior officers of the RAF, locked in discussion in the mess, had no such resources in the late Fifties, but did have the clout to instruct the Director of Flight Safety to conduct a study. Suffice to say, Tuttle was correct but while the reasons are interesting, the Director of Flight Safety, Air Commodore J C Millar revealed something most enlightening in his study.

Millar's staff examined the occurrence of F-Accidents, those occurring on take-off, landing or in the circuit for a flying boat, the Sunderland, and four land types: Avro Shackleton, Lockheed Neptune, Handley Page Hastings and Vickers Valetta. This allowed the comparison of four- and two-engined aircraft in the maritime and transport role. Given that the Sunderland had been in service for at least ten years more than the landplanes, an allowance was made for the annually improving trend in accident rate. This was typically taken to be

On the step and with a rooster tail in its wake – even the Shorts Shetland looks graceful at this point in take-off or landing.
Bombardier via Phil Butler

13% improvement per annum for a particular aircraft type as its time in service increased. Fortunately the study soon revealed that the Sunderland's F-Accident rate was so much in excess of the landplanes that this correction factor could be all but ignored, the other four types being of similar vintage.

The surprise revealed by Millar's study was that, despite operating aircraft in broadly similar classes, maritime patrol squadrons suffered more F-Accidents than transport squadrons by a factor of 2.5! This tweaked Millar's curiosity and so he and his staff began to delve into the problem. Firstly, it was apparent that maritime squadrons were suffering F-category accidents on a par with single-engine fighter squadrons! Secondly all landplanes were at the mercy of unreliable undercarriage mechanisms, which would affect both maritime and transport aircraft equally and could be ignored. Thirdly, both roles shared long sorties, be they patrols over the ocean or long stages on transport routes. So what was the key factor in the losses? It transpired that the transport squadrons carried out more continuation training than the maritime squadrons, with the transports conducting more training flights with 'circuits and bumps' and, with 10% of the transport fleet used for training, transport pilots spent more time in the flight regime that was under examination. Maritime squadrons habitually tacked such training onto the end of their sorties, but this appeared to be counter-productive as the pilots were fatigued after a long mission, thus compounding the problem.

The lessons learned had implications for Bomber, Coastal and Transport Commands and the Air Commodore's findings contributed to advice on training put forward to the respective commands. A significant contribution to flight safety derived from an argument between two men in a pub.

Missing the Boat

As for the flying boats, the British had a particular affection for them, possibly due to them being the aircraft that had linked the Empire since the early Twenties. It might even have been the image of the 'Flying Porcupine' battling Focke-Wulf Fw 200s and U-boats in the darkest days of the Battle of the Atlantic. The RAF still had many flying boats on its inventory at the end of World War Two but the big boats' days were numbered. Developments in large long-range landplanes and the proliferation of concrete runways at many a remote location across the world, all built to support the war effort, made such aircraft more practicable. In late 1946 the Air Staff Operational Requirement Branch opened discussions on a replacement for the Shorts Sunderland and the same company's S.35 Shetland that had been designed around Specification R.14/40 issued in 1940 for a Sunderland replacement. Shorts had beaten off the Saro S.41 in the R14/40 tender process and the first Shetland I flew in 1944, but when the war ended there was no pressing need for what was effectively an enlarged, uprated Sunderland and the Shetland II became a transport. Two Shetlands had been built with a further twelve laid down before the project was cancelled. The period immediately after the war saw very little work carried out on new British aircraft, the Labour Government was coming to terms with what actions were needed to set up the welfare state they had promised, not to mention rebuilding the coun-

A Shorts Sunderland, the type to be replaced by R.2/48, undergoes engine maintenance. Note the buoy, floating work platform and fuel scow, plus all the rigging involved. Such equipment needed skilful handling to avoid damaging the flying boat's hull. Precision work on a calm day was daunting enough, but on a rough day it could be very challenging. *Author's collection.*

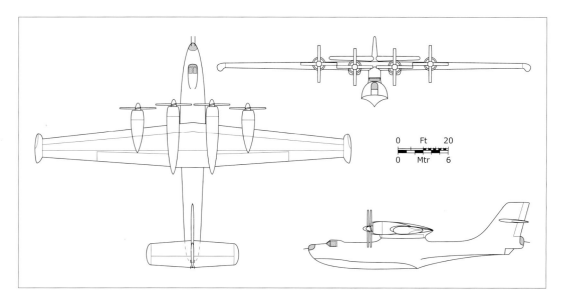

Saro's P.104/1 used the Saro narrow-beam hull, with a wing derived from the Bristol Britannia mounted on a Catalina-style pylon. Such designs allowed a smaller fuselage with engines mounted high and well clear of spray. Any of the late 1940s 'big engines' could be fitted.

try and converting the wartime industries to civilian-oriented production. Nor did the Government really know how new technologies such as the atom bomb and gas turbine would affect future developments in the aviation industry.

By November 1946 a draft requirement, OR.223 (later re-used for a Canberra variant), for a long-range reconnaissance flying boat was in circulation around the Air Staff with the minutes of the planning meeting stating '…this would be an easy aircraft on which to get the necessary action completed and a contract placed before the end of the financial year.' Famous last words, as the saga of the long-range flying boat to replace the Sunderland would continue until finally fizzling out a decade later. The rationale was that the use of the Sunderland V fleet could not be extended beyond 1952 as the availability of spares from America for their Pratt & Whitney Twin Wasp engines was in doubt. The OR.223 requirement that arose was for a six- or eight-engined aircraft with an all-up-weight (AUW) of 200,000 lb (90,700kg) powered by Bristol Centaurus engines. The Nomad, D. Napier and Son's compound diesel, had also been considered but dismissed due to 'the scanty information available' on this engine. The initial response came from Saunders-Roe in the form of the P.104/1. This used the wings, somewhat modified, from the Bristol Britannia and this with its Centaurus engines was mounted on a pylon above a narrow-beam hull, producing an aircraft not unlike an enlarged, four-engined Consolidated Catalina.

Further discussions revealed that Coastal Command considered a smaller aircraft preferable as the anti-submarine campaign in the Atlantic had shown that '…the best results were obtained in submarine hunting by having a large number of small boats or aircraft.' There was also the fact that more aircraft could be procured for the same funding and so the requirement changed to a smaller aircraft. This view also coincided with the debate on whether to replace the Sunderlands with a flying boat or procure land-based aircraft for the maritime role. On 21st June 1947 the Deputy Director of Operations (Flying) wrote that the intention was to continue to use Sunderlands for another seven years then '…to employ land based aircraft in all maritime roles from then on.'

By 10th January 1947 a firm requirement in the shape of OR.231, later to become Specification R.2/48, was issued to the various aircraft companies as the basis of design studies for a tender process. The basic outline of OR.231, as issued initially, was for an aircraft with an AUW of 80,000-100,000 lb (36,281-45,351kg) capable of flying a patrol of six to eight hours at a distance of 1,000nm (1,852km) from base with a weapons load of 8,000 lb (3,628kg) including Pentane torpedoes. With this new requirement, the Air Staff and Coastal Command went looking for a Sunderland replacement, only to be pulled up by Air Commodore Thomas Pike, Director of Operational Requirements. Pike advised on 26th June 1947 that the only firm in a position to build such a flying boat were Shorts but '…they are so occupied with the new bomber that they are unable to start serious work on it for about a year.' That new bomber was the Shorts Sperrin to meet OR.239 /Specification B.14/46.

By August 1947 Issue 2 of OR.231 was under discussion, outlining the need for a 90,000 lb (40,816kg) aircraft powered by four Napier

Nomad engines '…or should development of this prove unsuccessful, the Centaurus.' and a 4,000 lb (1,814 kg) weapon load plus 70 sonobuoys. This proved controversial; particularly the size of the aircraft, with Air Marshal Leonard Slatter, AOC Coastal Command, who considered this too small and that an aircraft in the 250,000-300,000 lb (113,378-136,054 kg) class would be required to meet this very long range (VLR) role. Pike responded that if Slatter insisted on this size of flying boat '…it might well be we should get no boat at all.' Director of Operations, Wing Commander L M Laws, added further to the AUW debate by pointing out that the Americans were performing similar operations with the twin-engined PBM-5 carrying 12,000 lb (5,442 kg) of weapons at an AUW of 58,000 lb (26,304 kg). In reality the Martin PBM-5 Mariner carried 4,000 lb (91,814 kg) of weaponry. The result of this sometimes heated debate was OR.231, issued in April 1948. It should be noted that by April 1948 the US Navy had contracted Convair to build the XP5Y patrol flying boat (although it entered service as the R3Y Tradewind transport) which was more or less in a similar class to the OR.231 aircraft. The MoS compared the XP5Y with their Nomad-powered OR.231 studies and concluded that the Allison T40 turboprops on the XP5Y used '…nearly *twice* the weight of fuel to do an air distance only *15%* more than OR.231 and at *lesser* speed.'

Having issued the requirement to the companies in April 1948, a year later the first tenders were coming in from Shorts, Saro and Supermarine. Interestingly, Tuttle was very wary of committing to the expense of the R.2/48 flying boat development and in a letter dated 16th May 1949 to Air Vice-Marshal Robert Foster,

ACAS(Projects), advised that the Air Staff should really '…consider flying boat policy to re-affirm the need for such an aircraft.' Tuttle's concern was that the R.2/48 was a white elephant because '…the land-based Shackleton will cover the Atlantic convoy routes…and it would appear that the boat is only needed…in the event of a submarine war in the Indian Ocean.'

By March 1950 Short Brothers, recently moved from Rochester to Belfast, were awarded a contract to build their PD.2 flying boat to meet the newly revamped R.2/48 which had been designated R.112 on 27th February 1950, (the designation R.2/48 will be used throughout this chapter unless in a quote) but that did not mean that Shorts could start building the PD.2. Why had it taken four years to award a development contract for a fairly straightforward large aircraft and why were the various members of the Air Staff, Air Council and Air Ministry still arguing about the pros and cons of flying boats almost a decade later? The Chief Scientist at the Air Ministry was concerned by the delay in making a decision and that this could cause the flying boat design teams to dissolve and Britain would lose valuable expertise meaning that none of the aircraft companies would be able to develop a new flying boat. Essentially the problem was that the air-minded fell into the two camps, those who considered flying boats an anachronism and those who considered them a useful type. The debate became most heated in the period from 1951 until 1955, when the perceived need to replace the Sunderland in 1960 prompted their utility to be questioned by the upper echelons of the RAF and Air Ministry. Chief of the Air Staff, Marshal of the RAF Sir John Slessor, in May 1951 asked 'Does our survival in the early

stage or our ultimate ability to win a future war depend on having a big long-range flying boat?' As for the Chief Scientist's concerns, Slessor assured him that there would be a decision by the end of the month. There wasn't, and the debate rumbled on.

An Embarrassment All Our Lives

Slessor took the view that flying boats were expensive to operate as they needed bases, slipways and marine craft (eighteen flying boats apparently needed 338 small boats to support them) and were only acceptable '…if nothing else will do the job and you couldn't get enough landplanes to do it.' On the costings front, Slessor pointed out that three units with five R.2/48 types each would cost £14.36m whereas three units of eight Shackleton MR.1 each would come in at £10.09m. The thinking was that eight Shackletons could be replaced by five R.2/48 aircraft, mainly due to R.2/48's greater endurance, being capable of a four hour patrol at 1,000nm (1,852km) from base while the Shackleton could only patrol for four hours at 800nm (1,482nm) from base. Operational experience soon showed that the Shackleton MR.1's radius of action for a four hour patrol was 630nm (1,167km) and could only reach the 1,000nm demanded for an R.2/48 type with a reduced warload which in turn meant that the 5:8 ratio of R.2/48 to Shackleton became 5:10.

At this point, in July 1951, Air Chief Marshal Ralph Cochrane, VCAS, weighed in with a note to the Air Council pointing out that flying boats would only really be useful outwith the area that would become the RAF's primary concern – the north east Atlantic. However, Cochrane pointed out that without R.2/48, more Shackletons would be required and Avro did not have the capacity to build them as they were committed to building Canberras (as were Shorts). The Air Staff were ostensibly in favour of the flying boat but Cochrane and some of the Air Staff officers had some misgivings on the type, being of the opinion that despite Shorts having been awarded a development contract for R.2/48 in 1949, it was Saunders-Roe that should develop the flying boat to meet R.2/48. The driving force behind Shorts being given the contract was the MoS who Cochrane thought were quite happy to see Saro go under. This is supported by a sentence in a memo from the Air Ministry's G S Whittock who writes '…the Ministry of Supply argues that the ultimate loss of Saunders-Roe capacity would not be a serious matter.' The MoS argued that Belfast was less vulnerable to enemy action than the Isle of Wight and that Shorts had room to expand. Whittock continues with the observation 'I suspect that the fact that the Government owns the majority of the shares in Shorts may have something to do with their preference for this firm.'

Cochrane appears to sit on the fence in the flying boat debate, having advised that Saro be awarded R.2/48 he then made the comment 'There remains the major question of policy, whether or not to discontinue further development and let the flying boat follow the airship.' Two years later the argument on flying boats was still ongoing and ACAS (OR) Air Vice-Mar-

Shorts PD.2 flying boat was selected for development despite the Air Staff's advice. The MoS were prepared to see Saro go out of business to enable Shorts to fill the R.2/48 contract. This image shows the later Shorts PD.2 (with smaller hull and Turbo-Griffon engines) dropping a pair of Pentane torpedoes. *Adrian Mann*

shal Tuttle waded in to say that he doubted that there were '…any good military reasons for acquiring a flying boat.' He went further and stated that '…flying boats are where airships were in 1933.' and that the protagonists of flying boats were forever providing examples of the usefulness of the flying boat, but ignored the fact that none of these were in the requirement. Tuttle concluded his note with the quote that appears at the start of this chapter.

Fairly soon after the discussions of 1951, the Air Staff had began to look at cancelling R.2/48 and replacing it with a new requirement based on a landplane, specifically the Bristol Britannia as the Type 175MR and Type 189 or the Avro 719. The DCAS, Air Marshal Sir Ronald Ivelaw-Chapman, in May 1953 stated that the role of the flying boat was in need of review and that a long-range maritime reconnaissance aircraft was required. This would be tailored for the new military need to concentrate force in north west Europe, the north east Atlantic and Mediteranean area, so an aircraft that could patrol for eight hours at 1,000nm (1,852km) and with a transit speed of 250kt (463km/hr) would be preferred, a requirement that a modified Britannia could meet. Apart from the eight hour patrols, the Avro 719 could meet this, but that limited time on patrol meant that, as with the R.2/48 case, more Avro 719s would be required. The DCAS went on to explain that developing a new flying boat would be an extremely expensive undertaking but if it was not developed, the flying boat industry would die. The UK had led flying boat development and the US was developing new flying boats for new roles, specifically the Martin P6M SeaMaster strategic bomber. DCAS concluded with 'In any event, if we propose to deal the British flying boat industry a death-blow, we shall certainly have to be ready to defend publicly our decision.'

Further to this comment by the DCAS, Air Vice-Marshal Tuttle, ACAS(OR) wrote that there were three options for the long-range maritime patrol aircraft: a developed Shackleton (the Type 719), a maritime Bristol Type 175 and the R.2/48 flying boat. Tuttle reiterated the case against the use of flying boats in the Atlantic and that the landplane would be a better option. Tuttle's unalloyed dislike of flying boats is summed up in his concluding paragraph: 'I can imagine nothing more embarrassing than to have one aircraft of this type needing all the resources that a flying boat needs and all the backing that one peculiar aircraft is bound to want. I would suggest that we make quite clear that we do not want anything to do with it

whatsoever, as it would be an embarrassment all our lives.'

Air Marshal Ivelaw-Chapman took the view that if a global war à la 1939 broke out, a flying boat would be required (that he uses 'required' rather than 'useful' suggests he was in the pro camp) but it was becoming obvious that in the current financial climate there would be no funds available for any new maritime patrol aircraft, never mind a flying boat. There was another reason for not developing a new maritime type at this time and in a report dated 28th May 1953 the DCAS explained it to the Air Council. 'There is no adequate technique in prospect for finding submerged submarines or for finding those snorting in rough seas. It would be premature to develop an aircraft which merely carries all the old and inadequate equipment slightly faster and further.' This was a prescient statement, as will be shown below when the procurement of the maritime Comet a decade later is described.

By July 1953 the advocates of the flying boat were fighting back. The MoS said that slender hull flying boats '…offered general improvements in performance and seaworthiness.' The MoS also voiced their concerns about loss of flying boat expertise, but it was the Chief Scientist (Air) who appeared to be most keen on flying boats and in a note to the Air Council on 17th September 1953 advised that '…for an aircraft with an AUW of 150,000lb or more, a flying boat can be developed with a slender hull to be equally as efficient as a landplane, so there's nothing technically between them.' The CS(A) continued that this new work had made the Saro SR.45 Princess obsolete and that research into slender-hulled flying boats should continue. Saro had by July 1953 prepared a report on the use of flying boats for ASW that described how by using the newly developed slender hull, a flying boat could operate in 'on the open ocean under a wide range of weather conditions.' This would allow a flying boat sortie to be extended by refuelling at sea and use dipping or towed sonar arrays which would allow the aircraft to detect and track '…deeply submerged and nearly stationary' submarines. Saro also pointed out that if the weather conditions prevented alighting on the sea and deploying the sonar array, the flying boat could still operate like a land-based aircraft. Oddly enough, thirty years on, such a dipping sonar system would be examined by Hawker Siddeley as the HS.828 described in Chapter Two. The flying boat that Saro described in that report was the Saro P.162, which is the only time this type is mentioned in a report or submission to the authorities in relation to R.2/48. Saro argued that the flying boat was the best anti-submarine platform available for the Fifties: it could operate on the sea with sensors as good as a ship and if the sea state prevented that, could still operate like a landplane.

The ACAS (OR) Air Vice-Marshal Tuttle countered this argument by stating that flying boats were on the way out and that there was no military reason for them and their continued development was a 'national issue' and that the only merit in a flying boat was that '…it could land in out-of-the-way places, but the new maritime focus is on North Atlantic.' Tuttle went on to recommend that '…the Air Council supports the view that no new flying boat be developed.' By the autumn of 1953 the Secretary of State for Air, Lord de l'Isle and Dudley, was agreeing with Tuttle and took the view that the RAF would need a replacement for its Shackletons and Neptunes and it would be a landplane. There was no point in buying two types and on 11th September 1953 OR.231 was cancelled.

The Naval and Air Staffs were by mid-1954 still prevaricating on the subject of flying boats and were reluctant to forego them. In late May 1954 ACAS(P) Air Vice-Marshal Edward Chilton presented the Air Council and Air Staff with some hard facts, specifically that the Shackletons *and* Sunderlands would need to be replaced no later than 1962 and that it would be prudent to start looking at that immediately. Chilton also advised that the overall research and development programme was oversubscribed and that there were no funds for a new maritime type or its Napier Nomad engine! The flying boat advocates fought back by informing all and sundry that a flying boat would be required if nuclear propulsion was to be developed and that if and when new anti-submarine kit was developed in the future, that might need to be carried by a flying boat. Those against flying boats pointed out that a flying boat was a nice to have type, but not a necessity, and that Canada was developing its own land-based maritime patrol aircraft based on the Bristol Britannia but this was proving very expensive. Canadair had by then proposed the CL-33, which was described as a 'Fat Lancaster' and powered by four R-3350-32W turbo-compound engines.

The arguments rumbled on until the 10th December 1954 when the Minister of Defence, none other than Harold Macmillan (with a brief to cut defence spending that would ultimately lead to the 1957 Defence White Paper), signed the death warrant of the flying boat in the RAF. Tuttle's view had prevailed and there would be no more flying boats for the RAF.

Another factor in the demise of the flying boat was progress made in the development of aircraft undercarriage and particularly the braking systems. As Air Commodore Millar's research had indicated, operating large aircraft was a somewhat dicey affair, with many aircraft lost due to running off the runway after brake failures or burst tyres caused by the wheels locking up under braking and the aircraft skidding out of control. As with any task, practice makes a difference and Millar had highlighted the difference in attitudes within the Commands to continuation training. One technology that revolutionised the operation of large aircraft were anti-lock brakes such as Dunlop's Maxaret, that allowed the aircraft to be controlled under braking conditions and made baulked take-offs and high AUW landings much safer than before. With much improved brakes and undercarriage, the large land-based aircraft became ascendant.

R.2/48 Flying Boat Design Conference

'This conference is of more than usual importance for upon its decisions rest not only the selection of the most promising tender, but the whole future of the flying boat.' H C B Thomas, AD/RDS, May 1949.

The Director (OR) asked for tenders for R.2/48 and received five, one each from Saro and Vickers Supermarine and three from Shorts. Three of the tenders were new designs, with the fourth being an upgraded Shorts Shetland and the fifth a re-engined Seaford. The Seaford was originally conceived as an interim type that gave the Sunderland a wider beam, lengthened hull and Bristol Hercules engines. Twelve Seafords were to replace Sunderlands until the Shetland was available and was therefore not considered a long-term prospect for R.2/48. One other reason for this was that to produce new Seafords, Shorts would need to close down their new Canberra production line and the first Seafords would not be available for at least two years from Instruction to Proceed (ITP). However, the Seaford formed the basis of the six Solent airliners for BOAC and it was these that the Air Staff and Shorts were looking at as a cheap, if not free, source of airframes for conversion to meet R.2/48 as the Shorts PD.3.

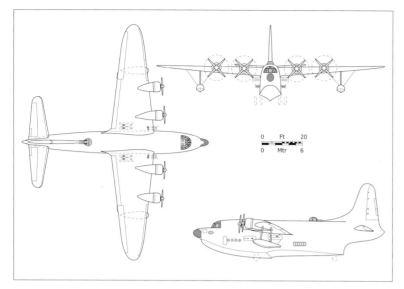

0 Ft 20
0 Mtr 6

Shorts' Seaford for R.2/48 was too small and offered little improvement over the Sunderland it was to replace. Nor could it carry the large scanner for the H2S Mk.9 ASV radar or the Pentane torpedo. Its only advantage was that it could be converted from ex-airline Solents that could be obtained free of charge.

ing engines, Rolls-Royce's Compound and Turbo Griffons. These were deemed to be 'very complicated and with many gadgets that were open to petty u/s components' by 'gadgets' the committee meant turbochargers and power recovery systems. Interestingly Rolls-Royce had ended development of the Compound-Griffon in March 1949 because they could see 'no useful application'. The Turbo-Griffon had shown better specific fuel consumption than the cancelled Compound-Griffon, but the committee observed that with the Turbo-Griffon '…too high a proportion of the total power available was used in cruising conditions. The engine would therefore tend to be unreliable.' The Air Staff were concerned that its lower power rating might require six to be used on the heavier versions of the R.2/48 designs. Then there was the Napier Nm.3 Nomad I, which shared the attributes of the Compound-Griffon. However, unlike the Compound-Griffon, the Nomad was based on an untried reciprocating design.

The powerplants were subject to great scrutiny, with the Air Staff being most concerned about reliability more than any other factor. At a pre-tender meeting on 25th July 1949, the various powerplants were assessed. Firstly the Bristol Proteus was considered and received a very favourable verdict in that it was relatively simple, was in a reasonable state of development and 'free from gadgetry'. Bristol's big radials, Centaurus and Hercules could not be fitted with a contraprop and so needed a larger diameter propeller and therefore a higher wing mounting to maintain propeller clearance. In fact the Bristol radials were '…considered to be a retrograde step', but were still looked upon more favourably than the other two reciprocat-

The Air Staff were also concerned about the Nomad's reliability, availability of diesel fuel and its potential for waxing in Arctic and Antarctic zones. Added to this was the complication of fitting reversible pitch propellers. Saro opted for fitting reversible propellers on the inboard engines only as experience had shown that this gave best manoeuvrability on the water. By applying power on an outboard engine and reverse pitch on the opposite inboard engine, the flying boat could make very sharp turns, even about a point within the plan of the hull.

The Seaford/PD.3 airframes would have been more or less free of charge but were not particularly suited to the R.2/48 specification. This was mainly because it was too small, lacked range and payload. Like the Sunderland it was to replace, it would have been just as obsolescent in 1960.
RAF Museum

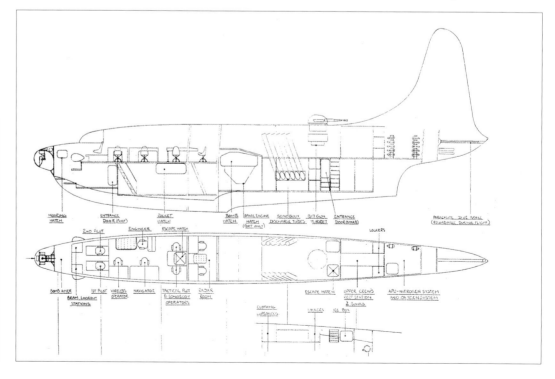

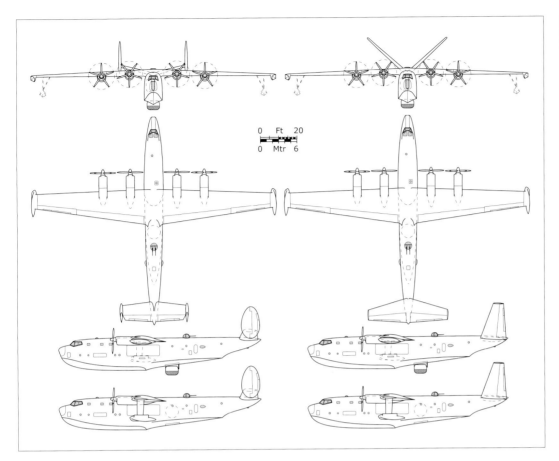

The Saro P.104/3 was an advanced design that incorporated the latest thinking on long, narrow hulls. The butterfly tail on the P.104/4 saved structural weight and reduced drag while keeping the tail surfaces clear of spray from the hull on take-off. The Air Staff deemed the P.104/3 the winner but the Ministry of Supply had other ideas.

The designs were scrutinised in depth, with Saro's praised for their butterfly-tailed P.104-4 and adoption of the most recent developments in hull designs, particularly the long/narrow hull shape and dead-rise that made for better high-speed planing ability, handling in choppy waters and much reduced spray at low speeds. Shorts' R.2/48 design was criticised for its lower length-to-beam (L/B) ratio of 7 (The Saro R.2/48 had a L/B of 9) with Shorts' response being that the narrow fuselage was unsuitable for conversion to passenger use. Summing up the hull designs, the committee stated that 'Saro have shown the most practical approach, Vickers (Supermarine) have shown little imagination but Shorts are still apparently working on the assumptions of ten years ago'. Saro were also commended for their wing structure that used a box beam rather than spars and was thus 500 lb (187kg) lighter than the sparred wings of Shorts and Vickers. Saro had obviously learned a great deal in the period since the A.33's failure.

Equipment-wise, the main concern was weapons load and the ASV radar. The weapons were to include a minimum of 4,000 lb (1,493kg) of mixed stores carried internally and comprising up to sixteen 500 lb (187kg) general purpose bombs, up to four Zeta torpedoes (14ft/4.26m long, 14in/35.6cm in diameter and weighing in at 2,000 lb/907kg). Up to 24 sonobuoys and 24 marine markers were to be carried. The ASV radar was to meet OR.3507 and probably based on the existing ASV.13 or more likely, the H2S Mk.9A, but it was the antenna size and placement that exercised the Air Staff who wanted a 6ft (1.8m) antenna in a 7ft (2.1m) radome. Saro and Vickers considered a retractable ventral radome fitted in a water-tight compartment aft of the step the best con-figuration, a view shared by the TRE. Shorts favoured a nose mounting as it allowed easier servicing and good coverage if the aircraft was in a dive towards a target. The lack of 360° coverage from the nose radome was to be addressed by installing a second rearward scanning antenna in the tail, but TRE foresaw problems synchronising the two sets. As with many of the UK's airborne early warning radar studies, the fore and aft scanner system was seen as an ideal solution to the airframe mask-ing problem.

As for the conversions of the existing Shorts types, the Shetland, at an AUW of 125,000 lb (56,689kg) was too big and, lacking the required endurance, needed extra fuel tank space that could have been used for carrying weapons. The modified Seaford was too slow and did not meet the range and payload

Shorts' novel 'flipping' radome for the H2S Mk.9 ASV radar. Based on the V-bomber radar, this possessed a capability on a par with the American AN/APS.20. The scanner was mounted on one side of a longitudinal shaft, with a hull planing surface on the other. The assembly rotated on the shaft to fit the hull shape for take-off, rotating to expose the radome in flight. *RAF Museum*

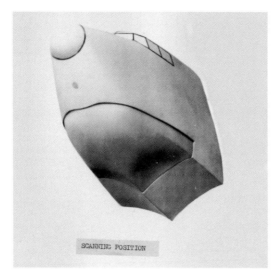

SCANNING POSITION

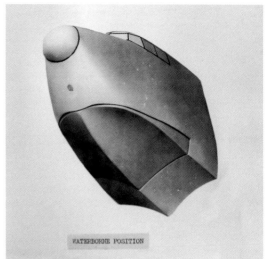

WATERBORNE POSITION

requirement, nor could it carry Pentane torpedoes internally. The Air Staff took the view that ordering a ten year-old design in 1950 would mean that it would be very out of date by 1963,

no matter what upgrades Shorts applied. The Shorts Shetland and Seaford were dismissed on the grounds that they did not meet the performance requirements, such as the 200kt (370km/h) transit speed and the carriage of the Pentane torpedoes under the engine nacelles rather than internally, which also failed to impress. Nor did fitting the radar scanner in a rotating bay in the bow of the planing bottom fill the Air Staff with confidence '…since heavy loads are imposed on the bow planing bottom in rough seas and in nose-down landings.' In fact the conclusions of the tender design conference state that the modified Shetland '…could never be more than an unsatisfactory makeshift.'

At the other end of the spectrum from the Seaford was the Shetland that Shorts tendered for R.2/48 as a cheap alternative to a clean-sheet design. The Air Staff said it was too big, too heavy and only carried two Pentane torpedoes (in the extended inboard engine nacelles) rather than the four required.

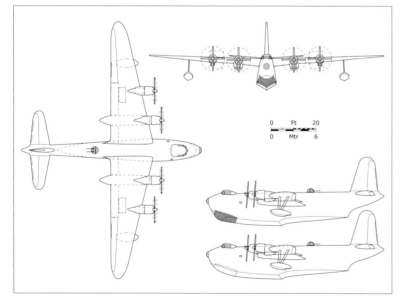

0 Ft 20
0 Mtr 6

The Shetland was too big and too expensive. Unlike the Seaford it could carry the H2S Mk.9 radar in the 'flippable' chin installation. The Air Staff disapproved of its carrying stores in the engine nacelles, a criticism that Shorts did not appear to address. *RAF Museum*

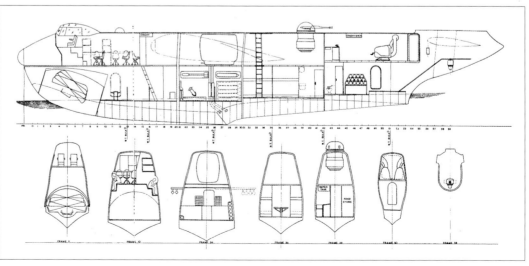

R.2/48 Tender Design Conference

The final design conference on 25th July 1949 concluded that the Saunders-Roe R.2/48 was the best design, if fitted initially with Centaurus engines and ultimately with Nomads, followed by the Vickers Supermarine R.2/48 (Type 524) and then the Shorts PD.2 to R.2/48. Shorts' modified Shetland and Seaford tenders were dismissed, mainly due to failing to meet the performance required, particularly transit speed and endurance. The Blackburn B-78 was not officially tendered for R.2/48 and is not mentioned in Air Staff or Air Ministry documents.

In a memo dated 6th July 1949 to the DOR, Squadron Leader M C Raban, DDOR1, summarised the findings of the conference. The Shorts PD.2 design study with Nomad or Centaurus weighed in at 105,000 lb (47,619 kg) and was criticised for its engine nacelle weapons stowage with respect to difficulty loading the four bays from a bomb scow (a shallow draught barge used to carry weapons to the boat) in adverse weather. The ASV scanner had limited coverage and its location in the bow of the planing bottom was seen as a potential problem. The Shorts 'Shetland Development' at 120,000 lb (54,422 kg) with Centaurus or Nomad engines, was too heavy and suffered from the same weapons stowage problems. As a result, it failed to meet the requirement, particularly in its range and endurance.

Vickers Supermarine Type 524, with an AUW of 103,200 lb (46,803 kg) when powered by Nomad, 106,400 lb (48,253 kg) with Centaurus, 117,500 lb (53,288 kg) with Proteus turboprop '…looks clean and attractive externally.' Oddly it was criticised for lacking a drying cupboard which Raban saw as '…a small point but I consider an important one when crews are expected to carry out long patrols and possibly be detached from their bases…' This might sound like a very odd criticism for an aircraft, but flying boats carried a great deal of equipment that was peculiar to their types. Mooring equipment and bilge pumps added weight and the messing facilities, crew rest bunks and 'settees' listed for

The obvious problem of dropping bombs from the water-tight hulls of flying boats was overcome with some innovative, and complicated, mechanisms. Supermarine utilised cradles on tracks, Saro used swing-out panels and Shorts ignored the Air Staff and fitted the Pentanes in the engine nacelles on the Shetland and PD.2. They were criticised in particular for the underwing bomb trolley on the Seaford that could not carry Pentane torpedoes.

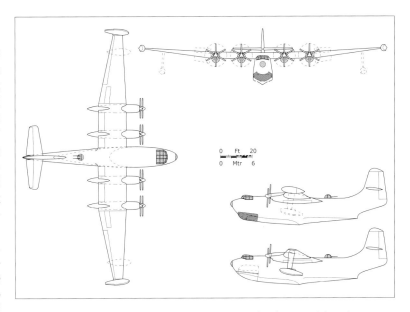

Shorts PD.2 to meet R.2/48 with Napier Nomads. Shorts' wish to have a flying boat that could be used as a passenger transport was its downfall as the Air Staff disapproved of the extra structural weight and wasted space in the deep hull. The Air Staff were particularly annoyed by Shorts ignoring their guidelines on internal weapons carriage.

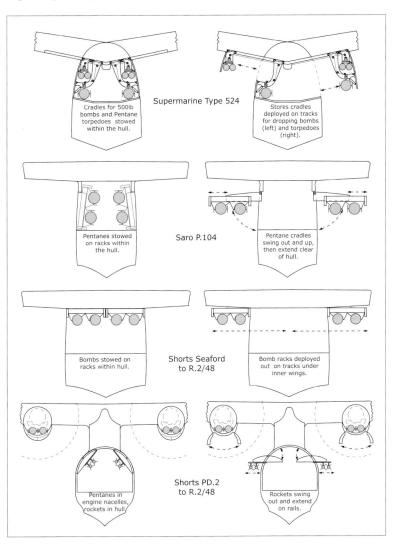

Supermarine artwork showing the Proteus-powered Type 524 in flight. A highly-streamlined design, it was fitted with a retractable turret, ASV scanner and tip floats. The Air Staff were impressed with the Type 524. Alternative powerplants included the Centaurus, Nomad and Turbo-Griffon. *RAF Museum*

the Supermarine Type 524 added 575 lb (261 kg). The equipment list did include a kitchen sink and, given the fate of a Shackleton MR.3 struck off charge due to rodent damage; the possibility of a ship's cat has to be considered.

The Saunders-Roe (Saro) P.104 was lightest at 89,900 lb (40,770 kg) and was to be powered by the Compound-Griffon, but 'The standard Griffon could be easily substituted but would result in a reduction in range.' It was, according to Squadron Leader Raban, an attractive design, but its handling on beaching gear was thought to be too restricted, with a 50ft (15.2m) minimum turning circle. Squadron Leader Raban advised 'I am in favour of adopting the Saunders-Roe design.' Raban went on to state that the Shetland '…is cheaper in the prototype stage, but cost of production is more expensive than any other design.' Raban fin-

ished by saying that the Supermarine and Saro designs were the most attractive '…and have chosen Saunders-Roe as it is by far the less expensive. Job done? Well, no, as three weeks later Raban was considering the firms' latest tenders, this time powered by Turbo-Griffons!

One change forced on all the firms by the Turbo-Griffon was a reduction in AUW (typically around 9%) due to the reduced engine power compared with the Nomad with the Shorts PD.2 changing from 105,000 lb (47,619 kg) to 96,000 lb (43,537 kg). This weight reduction was achieved by adopting a shallower hull with the wing being mounted on a pylon *à la* Consolidated Catalina. This also produced a single deck interior that used the space more efficiently. Unfortunately for Shorts, Squadron Leader Raban, had the same criticisms as for the earlier PD.2, and also commented on the lack of

Compared with the Shorts bids, the Type 524 hull appears cluttered but Supermarine used the available space much better than Shorts with the PD.2. This was because it was designed from scratch as a warplane rather than as a potential airliner. *RAF Museum*

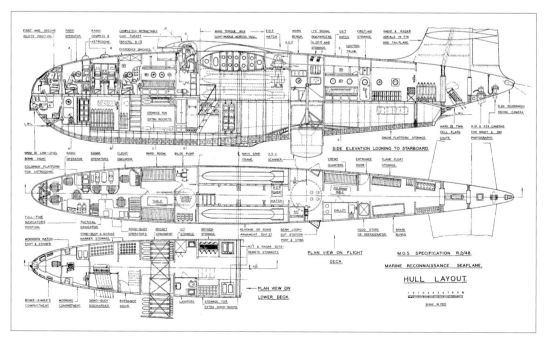

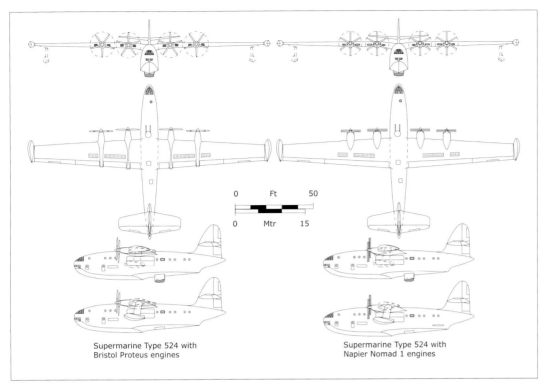

0   Ft   50

0   Mtr   15

Supermarine Type 524 with
Bristol Proteus engines

Supermarine Type 524 with
Napier Nomad 1 engines

Supermarine's Type 524 was a very clean design, complete with retractable radar and dorsal Bristol B17 turret. The 524's weapons carriage was quite ingenious, allowing different weapons to be deployed depending on the target. Shown here are the Nomad and Proteus-powered studies. There were also studies that used the Centaurus and of course, the Turbo-Griffon.

a drying cupboard. The Supermarine Type 524 now weighed in at 93,700 lb (44,126 kg) and was externally the same, but with all dimensions reduced by approximately 6%. Raban's earlier comments on the type were unchanged. The Saro type was *increased* in size, with the AUW rising to 94,250 lb (42,744 kg) and Raban was pleased to note that the radar scanner mechanism had been rearranged to allow it to be worked on within the hull.

As the years rolled by, the R.2/48 flying boat tender process continued. The discussion had by January 1951 moved to whether a smaller boat was better for operations from the UK and a bigger boat for overseas. In fact the argument was going against the flying boat as landplanes, such as the Shackleton, were gaining fans in Coastal Command (at least with the higher echelons, not the aircrews). Then in April 1951 BOAC announced they were selling their Shorts Solents and since they actually belonged to the Ministry of Aviation, the RAF had first refusal on them. Unfortunately these, which were based on the Seaford, were pretty much a short-term measure. 'Seaford is short range and short term, R.2/48 is long range and long term' was how it was summed up in a brief for the VCAS

When the type of reciprocating engines was changed, the airframe had to be tailored to suit the engine. This Saro drawing shows the changes forced upon the P.104 by changing from the Nomad to the Turbo-Griffon. *RAF Museum*

Saro artwork showing a P.104/3 with 'everything out' displaying the weapon racks deployed under the wings, 3in rocket rails on the forward hull and the stabilised, retractable radome for the H2S Mk.9 radar. For the R2/48 studies Saro applied the most up-to-date thinking on flying boats, but the big boats had had their day. *RAF Museum*

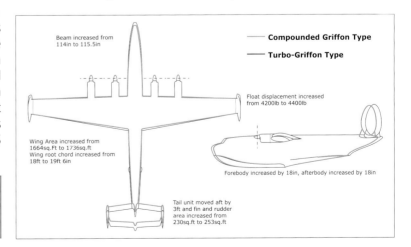

Beam increased from 114in to 115.5in

—— Compounded Griffon Type

—— Turbo-Griffon Type

Float displacement increased from 4200lb to 4400lb

Wing Area increased from 1664sq.Ft to 1736sq.ft
Wing root chord increased from 18ft to 19ft 6in

Forebody increased by 18in, afterbody increased by 18in

Tail unit moved aft by 3ft and fin and rudder area increased from 230sq.ft to 253sq.ft

Air Chief Marshal Sir Ralph Cochrane prior to a meeting on 2nd May 1951. During that meeting, Cochrane, evidently not a fan of the flying boat and usually associated with Bomber and Transport Commands, appeared to play devil's advocate and more or less demolished the case for the flying boat, asking at one point 'What is the present place of the flying boat within the structure of the Royal Air Force?' Further into the meeting, Cochrane sided with the flying boat protagonists and by the end had opted to allow 'the technique to survive'. But in reality the flying boat was viewed increasingly as an expensive anachronism associated with Empire at a time when the Empire itself was being viewed as anachronistic.

The End of the Flying Boat

The debate on the future of flying boats in the RAF, never mind whether a new type should be developed, continued into the late 1950s. Essentially only the Admiralty and the members of the General Staff concerned with operations in the Far East voiced any concern for the flying boat's demise. In late 1951 the Sea/Air Warfare Committee Policy and Plans sub-committee drew up a paper, *Replacement of the Sunderland Flying Boat*. The flying boat lobby pointed out that the Shackleton could not fill the gaps from island bases in the Indian Ocean and had a map to back that up. The landplane faction of the Air Staff cried 'Foul!' and had their own people examine the map and performance figures for evidence of 'creative accounting'. In a scribbled response to this paper addressed to the DDOR.1, the Air Staff's Director of Policy Air Commodore J G W Weston stated 'Please get your chaps to check the radius figures for me. Surely the Mk.2 Shackleton will do more than 650nm. What a horrid paper. I do hope we do

not subscribe to it?' Weston wanted his results within three days. A surviving copy of the paper contains a great deal of notes and underlining, in bright red pen. The last paragraph asks for recommendations on the flying boat matter, but under that portion of typewritten text the following is scribbled in red ink: 'A dishonest paper written by someone who is determined to have a flying boat regardless of fact.' By 1954 the thrust of that paper – coverage of the Indian Ocean and West Africa – was irrelevant as the focus of operations had moved from the Far East to Europe and the Mediterranean, where landplane operations were more promising.

The Air Staff in 1951 advised that there were sufficient Sunderlands to allow operations until 1955 at least and late 1958 if the reserve stock was utilised. The flying boat supporters stated that flying boats were necessary for operations off the African coast, to patrol the trade routes and the Indian Ocean/Far East regions to deal with any Chinese submarine threat. The ideal aircraft was, as laid out in R.2/48, essentially a flying boat with an AUW of around 130,000 lb (58,957kg) powered by four of the efficient Napier Nomad or Bristol Centaurus engines. When challenged on this and asked about alternatives, the VCAS stated that only a newly-developed type could meet the specification as the alternatives such as the Shorts Seaford, Martin Mariner or Marlin were too small. Oddly enough the contract for R.2/48 had been awarded to Shorts in 1950 for the PD.2, despite the Air Staff having concluded that the Saunders-Roe P.104 was a better design and the most efficient way to meet the task! Such are the machinations of the MoS, who took the view that Saro did not have the capacity to build the R.2/48 type.

By late 1951 the debate on flying boats operating around the UK was focused on the 'expensive runways' aspect of fielding land-based MR

The need for R.2/48 in the Far East and Africa is apparent from this map, as is the difference between the estimated and actual endurance of the Shackleton, just over 20%. The landplane advocates cried 'Foul!'

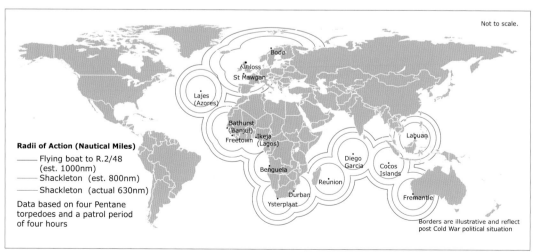

Radii of Action (Nautical Miles)

―― Flying boat to R.2/48 (est. 1000nm)
―― Shackleton (est. 800nm)
―― Shackleton (actual 630nm)

Data based on four Pentane torpedoes and a patrol period of four hours

Not to scale.

Borders are illustrative and reflect post Cold War political situation

aircraft and the impending acquisition of the Lockheed Neptune MR.1 from the USA. Coastal Command was expanding and the additional Shackletons this required could not be met by Avro, who were building Canberras and developing the Vulcan. At this time a major re-armament programme was under way, prompted by the Korean War and a burgeoning Soviet threat. That this threat was centred on the European theatre shifted the emphasis of Coastal Command operations onto the North Atlantic and the Mediterranean. This meant that land-based aircraft could meet the needs of Coastal Command and the 'expensive runways' argument was nullified because the forthcoming Valiant, Victor and Vulcan bombers tasked with carrying the deterrent would rely on runways and these would be built in a variety of locations around Britain to allow dispersal of the deterrent carriers. Interestingly the US Navy was at this time developing the Martin P6M SeaMaster, a four-jet strategic bomber and patrol aircraft to a naval requirement and on getting wind of this the flying boat supporters in October 1952 began discussing a jet-propelled replacement for the Sunderland. As it happened, Saro had such an aircraft on the drawing board in the shape of the P.131.

The Duchess for R.2/48

Saunders-Roe (Saro) proposed a version of their P.131 Duchess jet-powered flying boat airliner as an anti-submarine/transport, but with many, quite radical changes from the passenger aircraft. The original Duchess was a sleek swept-wing flying boat with an all-up-weight of 130,000 lb (58,967 kg) and powered a six de Havilland Ghost turbojets, rated at 5,000 lbf (22 kN) apiece. Aimed at airlines that were operating flying boats such as the Solent, Saro quickly realised that the airline market for flying boats was shrinking but the type was suitable for conversion to a military role.

Saro approached the Air Ministry in October 1951 with a scheme to modify the Duchess design in what could have been construed as modular form. This involved a pair of wing centre sections, the first echoed a development of the civil airliner being offered to Tasman Empire Airlines Limited and substituted the six DH Ghosts with the same number of Rolls-Royce AJ.65 Avons and replaced the conventional swept tail surfaces with an elegant butterfly tail. This was seen as an ideal type to use in the military transport role, effectively a minimal change version of the civil airliner. The RAF had

used the Consolidated PB2Y-3B as the Coronado GR.1 in a similar manner during World War Two, mainly because the Coronado's endurance in the anti-submarine role was poor in comparison with the Consolidated Catalina and Shorts Sunderland. This no doubt had some bearing in the formulation of the range requirement for R.2/48 and the development of the Napier Nomad engine.

The Avon Duchess could carry a maximum payload of 25,000 lb (11,340 kg) or 106 fully-equipped troops over a distance of 2,430 nm (4,500 km) which was a respectable range for a jet transport. The second proposal saw a new wing centre section carrying four Napier Nomad compound engines for maximum economy in the maritime patrol role. If fitted with radar, turrets, anti-submarine equipment and a weapons bay, the Nomad Duchess would be capable of patrols of up to twenty three hours, if a double crew was carried, or in the transport role the aforementioned payload for 5,700 nm (10,556 km). Saro pointed out that in its Nomad-powered guise the Duchess would meet R.2/48 with ease. The Air Ministry and MoS Directorate of Military Aircraft Research and Development (DMARD) had by January 1952 examined Saro's brochure on the Duchess developments and were less than impressed, their response being that 'The use of a flying boat for a military transport has been ruled out by the Air Staff'. As for its use for maritime

A thoroughly modern flying boat. A Martin P5M Marlin of VP-49 banks away from the camera to show its key features. These include high aspect ratio wings, curved step on the narrow beam hull, weapons bays in the engine nacelles and the T-tail. The MAD 'sting' is mounted at the top of the fin and the radome in the nose. The French *Aéronavale* replaced its Sunderlands with Marlins and Breguet proposed the similar Br.1250 for NBMR.2. *Terry Panopalis Collection*

The elegant Saro P.131 Duchess was a large airliner intended for Empire routes. The de Havilland Ghost turbojet was the original engine, with six installed, but by 1952 the Roll-Royce Avon was preferred as it was more powerful and more efficient. *Author's collection.*

By swapping its Avon turbojets for Nomad compound diesels, Saro considered it suitable for trooping duties and the R.2/48 specification. The Air Staff were looking for something a bit smaller and bespoke as they had already despaired of Shorts' determination to design the PD.2 type with an eye on passenger service.

patrol, DMARD stated that the '...future of flying boats is now under review by the Air Staff.' but that if a flying boat was required '...it will almost certainly be in the form of a fresh submission by Saunders-Roe.'

So in that concise, scribbled note appended to a brochure receipt, the Air Staff's thinking on flying boats was laid out. There would be no new boats, but that did not stop the companies attempting to develop one, nor did it stop the Air Staff from ordering its development.

The proponents of land-based MR aircraft were looking at replacing the Sunderlands and Shackletons with a single type based on the Britannia, Bristol's Type 175MR powered by Centaurus engines or the Nomad powered Type 189, with Avro looking at the Shackleton-derived Type 716 or 719. The Canadians were also looking at a modified Britannia for their ASW needs, a study that ultimately became the Canadair CL-28 Argus. In the light of this the

DCAS in May 1953 decided that R.2/48 needed revising to reflect current thinking and optimise it for the UK-oriented needs of Coastal Command and considered eight hour patrols at 1,000nm (1,852km) from base to be feasible, as was a 250kt (463km/h) transit speed. While the flying boats could meet the endurance requirement, the transit speed was unattainable without jet power. The Shackleton MR.2A (later renamed MR.3) could reach the 1,000nm (1,852km) patrol range but was limited in the time it could spend in the area, which in turn necessitated a larger Shackleton force. The case for a mixed MR force of flying boats and land-planes was becoming less attractive, particularly from the financial aspect, so the Air Staff began to consider the replacement for the Shackleton and Sunderland and that should happen before 1962.

By April 1954 the end of the flying boat was looking inevitable, with replacement by land-based aircraft on the cards. At this point the Admiralty got wind of the Air Staff's plan and were not at all pleased about it. At a Chiefs of Staff Committee meeting on 15th September 1954, the First Sea Lord, Sir Rhoderick McGrigor, said that '...the Admiralty was most anxious that flying boats should be kept in existence because of their great flexibility and most useful role in the maritime aspect of the Cold War'. In response to this, the Chief of the Air Staff, Marshal of the RAF Sir William Dickson said that he agreed with His Lordship, but explained how the RAF could not afford a flying boat development programme and added '...that in the Mediterranean and Atlantic, land-based aircraft were possibly superior since they had a better weapons carrying capacity.'

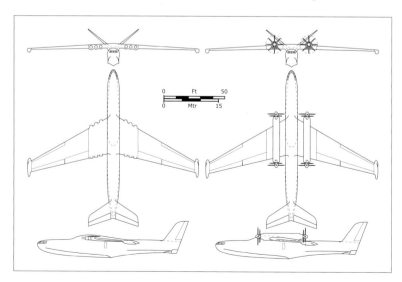

The Chief of the Imperial General Staff Field Marshal Sir John Harding added his thoughts on flying boats in general and said that he '…did not believe that the flying boat was of any great value in assisting army/air operations in the Cold War."

It had been noted that New Zealand had shown an interest in R.2/48 and the Air Staff's pro-flying boat faction also hoped to interest Australia but in reality the R.2/48 was too big for the RNZAF (who wanted a 75,000 lb/ 34,019kg aircraft) and the RAAF had little interest in flying boats. The hope was that the antipodeans would chip in and help fund development of R.2/48, with Defence Secretary Harold Macmillan on 4th December 1954 stating '…the question of asking them to make some contribution towards that cost of developing a new flying boat might be considered.' The question of costs was close to Macmillan's heart, tasked as he was with reducing the cost of the armed forces, a task that would lead, via the Suez Crisis, to that mass extinction of the UK aircraft industry at the hands of Duncan Sandys. Macmillan noted that the cost of defence research and development was rising annually at a rate of £5m per year to maintain the current level of activity. The level of R+D activity would need to rise as the complexity of weapons grew and as a service with an increasingly complicated inventory (supersonic fighters and bombers, high-power radars, SAGWs and ballistic missiles) the RAF's share of the defence budget had to rise just to maintain the 1955/56 levels. Using £4m of that budget to fund a flying boat, whose utility in any coming war was questionable, was not an option, there were other projects that had higher priority and hence funding allocations than a flying boat. In a move that foreshadowed the cancellations of April 1957, Macmillan wrote 'Though we would like to see flying boat development continued we feel that this is desirable rather than essential and we cannot, therefore, agree that such development should be supported from the Defence Budget.' Macmillan left the door open for such development to continue under civil funding, for a nuclear-powered aircraft or transport aircraft, but the military were out of the flying boat development business.

By mid-January 1955 the ACAS, Air Vice-Marshal Edward Chilton, in a brief for the Secretary of State for Air, exploded the myth that the Far East lacked airfields by writing 'Fairly evenly distributed over the Far Eastern area there are 12 airfields now suitable for either one or both variants of Shackleton and 2 more projected by 1956 and 1958 respectively. These permit a much more effective maritime patrol cover by landplanes…'

On 14th January 1955 the Cabinet reviewed the findings of the Chiefs of Staff Committee and heeded Macmillan's advice on R+D funding and the Cabinet minutes stated '…resources should not be applied from Defence Votes to the development of a new type of flying boat, whether for military or civil use.' Britain was out of the flying boat business for good and No.205 Squadron conducted the last operational sortie by an RAF Sunderland on 15th May 1959, from RAF Seletar in Singapore. The RAF's relationship with the flying boat was over. As of 2014 a handful of air arms operate amphibians (seaplanes with undercarriage that can alight on land or water) such as the Canadair CL-215 mainly in the firefighting role. The flying boat still exists in the Far East with the Japanese Maritime Self Defence Force and their Shin Meiwa PS-1 and US-1, direct descendants of the Kawanishi H8K *Emily* that was described as the most advanced flying boat of World War Two. The PS-1 is a flying boat with retractable beaching gear while the US-1 is an amphibian and both benefit from the advances in flying boats that arose in the immediate postwar era. The Chinese People's Liberation Army Air Force flies the Harbin SH-5, which while described as amphibious, has retractable beaching gear rather than true undercarriage. The Russian and Ukrainian Naval Aviation forces have a few Beriev Be-12 *Mail* amphibians and the Beriev A-40 *Mermaid* and its derivatives have been under development for years, but the true flying boat today is a very rare beast. The last of the wartime big boats, the gigantic Martin JRM-3 Mars, ended its long career fire-fighting in British Columbia.

What must be taken into consideration with any analysis of flying boats and seaplanes is the massive leap in the capability of landplanes in the period from 1939 to 1945. Before the war seaplanes held many of the range and speed records for aircraft with one reason for this being that by operating from water, they could have a much higher landing speed than landplanes. As a result seaplane wings could be optimised for higher speeds than landplanes whose wing designs were compromised by the need to have slower landing speeds to allow alighting on fixed length fields and further complicated by the poor efficiency of existing braking systems. By the end of the war, developments in high-lift devices, braking, tyres and of course, the proliferation of long runways around the world, saw the advantage turn to the landplane.

The 1950s should, or possibly even would, have been the heyday of the flying boat. Wartime work by the American companies such as Martin and Convair plus Saro in Britain had shown how the gas turbine, narrow beam hull and planing surface developments improved the efficiencies of flying boats. The new technologies that improved the safety of landplanes could have been applied to flying boats, but the ancillary services associated with them; tenders, quays and buoys ensured they could never prosper. As the Air Ministry's Chief Scientist feared, the British, then the Americans ultimately lost their expertise in seaplanes, leaving the Soviets, and subsequently the Russians, to carry on with their amphibians and applying that expertise to the *Ekranoplan*. In the West, save for floatplane conversions of small aircraft, the knowledge appeared to be lost by the mid Fifties. Then in 1954, Dornier resumed research and development of flying boats at Freidrichshafen, preserving the art of the flying boat designer. Dornier designs were considered the most beautiful of all flying boats and the sight of a Dornier flying boat on the step is indeed a joy to behold, a sight that can still be seen on Lake Constance in the shape of the Dornier Do 24ATT. The final chapter of the flying boat story was written in 1958 when Dornier, Breguet and Saro tendered the last flying boats for a Western military requirement, NATO Basic Military Requirement No.2 (NBMR.2). They were dismissed as anachronistic.

Having spent a decade searching for a Sunderland replacement, time had caught up with the Air Staff and by 1955 their sights were on a replacement for the Shackleton as well. This hunt would last almost as long.

Economy over the Ocean – Engines

The aircraft designers whose proposals were put forward for R.2/48 looked to Britain's aero-engine builders for their powerplants. Rolls-Royce was ready with the established Griffon, plus two new Griffon variants under development to provide the extra power and efficiency required while Bristol Engines pitched their proven Centaurus and the forthcoming Proteus turboprop. Then there was D. Napier and Son. Their Sabre had powered the Typhoon and Tempest fighters, but Napiers had other ideas in the propulsion field. The story of the R.2/48 flying boat is interwoven with that of large aero-engines as R.2/48 was issued at a time when high-power aero-engines could have taken any of three routes – reciprocating, gas turbine or compound.

Bristol Power for Flying Boats

Bristol's Centaurus was a powerful, 2,625hp (1,958kW), two-row, eighteen cylinder, sleeve-valve radial whose numerous variants powered aircraft as diverse as the Hawker Sea Fury fighter and Blackburn Beverley transport. The Centaurus followed an established design evolution from the Jupiter engine of 1918 through the Perseus and Hercules, all of which were designed by Roy Fedden.

Shorts were already flying the Centaurus in their Shetland and Solent flying boats, variants of both being proposed for R.2/48 while the PD.2 clean-sheet design studies were to be fitted with the Nomad or the Centaurus, with the brochures stating that the engines should be installed '…in the form of a complete power egg'. This method had become commonplace

The last of the true flying boats: the massive Martin JRM-1 Mars. Designed as a transport for the Pacific War, the Mars ended its days as a water-bomber fighting forest fires in North America. *Author's collection*

during World War Two and involved aero-engines and their ancillaries installed as a single unit that attached to a firewall on the airframe with a minimum of standardised connectors. This technique allowed for quick engine changes and opened the possibility of fitting alternative engines with little redesign of the nacelle. This desire to fit alternative engines arose quite early in the R.2/48 story when Air Commodore G W Tuttle, Director of Operational Requirements queried the MoS on their being '…anxious for us to authorise an increase of approximately 20,000 lb in the all-up-weight so that it can take the Napier "Nomad" engine'. So here was the MoS pushing the Air Staff, who wanted more numerous smaller flying boats, to specify a bigger boat to accommodate an engine whose future was, at the time of writing his letter on 20th January 1949, somewhat uncertain.

Interestingly Bristol Engines proposed their Proteus turboprop for R.2/48 types. A contemporary of the Nomad, its development had begun in September 1944 and was one of the first engines to use a free turbine. This referred to the main engine acting as a gas generator, with a separate turbine section on a second co-axial spool being connected to the propeller, thus the free-turbine and propeller were driven by gas flow from the gas generator rather than gearing.

Japan has a long and distinguished history in flying boats and continued this with the Shin Meiwa US-1. The US-1 used the latest technology: narrow hull, turboprop engines, boundary layer control and spray suppression. Shin Meiwa Industries arose from the Kawanishi Aircraft Company, famous for the H8K *Emily* flying boat and the N1K-J *George* fighter that was derived from the N1K *Rex* floatplane.
Author's collection

Engines for Flying Boats

Engine	Configuration	Capacity	Output
Merlin (620)	V-12	27L/1648 cu in	1,175hp/876kW)
Griffon (57)	V-12	37L/2258 cu in	1,960hp/1,460kW
Turbo-Griffon	V-12	37L/2258 cu in	2,850hp/2,125kW
Compound-Griffon	V-12	37L/2258 cu in	2,500hp/1,864kW
Eagle	H-24	46L/2807 cu in	3,200hp/ 2,387kW
'Falcon'	H-24	Est. 100L/6102 cu in	Not known
Centaurus	R-18	53.6L/3271 cu in	2,405hp/1,793kW
Hercules	R-14	38.7L/2362 cu in	1,356hp/1,012kW
Nomad I	H-12	41.1L/2508 cu in	3,000hp/2,200kW)
Nomad II	H-12	41.1L/2508 cu in	3,150hp/2,344kW)
R-3350	R-18	54.9L/3350 cu in	3,400hp/2,536kW
Proteus (705)	Turboprop	N/A	3,320hp/2,475kW plus 1,200 lbf (5.33kN) residual thrust giving 3,780eshp

The Last Rolls-Royce Recip.

Rolls-Royce had ended the war with a number of high-power reciprocating engines under development including the Eagle, a 46 litre 24-cylinder H-configuration, 3,200hp (2,387kW) monster used on the Westland Wyvern and the Crecy, a V-12 two-stroke 'sprint' engine for interceptors that lost out to the new gas turbines. One other Rolls-Royce engine that, like Eagle and Crecy, was beaten by the gas turbine was the Falcon, a massive 24-cylinder H-configuration 100 litre engine. The Falcon is briefly mentioned in the Air Staff documentation, and there may also have been an intermediate design that used Merlin 'pots' – cylinders and pistons – that would have had a capacity of around 54 litres, but evidence for either of these engines is sketchy. In fact the Turbo-Griffon is called Falcon in the minutes of the Tender Design Conference in July 1949 referring to it as 'Turbo-Griffon, now to be known as the Falcon'. Throughout the war, Rolls' bread and butter was the Merlin and latterly the Griffon. Developed at the request of the Fleet Air Arm for the forthcoming Fairey Firefly, the Griffon was adopted with much success by Supermarine for the Spitfire and Seafire. Most applications of the Griffon were single-engined fighters and attack aircraft, but the Griffon was also to be fitted to the R.5/46, the Avro Lincoln ASR.3 that ultimately became the Avro Shackleton GR.1/MR.1. The Griffon 57 was rated at 1,960hp (1,460kW) and was fitted with contra-rotating propellers.

By 1946 there was a perceived need for a more powerful and more efficient engine for patrol and transport aircraft and, having witnessed the success of the Pratt and Whitney R-3350 on bombers and transports, followed by the Turbo-Compound version for the Lockheed Constellation and Douglas DC-7, Rolls-Royce decided to develop similar technology based on the Griffon. The first was the Compound-Griffon, which is thought to be similar in operation to the Napier Nomad in combining the V-12 Griffon with axial turbomachinery, for an estimated power output at take off in the region of 2,500hp (1,864kW) using water/methanol injection. An axial compressor supercharged the Griffon, with the compressor driven by an axial turbine using the Griffon exhaust as its working fluid. Unlike Napier, Rolls-Royce saw the folly in such complex equipment, but recognising the potential in turboprops at an early stage, gave up on the Compound-Griffon. They did, however, see that an engine along the same lines as the Pratt & Whitney R-3350 that used power recovery turbines, could be useful for transport aircraft.

For the Turbo-Griffon, Rolls-Royce took the Griffon and increased that engine's output to 2,850hp (2,125kW) through the use of a power recovery turbine and the engine was being developed under contract for the Supermarine R.2/48 flying boat (Type 524) and the Avro Type 719 (original Shackleton Mk.3). Performance at low level was similar to the Griffon 57 and at low altitude the turbocharger was partially

Shorts PD.2 to meet R2/48 with Rolls-Royce Turbo-Griffons. Sent back to the drawing board with the Air Staff's criticism ringing in their ears, Shorts took the PD.2 and the Turbo-Griffon engine and designed a flying boat more suited to the Air Staff's needs. Shorts got the contract because the MoS wanted to preserve jobs in Belfast and were happy to see Saro go under.

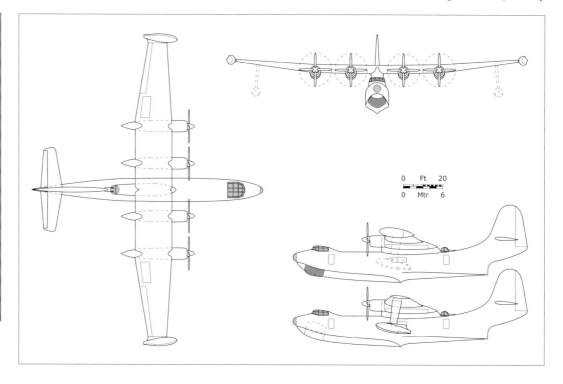

bypassed via a pressure-operated valve system (called a wastegate) that controlled the manifold boost pressure. Once the aircraft had climbed to an altitude at which the differential pressure fell, the wastegate closed to divert the entire exhaust gas stream into the turbocharger. The single turbine lay in a horizontal attitude at the rear of the rocker covers and took gas from each exhaust manifold. This gas drove the turbine, which turned a vertical shaft that fed power via gearing, into the crankshaft rather than boosting the inlet manifold pressure.

Meanwhile, at Acton in west London, a very unusual engine was being developed. It was an engine that has intrigued many in the aviation world, not only the aero-engine specialists, but anyone interested in the history of propulsion. Described as 'baroque', it was quite probably the most complicated engine ever built: the Napier Nomad.

Sandys, the Duke, the Marshal and the Nomad

'I am told there is a real danger that in the drive for economy, the Napier Nomad engine may be abandoned. I believe that would be not only the worst type of false economy but a disaster...' Marshal of the RAF, Sir John Slessor to R A (Rab) Butler, Chancellor of the Exchequer, 15 Sept. 1953.

What Slessor had heard, probably in the Napier chalet at the SBAC display at Farnborough, struck him as folly, but not because he saw the Nomad as the ideal engine for a Coastal Command flying boat. Slessor saw the Nomad as the key to a fleet of freight aircraft that would do for the air transport trade what the triple-expansion steam engine had done for the British merchant fleet. The Comet, Britannia and Viscount were, in Slessor's opinion, the glamorous equivalent of the RMS *Queen Mary* or SS *America*, but what created the wealth that made these possible in the first place was '...the vast fleet of merchantmen and tramps – sea-going freighters'. Not that Avro would have applied the description 'tramp' to their Nomad-powered Type 708 and 709 studies, the former a clean-sheet design and the latter a variant of the Type 689 Tudor 2, intended as long-range Empire transport aircraft.

Slessor was convinced that without 'export trade and economic strength' made possible by trade, the British lead in the glamorous aspects of British aviation – the jet transports, V-bombers and fighters – could not be maintained. There

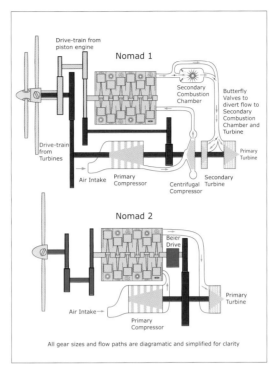

All gear sizes and flow paths are diagramatic and simplified for clarity

The Nomads, in diagrammatic form. Described as the most complicated aero-engine ever built, the Nomad I combined a contra-prop and sleeve-valve two-stroke diesel with axial and centrifugal gas turbines with reheat. The Nomad II addressed that complexity by removing the contra-prop, centrifugal compressor and reheat.

was a need for '...an economical and reliable power unit for freight and medium-speed passenger traffic...' and given that the 'normal reciprocating engine is dead, or anyway, moribund...' the answer was the compound engine. Slessor believed that the Napier Nomad was that answer and Britain, with its lead in gas turbine technology, could also attain the lead in economical propulsion with the Napier engine. The Chancellor replied to Slessor's letter pointing out that a decision on the Nomad hadn't been confirmed, but '...the fact is that our resources for research and development are very far from being unlimited.'

So, what was innovative about the Nomad and inspired such passion in the recently retired Chief of the Air Staff and former C-in-C of Coastal Command? The E.125 was conceived in 1944 by Napier as a 'completely new type of aircraft powerplant' with fuel consumption 30% lower than the then current reciprocating petrol engines and 60% less than the contemporary gas turbines. Research work began in earnest in 1946 and culminated in 1949 with a prototype engine that output 3,000hp (2,237kW) while running for 1,200 hours on a test rig and 100 hours on an Avro Lincoln B.4 (SX973) testbed aircraft. Renamed Nomad 1, the E.125 married a twelve cylinder horizontally-opposed, sleeve-valve, two-stroke diesel engine to a gas turbine with a twelve-stage compressor and a three-stage turbine based on Napier's Naiad. The diesel engine effectively became the combustion chamber for the gas turbine, with its axial compressor feeding a centrifugal compressor that further supercharged the reciprocating engine. The Nomad 1 was fitted with contra-rotating propellers, with the piston engine driving the forward three-bladed propeller and the gas turbine driving the aft three-bladed propeller. The centrifugal compressor could also be driven by the reciprocating engine's crankshaft to start up the gas turbine. To add to this complexity, the engine was reheated, as fuel could be injected into an extra combustion chamber to use up the remaining oxygen in the engine exhaust and provide extra power for take-off and climb.

The Nomad 1 was soon superseded by the (relatively) less complicated E.145 Nomad 2 that dispensed with the centrifugal compressor and the auxiliary combustion chamber. The combined output of the turbine shaft and diesel engine crankshaft was fed to the single propeller by connecting the two powertrains through an infinitely-variable gear train called Beier gears. This power transmission system used three shafts that carried a series of disks driven by the turbine that were arranged around a central set of disks mounted on an extension of the crankshaft. The outer and central disks were interlaced and because they were tapered across their radii, by moving the planetary disks in and out across the radius of the main disks, the viscous drag of fluid between the disk sets allowed the torque from the turbine to be transferred to the diesel's crankshaft. Water injection also increased the power output, with bench tests showing in increase of 600hp (447kW) with Napiers estimating that ultimately the Nomad would be good for 4,000hp (2,983kw). These changes reduced the length of the engine by 3ft 6in (1.1m) and reduced its weight by 1,200lb (544kg). The MoS gave the instruction to proceed with Nomad 2 in October 1951 and the

first engine ran in January 1953. However, the knives were out for the Nomad.

By 1953 the way ahead was becoming clearer: piston engines lacked the performance for the new generation of aircraft and the thirsty turbojet was being superseded by the second generation 'propeller-turbine' following on from the Bristol Theseus. Bristol Engines, learning from the Theseus, were developing their Proteus turboprop and Rolls-Royce had embarked on the RB.109 Tyne having cut their teeth on the Clyde. From its inception, the Nomad had been intended to power two aircraft: the R.2/48 flying boat and the C.3/46 transport, ultimately the Blackburn Beverley. It was this transport that Slessor saw as the new 'tramp steamer' of the skies while the Nomad's efficiency would provide the endurance Coastal Command required.

By 1953 the Nomad project had consumed £5.1m and Napiers had disclosed that to bring the Nomad 2 to production standard, its development would cost £1m per year for the next three years. This rang alarm bells at the Ministry of Supply, with Duncan Sandys being the minister in charge. This was a time when the UK armed forces were undergoing what could be described as a complete re-armament following stagnation between 1945 and 1951 when the Korean War had shown that new weapons were required. Research and development resources were tight and had to be prioritised, specifically to the atomic bomb, the V-bombers plus the Surface-to-Air Guided Weapons (SAGW) and fighter aircraft to defend the deterrent.

D. Napier and Son was the engine subsidiary of the English Electric Company Ltd (EECo) the industrial group that also controlled English Electric Aircraft and The Marconi Company, thus providing the group with the range of expertise to tender for government contracts. So, when the chairman of EECo, Sir George Nelson got wind of the withdrawal of funding from the Nomad, like Slessor, he wrote to the men at the top. Sandys was in New Zealand when Slessor's original letter arrived, but over the next year letters were exchanged, posing the pros and cons of the Nomad. The Private Secretary to the Chancellor, Ian Bancroft, appears to be handling much of this correspondence. However, on 30th August 1954 as the letter was about to be sent to Sir George informing him of the removal of Government support for Nomad, Bancroft scribbled a note to Butler. The note said 'Ministry of Supply rang to say would we hold the reply for a few more days. Sir G Nelson has now roped in the Duke of Edinburgh.' The Duke's views on

The Nomad 1 installation on a Shorts R.2/48 (PD.2). The installation was very neat and had fairly low drag compared with its rivals the Centaurus and Turbo-Griffon. Note also the weapons bay in the engine nacelle, something the Air Staff would not entertain due to loading problems. *RAF Museum*

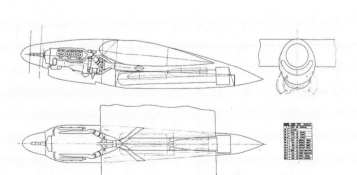

FLYING BOAT — SPECIFICATION R 2/48 – 4 NOMAD COMPOUND ENGINES
ENGINE & BOMB ARRANGEMENT

the matter are not recorded, but Prince Philip was a great advocate of Britain's aviation industry and his involvement might well have swayed a lesser politician than Sandys.

Sir George Nelson kept up the pressure to continue the Nomad's development, with letters direct to R A Butler. Nelson maintained the line that the Nomad would be a boon to trade 'I would like to emphasise that my letter was not a plea for the development of an engine as such but on in the much wider interests of earning foreign exchange…'. With the burden placed on the economy by the need to re-arm and develop new advanced aircraft to catch up on almost a decade of torpor, the Chancellor had to act on the advice of his Ministers of Supply and Defence. At this time military projects were assigned a priority, with the atomic bomb and SAGWs being assigned 'Super Priority' and therefore given first call on funding. An engine whose development period was uncertain, application subject to endless debate and requiring not inconsiderable funding was undoubtedly a long way down the priority list.

Duncan Sandys on 9th September 1954 wrote to Sir George Nelson. 'It is with considerable regret that we have decided that we cannot continue to support the Nomad engine…to concentrate our resources on those lines of development which, all things considered, appear to be of first importance.' Sir George had asked if there might be a 'special financial allocation' but Sandys stated that '…if additional money could be provided, the Nomad is not the only project on which it could be usefully spent. By October 1954, government funding for the Napier Nomad was withdrawn.

Although both the Nomad and the R.2/48 had their origins in 1946, neither had kept pace with developments elsewhere. An ideal application for such an engine, the flying boat, was the subject of a protracted debate for the next decade, a period during which conventional engines, gas turbines especially, improved and became the choice for long-range aircraft. Saro, Supermarine and Shorts all drew up studies for their R.2/48 types to be powered by Nomads, but as the tenders went to the Ministries, alternative engines were proposed. Initially the Compound Griffin was proposed for the Saro P.104 tendered in 1949, but as the more powerful Turbo-Griffon was being developed, the Air Staff opted for that rather than the Nomad. Shorts tendered their four design studies for R.2/48, including the modified Shetland, with Bristol Centaurus and the Nomad as an alternative '…when they become available.' Shorts also became more inclined towards the Turbo-Griffon. Supermarine's Type 524 was intended to use the Bristol Centaurus, with that company's Proteus turboprop as alternative. Supermarine hedged their bets by stating 'It is felt to be undesirable to build a completely new flying boat which can accommodate one type of engine only' and so designed the Type 524 to take any of the three engines.

The flying boat companies appear to have included the Nomad in their plans with many caveats – typically appending 'when available' to the name Nomad. Addenda to the tender brochures for Saro's P.104 and Supermarine's Type 524 examine powering the R.2/48 with Rolls-Royce Turbo-Griffon engines. Supermarine stated that '…a higher performance and lighter aircraft can be obtained by contracting the design to suit the 'Turbo-Griffon' powerplant. This is considered a better course…' The Nomad forced a heavier all-up-weight and a larger airframe. This may be another area where the Nomad, or any other reciprocating engine for that matter, was inferior to the gas turbine – piston engines do not 'scale' in the same manner as gas turbines. To match the Nomad-powered R.2/48 airframe to the Turbo-Griffon engine, Supermarine had to reduce and Saro had to increase the size of the airframe. Aircraft designers fell for the gas turbine because these could be scaled to match the airframe and on the drawings for many design studies the engine is described as being an current engine 'scaled to 0.8'. This airframe scaling effect can be seen in the 'outline comparison' drawing in the Saro P.104 brochure showing the difference that 350hp (261kW) per engine made to the airframe.

The trouble with the Nomad was that it was a niche engine intended for a niche market – long endurance at medium speed. It wasn't scale-able in the same sense that a turboprop was and suffered from the age-old problem with reciprocating engines – many moving parts and 'gadgets'. Another consideration with the Nomad was that while it could run on kerosene, it performed best on diesel, adding a third fuel to the RAF's logistics burden. This had been listed as one of three perceived problems with the Nomad in a letter dated 8th June 1948 from Air Commodore Pike, DOR, to Captain M Luby Director of Engineering R+D at the MoS. At a meeting, Pike had been asked if '…the Air Staff have any other objection to the Nomad other than that of having faith in the ability of the engine designers.' Pike replied 'yes we have' and listed three. The first was that the

MoS estimated that the Nomad added 2,000 lb (907kg) in weight to each aircraft which Pike said was 'Very distasteful.' Secondly the fact that 'The Nomad requires special fuel and therefore supplies will have to be put down all over the world just for this one aircraft.' Thirdly Pike was alarmed that 'the fuel itself tends to solidify at low temperatures…' adding 'This is very embarrassing.'

In response to this, on 4th August 1948, Engineering R+D1, delivered their findings to Pike and explained that the E.125 Nomad ran on a fuel called *Diesoleum*. This was the trade name for a British Petroleum (BP) gas oil of 0.848 specific gravity and having a calorific value of 10,220CHU/lb (42.7MJ/kg). The Nomad would also 'run equally well on aviation kerosene' with a specific gravity of 0.810sg and a calorific value of 10,320CHU/lb (43.1MJ/kg). Given Pike's concern about 'waxing' in cold weather'…it has been decided to therefore to do all further development running using this fuel.' This use of standard aviation fuel would also allow the performance of all gas turbine engines to be compared.

The Nomad was top of the agenda at the first R.2/48 meeting in Thames House on 27th January 1949, D. Napier and Son's representative, Mr Sammons, had pointed out that if the AUW of the R.2/48 was to be kept below 100,000 lb (45,351kg), '…the Nomad engine would be less attractive than the Centaurus and the Griffon.' Sammons had asked how firm the 100,000 lb AUW was. He was informed that the R.2/48 was '…the only aircraft planned for Nomad and therefore provided the sole avenue for its development.' However, to achieve the required endurance, the Nomad would be best suited and that the AUW was selected purely on economic grounds, 'There was therefore nothing absolute in the figure of 100,000 lb' Then Mr J E Serby, Director, Military Aircraft Research and Development – Ministry of Supply (DMARD), indicated that the Centaurus would need to be used, given the time being taken to develop the Nomad, and that there was '…little to choose between the aircraft performance with Nomad and with Centaurus.'

Nomad by Name, Nomad by Nature

References to the Specification R.2/48 first appeared in Air Staff and Air Ministry documents from September 1948 with the Spec. itself issued on 6th October 1948 to invite tenders for a 'Marine Reconnaissance Flying Boat'. By 1954 the very existence of the flying boat was in doubt and with its demise, also went the Nomad. In short, in times of stress, the niche-dwellers become extinct and in the UK in 1953, the economy and armed forces were under stress. Oddly enough development of the Griffon variants was terminated in 1949, long before the Nomad, victims of Rolls-Royce's 'dash for gas (turbines)' that saw their piston engine work shrink to merely supporting the Griffon by the end of the 1950s. Likewise Bristol Engines did not develop any piston engines after the Centaurus, preferring to concentrate on the Proteus turboprop and other gas turbines. The Proteus was fitted to the Saro Princess, hovercraft and, most famously, the Bristol Britannia which served from 1957 until ending its days plying cargo routes around Zaire in the 1990s and this author recalls seeing the last Britannias at Kinshasa's N'jdili Airport in the late 1980s.

The Nomad could have prospered but its complication, lack of a suitable aircraft – it had no home – and length of time it was in development meant that like a lot of British innovations, it had been outstripped by a simpler, if less efficient, class of engine – the turboprop. That engine was the Rolls-Royce Tyne and it would play a major role in the development story of maritime patrol aircraft, but not in the UK.

The Nomads today: Engineering curiosities on display. Uppermost is the Nomad 1, lovingly restored by Mr Bob Thomson of the Aviation Preservation Society of Scotland and below, the Nomad 2 on display at the Stephen F. Udvar-Hazy Center of the National Air and Space Museum, Chantilly, Virginia USA. Their appearances are deceptive, as the complexity is not apparent to the casual observer.

4 The Kipper Fleet: Short and Medium Range MR

'Access to the cockpit is difficult. It should be made impossible.' Test pilot's report on the Shorts Seamew.

The Lords of the Admiralty in 1950 were becoming increasingly exercised by the rising Soviet threat to inshore waters from mines, which prompted the preparation of a short report. While their Lordships were content with the state of affairs in the blue water navy and their defence of oceanic lines of communication, they were less content with the defence of the coastal and cross-Channel traffic. This more visible aspect of Coastal Command's duties had led to the entire organisation being called 'The Kipper Fleet', a typically British disparaging name that ignored the increasingly important oceanic anti-submarine role of the Command being conducted over the horizon in the Atlantic. Their Lordships wanted more emphasis placed '…on training for the defence of our coastal shipping' and saw the main threats to be mines and attacks from enemy aircraft, surface ships and submarines. The Admiralty saw mining as the most serious of all,

particularly mines laid by submarines, and to combat this threat, the Sea Lords requested increased co-operation with RAF Fighter and Coastal Commands in anti-submarine and anti-minelaying operations.

Their Lordships saw the threat from mining as grave and expected an initial Soviet campaign to comprise the laying of 1,200 mines each month, exacting a toll of 50 ships per month. Operational research had shown that should bases in the Low Countries and France be obtained by the Soviets, the number of mines would rise to 10,000 per month, with a corresponding increase in shipping losses, and that would be untenable. Group Captain N V Moreton DD Ops (Maritime) at the Air Ministry passed this to Group Captain M F D Williams at HQ, Coastal Command. Their Research Branch in turn examined the problem and set about developing a plan to provide cover for coastal convoys and to prevent Soviet mining operations, concluding that by using three bases, 2,200 sorties would be required to escort the coastal convoys or 1,900 if five bases were used and would require 190 aircraft. As for the types

Banking to port high above the cloud, Vickers Viscount VX211 was a Ministry of Supply aircraft finished in RAF livery. Vickers hoped to interest the Air Staff and Air Ministry in a maritime patrol version for short and medium range maritime patrol. *via Phil Butler*

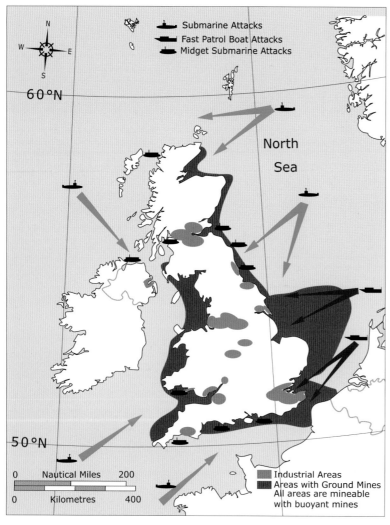

The Admiralty expected Soviet Naval forces to attack the UK's shipping by a variety of means and the Air Staff wanted the medium MR type to counter these. Fast attack craft would attack the ports and shipping off East Anglia, midget submarines the ports around the country and submarines could attack anywhere. Shallow and coastal waters were vulnerable to ground and tethered mines.

involved, large aircraft like the Shackleton would be capable of maintaining these patrols but would be uneconomic, although if they were fast, they could operate on a scramble basis from a number of coastal bases. Given that since the Great War the first rule of airborne ASW operations was that the mere presence of an aircraft deterred submarines, this scramble tactic was a non-starter. The next suggestion was to use a smaller type, along the lines of the de Havilland Tiger Moths and Airspeed Oxfords of the early war years, acting as an armed 'scarecrow' to keep the enemy's heads down.

Air Commodore Harold Satterly, Director of Operational Requirements, in April 1951 produced a memo in reply to the Director of Operations who had suggested that civil aircraft be used for coastal ASW. Satterly suggested that the MoS, rather than the Air Staff, should '…examine ALL non-operational aircraft – service and civilian – to see if they can be modified to do this coastal anti-submarine work…' Satterly saw this as a massive task and produced a list of seven questions beginning with 'What is your minimum crew – should MoS look at single-seaters upwards or what?' Satterly went on, asking about ASV, weapons sighting, day or night or both and his final point at the end of the list was 'etc. etc.' The memo concluded with the need for clarification on a minimum crew and warload 'You could then decide roughly what minimum payload you needed and this should help you narrow the field still more.'

Sense prevailed; Satterly obviously considered the prospect of civil aircraft lugging depth charges 'round the bay' with horror, and opted to look into using RAF and RN communication

The second prototype Fairey Gannet showing its capacious weapons bay for a Pentane torpedo and other ASW weaponry. Fairey proposed a 'lightweight' Gannet with a single engine for use by the RAF for its coastal MR task. *Terry Panopalis Collection*

aircraft for the task. Wing Commander J D E Hughes, OR2, advised 'There are about 200 Oxfords in reserve stock at the moment.' Hughes continued, saying that the MoS and Airspeed had examined the carriage of stores on the existing strong-points under the fuselage and that if the fuselage was '…reinforced by local spreading plates' two depth charges could be carried. Next suggestion was the Avro Anson, which was quite ironic as it had been designed as a maritime patrol aircraft back in 1935. The Anson already had fittings for bombs, with a warload of 360 lb (163kg) and would require no major modification for the ASW role. The Percival Prince and Pembroke were by 1951 under development for service entry in 1953 as communications aircraft for the RN and RAF respectively. Like the Anson, these were to be fitted with strong points and could be converted to carry depth charges.

Satterley's question about these aircraft using ASV radar seems to have prompted the realisation that radar might be a good idea for the coastal surveillance role, while a useful warload would be desirable. Never ones to take on a naval aircraft through choice, the Air Staff soon realised that the FAA might have an answer. The GR.17/45 (Fairey Gannet or Blackburn B-54), with a four and a half hour endurance and more than adequate weapons-carrying capacity, was deemed suitable, but these were designed for fleet carrier operations and thus carried a great deal of superfluous equipment. As luck would have it, another carrier-borne ASW aircraft was being studied, aimed at the Royal Navy Volunteer Reserve to equip the smaller escort carriers that had multiplied during the last three years of the war. This gave rise to Admiralty requirement NR/A.32 and Specification M.123 in 1951 for a lightweight ASW aircraft that could operate without the need for arrester or catapult gear from aircraft carriers. Shorts, Westland and Fairey tendered for this, with Fairey offering a single-engined GR.17/45 Gannet, which was deemed too elaborate for the task. Westland offered a T-tailed, mid-wing type with a narrow fuselage fitted with a single ASM Mamba turboprop in the nose and a tandem cockpit, with the entire airframe dominated by the large ventral bay that carried an ASW torpedo or depth charges. Neither of these reached prototype stage as the Shorts proposal, the PD.4/SB.3 design, was adopted by the Fleet Air Arm in 1951.

This odd looking aircraft was called the Seamew AS.1 and had its tandem cockpit mounted above the single ASM Mamba turboprop in the nose of the narrow, but deep, fuse-

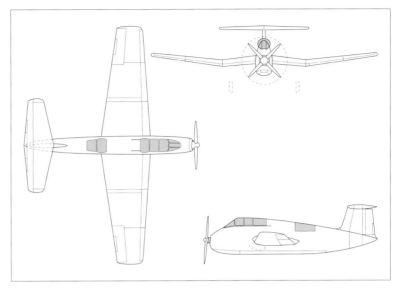

lage. A broad, thick wing provided the low wing loading of 26 lb/ft² (127kg/m²) required for unassisted take-off and landing on carriers, but flight trials soon revealed vicious handling characteristics, prompting the quote at the start of this chapter. Coastal Command was to acquire the Seamew, with reluctance, and designated it the MR.2 and these were ordered in 1955. They were to operate from short, rudimentary airstrips around the coast and patrol inshore waters with their ASV.19 radar and sonobuoys but the type could only carry the RAF's Pentane torpedo if the radar was removed. The lightweight ASW aircraft for Coastal Command was but a passing fancy, as a more capable type was preferred and the Seamew project was cancelled in its entirety in the 1957 Sandys Defence White Paper. Many were glad to see the back of the Seamew, including the Fleet Air Arm, which had never been keen on the type. The entire programme had, not for the last time, been drawn up by the Ministry of Supply in an attempt to keep Shorts' Belfast workforce in employment.

Fairey/Westland M.114, a study for the lightweight MR requirement. This used a single AS Mamba and was based on a modified Gannet airframe. Seamew was preferred, possibly because it was under development.

A folded Seamew shows its large weapons bay, with the ASV.19 radar housed in the forward portion. The RAF Coastal Command MR.2 version lost the powered wing folding and had larger low-pressure tyres.
Author's collection

A Vickers Varsity
T.1 shown in its
secondary role as a
maritime patrol
aircraft. In reality
such patrols were a
throwback to the
Scarecrow patrols
in the Great War
when it was
discovered that the
appearance of any
aircraft, no matter
what type,
hampered U-boat
operations.
Author's collection

As noted previously, the flying boat in the RAF's order of battle was fading into history and by 1950 the Shackleton was under development, for service entry in 1951. Having relied on flying boats such as the Sunderland for its long-range maritime reconnaissance needs, HQ Coastal Command was somewhat concerned that the Shackletons could not be based in the correct places to cover the North Atlantic. Essentially, all the long runways suitable for large landplanes had been built on the east coast and, apart from Prestwick that had been used as a transit and refuelling stop for transatlantic ferry flights, the west coast wasn't well served with long runways. This meant that the South West and North West Approaches were lacking in long-range aircraft cover, but in reality a smaller type could cover these near-shore areas where convoys would gather before departure and these could operate from smaller airfields.

On the east coast, where the suitable runways were, Coastal Command did not require an aircraft with the range and endurance of the Shackleton and so in May 1950, ACAS (OR) Air Vice-Marshal Claude Pelly asked ACAS (Policy) Air Vice-Marshal Douglas McFadyen if a smaller aircraft, such as the GR.17/45 would be suitable for this short-range reconnaissance task. In a change from earlier thinking, McFadyen advised that the type was too short-ranged and incapable of carrying a suitable warload, but he considered an aircraft such as the Avro York, Handley Page Hastings or 'a variant of the Vick-

ers Varsity' would be 'capable of both anti-submarine work and attacking light surface craft.' Air Marshal Charles Steele, C-in-C Coastal Command was very keen on this and stated that the aircraft should have an attack team of four, an ASV radar, carry two Dealer or Pentane torpedoes, have an endurance of 4-5 hours (transit plus two and a half hours patrol) and be 'cheap and numerous'. Steele and McFadyen's thoughts were summed up in a briefing by the Air Ministry's R W Simmons under the title '*Short-Range Maritime Reconnaissance Aircraft for Protection of Coastal Convoys and Shipping in Focal Areas*'. The Hastings and York were dismissed due to their need for too many modifications to enable defensive armament and anti-submarine weapons to be carried, which left the Varsity. The briefing paper was passed to Vickers, who very quickly advised that the Varsity T.1 would be an ideal basis for this type and submitted a brochure in October 1950 outlining their proposal.

Vickers' proposal for the short-range maritime reconnaissance (SMR) Varsity weighed in at 41,000 lb (18,597kg) and was powered by a pair of Bristol Hercules 230 engines rated at 1,975hp (1,473kW), later replaced with Hercules 274 engines rated at 2,200hp (1,640kW). Vickers advised that the Varsity could patrol for two and a half hours at a distance of 550nm (1,019km) which was less than Coastal Command wanted. Vickers advised that the fuselage could be sound-proofed to '…a noise level consistent with the best civil aircraft…' and Vickers claimed it was ideal for fitting an anti-submarine attack station. Detection equipment included a retractable ventral radome housing an ASV.13B radar, as used on the Bristol Brigand and de Havilland Mosquito, but this was not capable of providing 360° coverage without modification. Vickers were put right on the radar matter and it was the ASV.13, as used on the Shackleton GR.1, that was mentioned throughout later documents. Up to six directional sonobuoys and marine markers were to be carried, in addition to twelve non-directional sonobuoys.

Vickers' initial drawing showed the Varsity SMR as a straightforward conversion of the Varsity T.1 with the nose modified to carry a radome for an ASV.13 antenna and the ventral pannier extended to house a longer weapons bay. No defensive armament was fitted to this minimal-change version but for the brochure submitted to the Air Ministry, a single Bristol Type 17 Mk.6 dorsal turret that mounted a pair of 20mm Hispano cannon was added. The Varsity SMR as laid out in the brochure was fitted with an extended nose with a glazed bomb aimer/lookout position. The antenna for the ASV.13 was installed in a retractable ventral radome that was to be raised into an underfloor compartment aft of the nosewheel bay. This version also dispensed with the ventral pannier, as a weapons bay was installed under the cabin floor. More capacious than the original pannier, this could carry longer stores such as the 2,000 lb (907kg) Pentane or two 665 lb (301kg) Dealer torpedoes in addition to directional sonobuoys and marine markers. The final version featured the same weapons bay, but moved the ASV.13 scanner to a chin position and shortened the fuselage. The Bristol turret

was replaced by a Boulton and Paul Type T model fitted just behind the cockpit and armed with a pair of 0.50-calibre machine guns with 1,000 rounds per gun. This change of defensive armament worried the Air Staff somewhat, prompting a response that laid out the reasons for their concern, not least that the 0.50-calibre machine gun wasn't up to the job.

In a memo dated 26th October 1950, the Deputy Director (OR) Group Captain S R Ubee voiced concern that since the type would be operating off the east coast of the UK, there was a distinct possibility of '…considerable enemy interference over the North Sea…' and that since the turret would also be used for flak suppression, '…it should therefore have guns of at least 20mm calibre.' As for the rest of the warload, the main offensive weapon was to be a single Dealer torpedo and four depth charges, with the capability of carrying eight 25 lb rocket projectiles '…capable of underwater travel' for use against surface units. In the air-sea rescue role, a set of Lindholme gear and four smoke floats were to be carried. Coastal Command were keen on the Varsity SMR, but the original inclusion of the Hastings and York suggests that they sought a more capable aircraft, especially if it could carry two Pentane torpedoes. With the Varsity SMR, Vickers had been quick to respond to the Coastal Command request, but given a bit more time, they were able to apply the RAF's responses to the Varsity proposals with a view to carrying more weaponry and more detection gear, all of which amounted to and AUW increase of 2,000 lb (907kg) to 42,000 lb (19,048kg).

As the OR branch examined the Varsity proposal, a sharp-eyed officer, Squadron Leader Chandler, drew the Deputy Director OR1's

The initial version of the Varsity MR housed the ASV.13 radar antenna in a chin radome but the version shown on the right saw this moved behind the nosewheel. The gun turret was placed as far forward as possible to ensure the widest field of fire but when trained forward for use against surface targets, the blast and flash would severely affect the cockpit and pilots. The underfloor baggage hold was converted to a weapons bay. The type in the centre was a minimum change version with no guns, much to the Air Staff's annoyance.

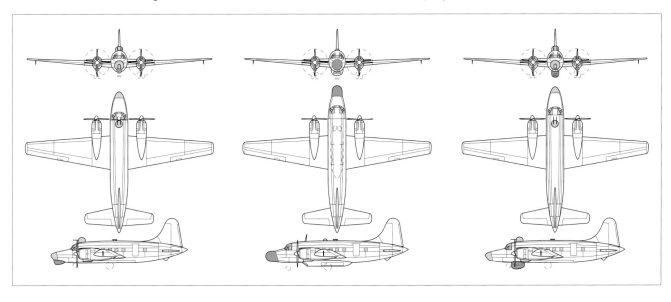

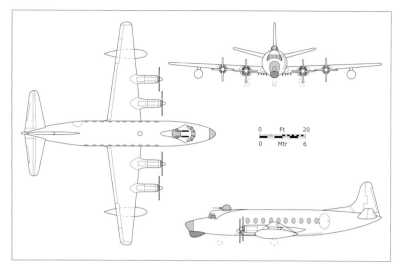

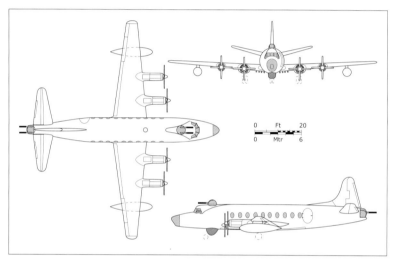

attention to a rather alarming discovery: due to undercarriage weight limits, the Varsity SMR would be unable to land with its full warload and would need to dispose of some of its stores before landing! In his memo, Chandler showed his working and it transpired that 'With a full crew and full stores load it would therefore be safe to land with 28 lb of fuel (or about 3½ gal-

lons)'. Chandler stated that he had contacted Vickers about beefing-up the Varsity SMR undercarriage to allow operations at higher weights. Chandler also pointed out that while the brochure described a crew of six, the drawings showed that there were nine crew stations and that these would all need to be manned during an attack on a submarine. He also mentioned that Vickers would not have the capacity to produce the Varsity SMR as it was committed to the 'B.9' (the Vickers Valiant to meet B.9/48).

The Varsity's main drawback as far as the RAF were concerned was that it was just too small to carry the necessary payloads for the required endurance and to address this; Vickers' George Edwards had been to see Bristol Engines and Rotol about the possibilities of fitting improved Hercules engines with new propellers. Edwards in mid-October 1950 also stated that if HQ Coastal Command '…insist on these requirements going in, we (Vickers) will have to re-engine with Centaurus.', which was the Hercules' more powerful sibling. In reality, the Varsity was too small and certainly did not meet the endurance HQ Coastal Command required and therefore Coastal Command considered a medium-range aircraft to be preferable and so turned to a project that had been spurned previously by the RAF.

By the end of October Vickers had also submitted a brochure for a short-range maritime version of the Viscount that addressed the shortcomings of the Varsity. Vickers had proposed a variant of the Viscount for the long-range maritime patrol role, but this had been rejected on 3rd October 1950. The Viscount carried a heavier warload of two Pentane torpedoes in tandem within a longer bomb bay and the eight rockets on rails under the inner wing. Defensive armament still comprised a single Bristol B17 Mk.6 turret behind the cockpit, carrying a pair of 20mm Hispano cannon with 360 rounds per gun covering 360° in azimuth and -10° to +45° in elevation. A later version of the short-range maritime Viscount was to be fitted with a Boulton and Paul Type D tail turret with a pair of 0.50-calibre guns and 500 rounds per gun. The tail turret covered 158° in azimuth and +45° to -45° in elevation. Sensor-wise, the Viscount Short Range MR carried a modified ASV.13 in a radome just aft of the nosewheel bay and could carry up to sixteen directional sonobuoys, 24 non-directional sonobuoys, 20 large marine markers and 24 of the smaller No.2 marine markers.

Then, in the first week of January 1951 while he was attempting to put together an appropriate operational requirement, a brochure for another contender for the Short Range Maritime Patrol Aircraft landed on Group Captain Ubee's desk, proposing a maritime variant of the Airspeed AS.57 Ambassador. Airspeed had been formed in 1931 by Hessel Tiltman, an aspiring aircraft designer, and Neville Shute Norway, a structures man and future novelist, who had met while working on the Vickers R100 airship. They formed Airspeed Limited at York in 1931, moved the company to Portsmouth in 1933 and designed the AS.10 Oxford twin-engined trainer and, most famously, the AS.51 and 58 Horsa gliders. The company became part of de Havilland Aircraft in 1940. In 1943, responding to the Brabazon Committee's specification 25/43, Tiltman drew up a twin Hercules design with an AUW of 32,500 lb (14,740kg) which soon grew into a larger, 53,800 lb (24,380kg) pressurised type powered by two Bristol Centaurus 661 engines rated at 2,625hp (1,957kW). The AS.57 prototype, had its first flight on 10th July 1947 and, called the Ambassador, entered service with British European Airways (BEA) in 1952. The second prototype was used as a flying testbed for various turboprop engines including the Bristol Proteus, Rolls-Royce Tyne, while D Napier and Son used the first prototype to test their Eland engine.

Airspeed's proposal, designated AS.69, was around 30% bigger than the Varsity with an AUW of 60,000 lb (27,216kg) and correspondingly greater endurance and warload, 7,240 lb (3,284kg) against the Varsity's 4,182 lb (1,897kg). It carried a crew of seven rather than the six of the Varsity, with the seventh crew member being a gunner in a second upper turret, both of which carried a pair of Hispano 20mm cannon with 350 round per gun while eight 60 lb (27kg) rockets were mounted on

rails on the fuselage sides, forward of the bomb doors. The AS.69 would require a complete re-design of the fuselage with the two gun turrets and a repositioned wing prompted by a change in the centre of gravity due to the location of the weapons bay. This last factor saw two proposals put forward. The first with the existing high wing and a second with the wing and main undercarriage lowered to improve the ground clearance for loading stores and servicing the ASV.13 radar. The turrets on the AS.69 were initially mounted in nose and tail, thus providing excellent fields of fire, although there were concerns about flash and blast effects on the cockpit glazing and the pilots' night vision. These, when combined with the structural changes to the nose and tail, prompted Airspeed to move the turrets to dorsal positions on the fuselage. The forward Bristol B.17 turret was installed just aft of the cockpit, far enough back to allow the guns to be fired forward without affecting the pilots. The second turret was fitted forward of the tail assembly at a point where 'no-fire' zones around the empennage were minimised.

As January became February yet another contender appeared in the somewhat portly form of the Bristol Type 170 Freighter. From the off the Bristol Freighter was a rank outsider for the role with the Director, OR2, stating 'The Bristol Freighter is not likely to match the performance of the Varsity in this role, and the Varsity does not entirely meet the requirement.' Both types had similar AUW, but the Freighter had 320hp (239kW) less power than the Varsity and its greater profile drag would affect its range performance. OR2 went on to list the cons of the Freighter, specifically the nose needing much modification to fit a bomb aimer's position and forward firing guns, the fixed undercarriage producing radar blind spots and

Opposite page: The initial Viscount MR carried similar modifications to the Varsity, chin radome and dorsal turret behind the cockpit. Long-range tanks were to be fitted to improve the type's endurance.

The Air Staff's concern about enemy fighters saw the Viscount MR's defensive armament beefed-up with a tail turret. Again the dorsal turret is placed behind the cockpit, but with eight 3in (7.6cm) RPs under the wing roots, the guns would be less likely to be used on surface targets.

Airspeed's AS.69 maritime patrol aircraft was a much-modified Airspeed Ambassador airliner. Aside from turrets and weapons bay, the AS.69 was mid-winged unlike the high-wing airliner. Essentially only the wings and tail unit were unchanged. *David Cook*

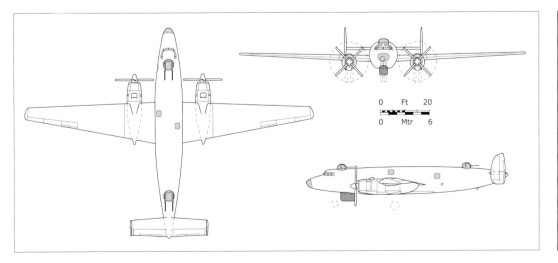

The Airspeed AS.69 was fitted with a pair of Bristol B.17 turrets with 20mm Hispano cannon. The Air Staff required this armament for the medium MR type because it was to operate mainly in the North Sea and thus be vulnerable to Soviet fighter aircraft operating from captured bases in Holland and Denmark.

0   Ft   20
0   Mtr   6

The portly Bristol Freighter was suggested as a short/medium-range MR type. The Air Staff considered it to be too slow and could not provide adequate radar coverage due to its fixed undercarriage. Defensive armament would also be problematic. *Terry Panopalis Collection*

the tail turret which '…would not improve the Freighter's handling' while the mid-upper turret's field of fire would be severely restricted. Lastly and probably the most apposite to the Freighter, 'Coastal Command would like an aircraft which is moderately nippy for RP attacks.' OR2 summarised the Freighter with 'This proposal does not look promising and if Bristols are persistent they should presumably make a proper study and have it vetted by the MoS.'

Mr A S Crouch, Director, R+D Projects, on 17th February 1951 prepared a brief comparison of the Ambassador and Varsity and stated that the Varsity SMR could be in service much sooner than the AS.69 Ambassador and 'If therefore the payload and range is adequate this would be the aircraft to go for.' Crouch concluded by pointing out that the Ambassador could carry the same weapon load for twice the range or a 50% greater weapons load for a greater range than the Varsity and that 'it is for the Air Staff to say whether they would prefer this and accept a later date for Service aircraft'. However, a week later Mr R W Symmons, Assistant Director, RDL, advised that this was only really the case if the Ambassador's configuration was to be changed to the mid-wing variant. Symmons also took the view that Vickers were much more heavily loaded with the production of the Valiant and that if the PR version to meet OR.279 went ahead, this would only add to delays to the Varsity SMR. In fact Vickers would have struggled to fulfil the Varsity MR order, its manufacturing capacity being taken up by a variety of programmes.

By August 1950 the RAF had been told that the Lockheed P2V Neptunes that the US Navy were intending to bring over to Britain for sonobuoy trials were needed in Korea and therefore were unavailable boosting hopes for a UK short-range MR type. This non-availability was short lived, as ultimately the role intended for the Varsity, Viscount and Ambassador was filled by US Navy-surplus Neptunes, with 52 being supplied in 1952.

Neptune – No Better Than Southamptons?

Air Vice-Marshal Robert Ragg, AOC No.18 Group Coastal Command on 12th October 1953 wrote to Air Marshal Alick Stevens, C-in-C Coastal Command on the subject of the performance of AN/APS.20 radar and the Neptune in general. Ragg informed his C-in-C that he had been on Exercise *Mariner* at RAF Kinloss and commented on the sheer number of contacts that the AN/APS.20 was displaying on the PPI, after bemoaning the lack of a camera to photograph the displays, he described the display. Ragg started off with 'At first sight the number of contacts may not appear to be very great; but when you come to count them up it is surprising how many there actually are.'

While this alluded to the sensitivity of the AN/APS.20 radar, it must borne in mind that it was developed as a 'snort detector' and the plethora of contacts could not possibly be periscopes and snorkels. Ragg continues '…if a chap is to investigate every contact, we couldn't possibly count on him covering the sort of area we have been giving him. Indeed, he will probably be able to cover an area no larger than we used to in the old "Southampton" days before there was any radar!' Ragg also points out that in wartime there would be a lot fewer fishing vessels on the sea to provide contacts but 'The only frightful thought is the possibility of hundreds of radar reflectors being spread about the ocean by the enemy.' An interesting concept indeed and suggests that a form of 'Window' for obscuring maritime operations had been considered at some point in the past.

Stevens passed Ragg's letter on to his Chief of Staff, Air Vice-Marshal Cracroft who agreed with Ragg and offered his opinion on the contacts problem and on maritime patrol in general 'For years we have been bellyaching (for want of a better word) for long-range maritime aircraft and now that we have got them we employ them on inshore operations, therefore this is exactly the difficulty one could expect.' Cracroft goes on to suggest that the use of a long-range maritime patrol aircraft fitted with a good radar such as the AN/APS.20 on inshore operations was a waste of its capability, concluding with '...there is obviously an urgent requirement for a medium-range aircraft, cheap and simple, for inshore operations.' Before signing off, Cracroft asked Ragg to forgive his expression of personal views, but he found '...the official views are so nebulous that I cannot state them with any degree of certainty.'

From this correspondence it appears that the upper echelons of Coastal Command, while appreciative of the Neptune and its radar, considered its use on the coastal convoy and focal point operations the Admiralty wanted, to be a waste of their capability. Why was the RAF using the Lockheed P2V Neptune in the first place? Despite the austerity of the postwar years, Britain still had a large aircraft industry that could design and build a maritime patrol aircraft, best demonstrated by the Avro Shackleton and the ongoing debate on flying boats. Suffice to say, the reason, as ever, was money.

On 6th October 1949, US President Harry S Truman, in an effort to bolster European resolve against the Soviet Union, signed the Mutual Defense Assistance Act allowing American military aid to European member countries of the newly-created North Atlantic Treaty Organisation (NATO) and to friendly non-NATO countries. This became the Mutual Defense Aid Program and led to the direct transfer of US military hardware and US sponsorship of British research projects, with an early beneficiary being the RAF. From 1951 the RAF received Canadair Sabres and the Lockheed Neptune to fill gaps in the RAF inventory caused by lack of investment in the immediate postwar years and the aforementioned austerity (See this author's

Battle Flight for details on the lack of a swept-wing or supersonic fighter for the RAF). The main constraint on materiel supplied under MDAP was that it could only be used in the NATO area of operations and the equipment's '...use was subject to American surveillance' by the Military Assistance Advisory Group.

The Neptunes were intended to replace Lancaster GR.3s that had been in service since the war's end and the RAF were to take delivery of the Lockheed P2V-5 Neptune, to be called the Neptune MR.1, which came with standard US Navy ASW kit from existing stock. The first two Neptunes were to be utilised 'for evaluation purposes' (smacked of looking a gift horse in the mouth, but no point having useless kit) in the third quarter of 1951 and be based at RAF Kinloss in Morayshire. This was changed to the Anti-Submarine Warfare Development Unit (ASWDU) at St Mawgan in Cornwall, with Kinloss being the base for the operational conversion unit operating six aircraft, with a second squadron based at RAF Topcliffe in Yorkshire.

The planned procurement was for 52 aircraft: two for evaluation and 50 for operations, of which four would be modified for airborne early warning trials with Fighter Command's Vanguard Flight. They would be delivered at a rate of five aircraft a month with the first fully-operational squadron formed in September

A RAF Neptune MR.1 banks away from the camera showing its AN/APS.20 radome and nose and tail guns. This is one of the earlier variants based on the P2V-4 Neptunes transferred to the RAF as the later arrivals had the tail armament replaced with a MAD sting. *Author's collection*

Most of the RAF's Neptune fleet were the P2V-5 model that lost their nose and tail turrets but retained the dorsal turret. The P2V-4's wingtip pods carried a searchlight in the starboard and an AN/APS.8 short-range radar in the port, but in this P2V-5 example with a MAD boom, the AN/APS.8 radar is in the same pod as the searchlight. *Terry Panopalis Collection*

Hawker Siddeley drew up the HS.748MR study in 1961, long after the medium MR type had disappeared and long before the need arose to patrol Exclusive Economic Zones. It elicited no interest.

1952. Disbanded in 1945, 217 Sqn had been re-formed in January 1952, with many of the crew transferring from squadrons operating Lancaster GR.3s. The squadron commenced initial operations under the auspices of the ASWDU, whose two aircraft had failed to appear due to problems with delivery.

The warload for the Neptunes was to be up to four Dealer B torpedoes, two Dealer B and five 250 lb (113kg) anti-submarine bombs or ten anti-submarine bombs. If Dealer B was carried, a heating system had to be used to keep their electronics at optimum temperature. There was concern that the Dealer B might not fit in the Neptune's weapons bay, which was more tailored to the smaller US Mk.44 torpedo. The four 20mm cannons could use British ammunition, but would need to use American links and loading systems. Later RAF Neptunes, built to P2V-7 specifications dispensed with the nose and tail turrets, which were replaced with glazed nose and a MAD sensor respectively. The

dorsal turret with a pair of 0.50-calibre Brownings was retained. Coastal Command were also uneasy with the performance of the AN/APS.20 radar, which initially at least proved almost useless against snorts in moderate seas and above, a moderate sea state having wave heights of 4-8ft (1.2-2.5m), but this was pretty much the norm for the ASV radars of the era.

The Neptunes served from 1952 until replaced by Shackletons in March 1957. The surviving Neptunes were moved to RAF Silloth and put up for sale, with eight being sold to Argentina. The Neptune MR.1s, which were supplied at minimal cost, meant that a home-grown aircraft based on the Ambassador, Varsity or Viscount failed to prosper and thus faded into history, bleached-out drawings and foxed brochures in archives being the only evidence of their existence. The Air Staff wanted a full-capability, four-engined oceanic patrol aircraft and, in the late Forties and early Fifties, that could only be the Avro Shackleton.

Medium MR Postscript

Having seen the Air Staff more or less dismiss the medium maritime aircraft in favour of the Shackleton and other four-engined oceanic patrol aircraft, Hawker Siddeley at Woodford looked into the market for a medium MR type. Satisfied that there was a market for such a type they drew up in 1961 a design study called HS.748MR. This had a strengthened wing with tip-tanks to increase endurance and pinion pods on the outer wings to carry weaponry and ASW stores. Under the rear fuselage was a retractable dustbin radome for an ASV.21 radar similar to the Shackleton MR.3. The type attracted no orders and the concept would resurface twenty years on, but in 1961, nobody wanted a small maritime aircraft.

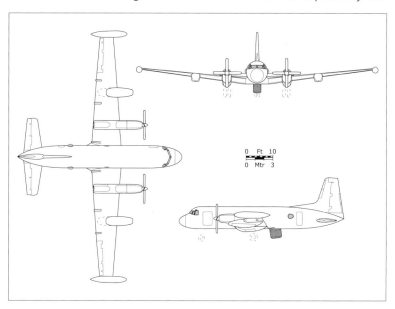

5 Avro's Bespoke Maritime Landplane

'The performance and operational characteristics of the Shackleton have not proved entirely satisfactory. An early replacement of this aircraft is therefore most desirable.' Mr A E Woodward-Nutt, DMARD, MoS, October 1951

Interspersed with the clatter of Super Pumas, Sikorsky S-61s and Boeing 234 Chinooks that contributed to the soundscape of north east Scotland in the Eighties, one sound could set this author scanning the skies over Aberdeen: the growl of four Rolls-Royce Griffons. The source was, of course, a Shackleton AEW.2 performing a practice ILS approach into Dyce Airport but the sound was one that characterised the 'Growler' wherever it served from first flight in 1949 until retirement. From St Mawgan in the south west to Ballykelly in Northern Ireland and Kinloss and Lossiemouth in northern Scotland, there was no mistaking what aircraft was operating from these bases. Apart from the Avro Anson, it was the only British landplane

designed from scratch as a maritime patrol aircraft and Avro's Shackleton, in its various forms, has been covered at length in a number of excellent books. Therefore only a summary of those Shackletons variants that entered service is provided here and the main thrust of what follows examines the larger 'Mk.4' Shackleton studies and the alternatives that were proposed as the first Shackleton replacement.

Throughout the Second World War, Coastal Command became increasingly reliant on landplanes for its oceanic operations, particularly American Consolidated Liberators and Boeing Fortresses, with British types such the Vickers Wellington and Warwick operating at shorter ranges. By late 1942 maritime patrol variants of the Handley Page Halifax and later, in 1943, the Avro Lancaster were conducting missions over the Atlantic and Bay of Biscay. The prospect of very long-range missions against Japan prompted the issue of OR.188 based on the Lancaster III, which became the Lancaster GR.3. The Lancaster, like all the RAF's 'heavies' from

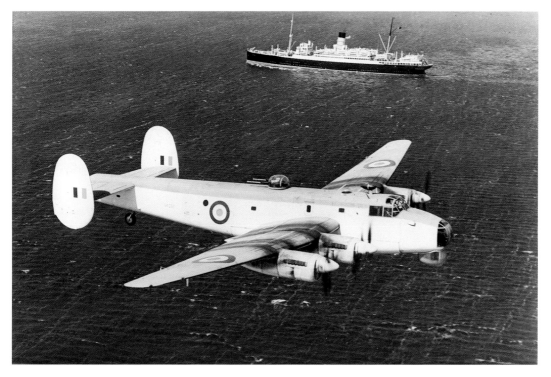

Shackleton MR.1 VP256 of 269 Sqn keeps an eye on marine traffic in the east Atlantic. The Shackleton MR.1 was an improvement on the Lancasters operated by Coastal Command, but conditions for the crew were not exactly palatial and noise levels caused crew fatigue. All this had been discovered by the time the early aircraft had undergone trials at the Aircraft and Armament Experimental Establishment. *via Tony Buttler*

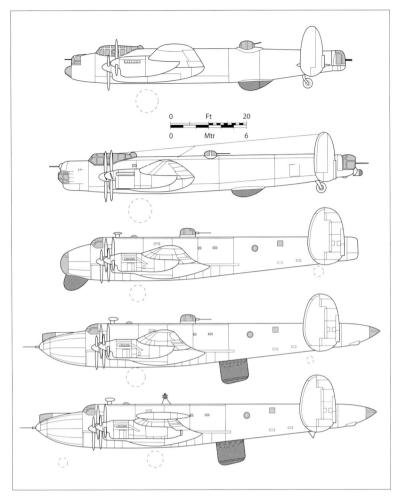

0   Ft   20

0   Mtr   6

How the Lancaster evolved into the Shackleton MR.3 is a cliché for British aircraft development in the postwar years. In fact the Shackleton borrowed more from the postwar Tudor airliner. The drawing here shows from, top to bottom, Lancaster, Lincoln, Shackleton MR.1, Shackleton MR.2 and the definitive Shackleton MR.3 Phase III.

was a twelve passenger transatlantic aircraft for which the Air Ministry issued Spec. 29/43 and OR.152. Avro's response to this was the Type 687, also known as the Avro XX, based on the Lincoln bomber, but this was soon superseded by the Type 688 to meet the needs of the British Overseas Airways Corporation (BOAC) and called the Tudor. This married the Lincoln wing and Rolls-Royce Merlin 102 engines with the first pressurised fuselage on a British airliner and first flew in June 1945. Ultimately undermined by BOAC's constant changes to the specification and the start of a long-term infatuation with American aircraft, at this point the Lockheed Constellation and the Boeing Stratocruiser, the Tudor did not prosper. It was the crash of a Tudor on a test flight on 23rd August 1947 that took the lives of four crew, including Avro's remarkable Chief Designer Roy Chadwick.

Fourteen months after the Lincoln's first flight the war in the Pacific was brought to an end by the dropping of the two atomic bombs that prompted the surrender of Japan. With the war's end and the withdrawal from service of the American types acquired under Lend-Lease, the burden fell on the Lancaster and Lincoln GR types to fill the gap. This reliance on landplanes, essentially converted bombers, revealed a major disadvantage that these types had compared with the flying boats used for long-endurance missions – crew fatigue and lack of crew rest areas. Flying boats such as the Shorts Sunderland and Consolidated Catalina (only eleven of the RAF's 600-plus Catalinas were amphibians) possessed the internal volume to allow the crew to move around and be fitted with a galley and rest bunks, whereas the cramped bombers lacked such facilities.

The ongoing debate on whether or not to replace the Sunderlands became somewhat heated in 1946 and ultimately led to the issue of requirement R.2/48, via R.8/42, for a long-range flying boat, but in reality the Consolidated Liberators had shown the way ahead for maritime patrol. Not that the advocates of flying boats saw it that way, as it would be another decade before they finally submitted to the landplane. To address this need for a British 'general reconnaissance' type, the Air Ministry issued Operational Requirement 191 and Specification B.5/46 for conversion of the Lincoln for general reconnaissance duties. Avro's Chief Designer Roy Chadwick, took the inner wings and tail of the Lincoln, the outer wings from the Avro Type 688 Tudor transport and married these to a new fuselage with more room for the specialised anti-submarine equipment and the

the early Forties was designed for the European theatre whereas the Far East demanded aircraft of much greater range, thus the Air Ministry issued Spec. B.13/43. The longer endurance needed for the war against Japan had driven the design of the Avro Type 683 called the Lancaster IV. With extended outer panels giving a wingspan of 120ft (36.6m) and an 8ft (2.4m) fuselage stretch, combined with improved construction techniques, a somewhat different shape of Lancaster appeared and so as its development progressed, the Lancaster IV was redesignated as the Type 694 Lincoln bomber. The prototype Lincoln flew in June 1944 and was intended to equip the Tiger Force for operations against Japan. As with the original Lancaster, a general reconnaissance version to replace the respective Lancaster variants was being worked on and by late November 1944 the Lincoln GR.III was the preferred choice for a long-range reconnaissance aircraft.

Contemporaneous with the issue of requirements for what became the Lincoln was the publication of the Brabazon Committee's findings on the air transport needs of postwar Britain. One of the aircraft types identified by Brabazon

crew to work it. This became the Avro Type 696, called the Lincoln GR.III, but this aircraft would become better known as the Shackleton.

The initial drafts of the design saw the Type 696 fitted with four Rolls-Royce Merlin 85 engines rated at 1,635hp (1,219kW) but these were soon replaced by four Rolls-Royce Griffon 37 rated at 2,055hp (1,510kW) that had been developed for the Fairey Barracuda V specifically for low-altitude performance. However, this engine never reached production and by October 1946 the Griffon 57 with contra-rotating airscrews had been adopted for the Type 696. The Griffon was preferred over the Merlin because the former produced the same power output at lower rpm than the latter, which improved endurance. As the design process continued the Lincoln GR.III looked less and less like a Lincoln, particularly with its portly fuselage and broad Liberator-style tail fins. The aircraft was evolving into a completely new type, upon which the name Shackleton GR.I was bestowed in October 1946. The specification also changed, becoming the R.5/46 and OR.200, in March 1947 reflecting its reconnaissance rather than bomber task. Development and construction of a prototype progressed rapidly, in part due to the type's use of existing Lincoln and Tudor components. The Type 696 prototype, VW126, made its first flight on a rather dismal 9th March 1949, the weather being apt for an aircraft that would serve over the north Atlantic.

Shackleton GR.1, Latterly MR.1

Originally the aircraft that became the Shackleton carried the General Reconnaissance (GR) designation but as the role had since become more focussed on anti-submarine and shipping surveillance, the designation Maritime Reconnaissance (MR) was adopted in 1949 and the Shackleton entered service as the MR.1. The prototype Shackleton had shown its wartime pedigree in the form of the defensive gun turrets in dorsal and tail positions, sporting the 20mm Hispano or Browning 0.50-calibre rather than 0.303in Brownings, with Hispano cannon carried in barbettes on the side of the nose. On delivery to RAF Coastal Command service in April 1951, the nose and tail guns had been removed from production versions of the MR.1. Veterans of the latter half of the Battle of the Atlantic would have felt quite at home in the Shackleton MR.1, although the cabin was a bit more comfortable than the Halifax and Lancaster and it did have a purpose-built galley and rest bunks.

One feature of the prototype that, had it been fitted to production aircraft, would have been completely new to veteran crews was the in-flight refuelling receiver fitted on the port side of the fuselage. This was to use Flight Refuelling Ltd.'s 'looped hose' method, a technique that was not adopted by the RAF and was superseded by the 'probe and drogue' systems commonplace today. Only the first prototype VW126 was fitted with this equipment. One other change implemented before production ramped up was changing the outboard engines from Griffon 57s to Griffon 57As that had a different air filter configuration. The first of what was known as the MR.1A flew in August 1951 and was delivered in January 1952.

As the MR.2 became available, the need arose for a replacement for the Lancaster GR.3s for ASW aircrew training and the MR.1s made surplus by the incoming MR.2s were converted to flying classrooms. This involved removing the dorsal turret and additional equipment for training, with the radar and systems from the later MR.1s and the MR.2 being fitted and this

The prototype Type 696 showing the Griffons with annular radiators and six-bladed contraprops. Also visible are the chin radome for ASV.13 and the cheek blisters for the 20mm Hispano cannon. This became the Shackleton GR.1, which was later amended to MR.1. *via Phil Butler*

became the Shackleton T.4. While by no means the perfect solution, the Shackleton MR.1 and 1A arrived on the scene at the right time, but in reality was a gap-filler, the term 'interim' could be applied to the MR.1 and 1A, something that always seems to be applied to aircraft whose requirement has been overtaken by events and technology. That term would crop up again and again in the procurement of maritime reconnaissance types.

Shackleton MR.2

Even before the first Shackleton MR.1 had been delivered to Coastal Command, aircrew visiting Woodford to inspect the Shackleton prototype had pointed out the high noise levels that would be the Shackleton's signature for its entire career. Later on, as trials progressed, comments from the A&AEE (Aircraft and Armament Experimental Establishment, renamed Aircraft and Armament Evaluation Establishment in 1992) were prompting the Air Staff Requirements Branch to re-assess the specification. The intention was to increase the range of what was to become the Shackleton MR.2 by reducing drag through fitting a new streamlined tail (as used on the Avro Lancastrian) and nose, a faired-in retractable radome, retractable tailwheel, lowering the turret height and removing the astrodome. The DF loop was

to be removed and replaced with an X303 radio compass. Other changes to increase range included fitting wingtip fuel tanks, which were deemed 'structurally difficult' and thus forgotten although provision was made for fitting a 400 Imperial gallon (1,818 litre) capacity fuel tank in the bomb bay, which would result in a reduction in the weapons load.

The main problem with the Shackleton MR.1 was noise and, to address this, a number of changes would need to be made. Noise in the Shackleton would be a perennial problem and was noted at an advisory design conference held on 9th May 1950. The DDOR1, Group Captain S R Ubee considered noise reduction a 'primary and urgent requirement both in the MR.1 and MR.2 Aircraft.' Tests had shown that the main noise problem was associated with 'high frequency resonance' and that the noise levels at '...the pilot's station gave cause for concern and was definitely worse than the Lincoln.' Of course, the source was the powerplant; the engine exhausts and the contra-rotating propellers. One reason the pilots' stations suffered from high noise levels was the relative position of the cockpit and the propellers, more or less in line with each other. The cabin noise was mainly due to the inboard engines' exhaust venting onto the fuselage sides while the source of the vibration was what was described as the 'order of rotation' and

This pair of Shackleton MR.2s of 42 Sqn showing the ASV.13 'dustbin' in its semi-extended 'patrol' position on WL800 (A) and retracted for take-off, landing and transit on the far aircraft, WL754 (F). *Author's collection*

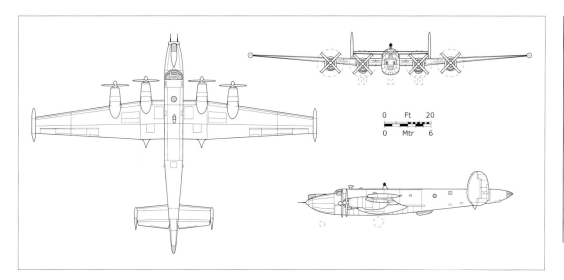

Concerns about the longevity and reliability of the Griffon prompted this interesting proposal from Avro to re-engine the Shackleton MR3 with the Napier Eland turboprop. This provisional drawing based on archive data shows how that type might have looked.

could be attenuated, but not resolved, by synchronising the engines.

The exhaust noise could be solved by fitting what Rolls-Royce described as a 'cross-over exhaust' whereby the exhaust manifolds on the inboard side of the engines were fitted with pipes that passed over the top of the engines to the outboard side. Such exhausts had been briefly fitted to Merlin engines during the war, but postwar they were used on the Canadair North Star airliner to reduce cabin noise. Unfortunately, given the time the cross-over pipes would take to develop, a quicker solution would be to shorten the exhaust pipes and fit thicker Plexiglas in the side windows of the Shackleton cockpit. As for the propellers, de Havilland was aware of the problem and advised that the roots of the propeller blades should be 'thickened'. Longer term, DH considered discarding the contraprops altogether but this would result in a reduction of cruise efficiency that the Air Staff would probably not allow. New four-bladed single rotation propellers could be developed but

these would, according to DH, need a new mark of Griffon engine to replace the Series 57s and by this time Rolls-Royce had little interest in reciprocating engines. The noise problem was so severe that Issue 2 of specification R.5/46 included a paragraph stating 'Noise is to be reduced so that normal speech is possible without raising the voice in the sound-proofed crew stations.' Of course this applied to the sonobuoy and radar operators, the pilots being by necessity exposed to the powerplant at their left and right shoulders.

By May 1950 the Air Staff were hoping to make the Shackleton Mk.2 a production continuation of the Mk.1, with modifications that could be applied to existing Mk.1s. Therefore the powerplant change was ruled out and the specification that resulted, R.5/46 Issue 2, was more or less '…intended to cover the production of a "cleaned up" Shackleton with all the modifications…' Other improvements that were to be included was the fitting of light-alloy radiators and oil coolers, which would save

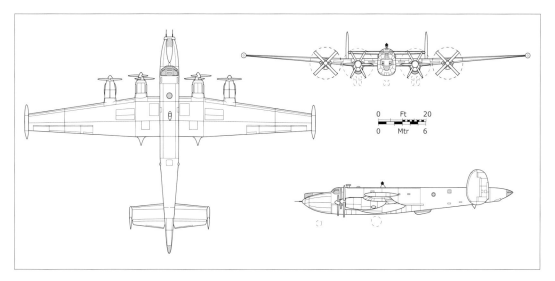

This interesting Avro study shows the variant proposed with two Griffons and two Nomads. All four engines would be used for take-off and transit, with the Griffons shutting down for extended range on patrols. Mixed powerplants have never been popular with any aircraft operator.

800 lb (363kg) and allow an additional 800 lb (363kg) of fuel to be carried, in the form of jettisonable bomb bay tanks. One item on the Air Staff's wishlist was that the engines should be interchangeable, but Rolls-Royce pointed out that this would not be possible if the cross-over exhausts were fitted.

The longevity, reliability and efficiency of the Griffon were of great concern to Coastal Command and the Air Staff as neither saw any future in the reciprocating engine. To address the reliability and longevity question, Avro suggested fitting the Shackleton with four Napier Eland turboprops, which might have changed the fortunes of D. Napier and Sons had the entire fleet been re-engined. On the efficiency front, Avro proposed replacing the outboard Griffons with Napier Nomad E.145 engines. The Nomad could run on pretty much any fuel, so using the same fuel as the Griffons would pose little problem and the efficiency of the Nomad would improve the endurance of the Shackleton. This of course, would rely on the Griffons being shut down and the patrol conducted on Nomad power. As for reliability, the question of whether the 'gadgets' on the Nomad would prove even less reliable than the Griffon was never answered.

Another factor that prevented the change from building MR.1s to MR.2s at an early stage in production was the need to replace the Lancaster GR.3s that Coastal Command had been using as a stop-gap after US aircraft such as the Consolidated Liberator had to be withdrawn under Lend-Lease regulations. By the end of 1950 the first MR.1 prototype had been converted to act as a MR.2 trials airframe but by that time the first production MR.1 had entered service at RAF Kinloss to replace 120 Sqn's Lancaster GR.3s.

Faster, Further, Heavier – The Original Shackleton Mk.3

Tropical trials in October 1950 of Shackleton MR.1 VW131 had shown that the aircraft's primary sensor, the ASV.13 radar with its antenna located in a chin radome, was susceptible to damage from bird-strikes. The A&AEE took the view that the antenna also suffered from airframe blanking and should be moved to a position under the rear fuselage in a retractable 'dustbin'. This was incorporated; along with existing improvements such as revised controls, 20mm Hispano cannon in the nose and retractable twin tailwheels, in the revised Issue 2 of R.5/46 that had been submitted to Avro in July 1950.

This gave rise to the Shackleton MR.2 that first flew on 17th June 1952 and entered squadron service seven months later. The MR.2 was operated alongside the MR.1s and did little to address the crew complaints about the MR.1's lack of sound-proofing and general discomfort on long patrols, with the levels of comfort not much better than those of the Lancasters and Lincolns that the Shackleton had replaced. Such was the level of complaint from the aircrews that RAE Farnborough became involved and to address the criticism, embarked on what amounted to a redesign of the Shackleton and this led to the issue on 18th November 1953 of R.5/46 Issue 3. Work on what became the Shackleton MR.2A, but ultimately MR.3, commenced in early 1955 and produced what at first glance looked like a MR.2 with wingtip fuel tanks. However, the changes were many and varied and in reality dated back as far as 1947 when OR.320 was issued in draft form for discussion.

The most obvious difference was the change from tail-dragger to tricycle undercarriage, with

The Avro Shackleton GR.1 prototype VW131 shows many of the reasons for the upgrade, if not outright replacement of the early Shackletons. These included the Perspex radome on the chin and high noise levels due to the propeller's proximity to the cockpit.
via Phil Butler

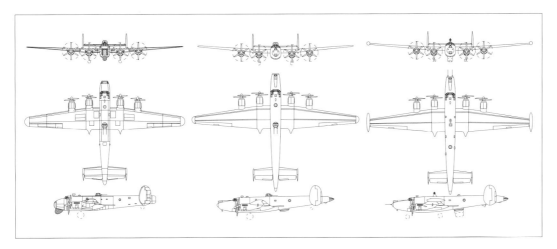

RAF Shackletons, left to right: The MR.1, MR.2, MR3. The MR.3, when fitted with the Viper turbojets, was the definitive type and the most capable. Former crew suggest it was superior to the Nimrod MR.1 in the ASW role.

twin wheels all round. The fuselage was lagged with fibreglass to provide better sound insulation, the exhausts were re-routed and large comfortable chairs installed at the crew workstations and look-out points. The dorsal turret was removed and a crew rest area with galley and bunks installed. The intention was to provide the operators with a comfort level on a par with contemporary airliners. A new canopy was fitted, replacing the heavily-framed canopy of the MR.1 and MR.2, which provided much improved vision for the pilots. Sensor-wise the new ASV.21 radar was installed, with the antenna in the same retractable ventral dustbin. The first prototype MR.3 flew in early September 1955 with the first Phase 1 examples entering service with No.220 Squadron in late August 1957. Almost as soon as it entered service the MR.3 began to be upgraded, initially with new equipment, but this second phase saw relocation of the HF detection kit to accommodate new ECM equipment such as Orange Harvest and Phase 2 became operational from 1961.

Phase 3 introduced most changes to the MR.3, to address the effect on aircraft performance of all the Phase 2 kit, with the AUW rising to 108,000 lb (48,988 kg). These modifications involved strengthening of the main wing spars to cope with the installation of a Bristol Siddeley Viper 203 turbojet in the rear of each outboard engine nacelle making the Shackleton MR.3 Phase 3 the RAF's only six-engined aircraft. The Vipers were only to be used to boost take-off performance and while two minutes at maximum thrust was the recommended limit, the Vipers could be run for five minutes at a push. These Phase 3 aircraft reached squadron service in 1965, but it was 1967 before all the Viper-equipped MR.3s were in service. With the Nimrod having been selected to replace the Shackletons and due to

enter service in 1970, the MR.3s were retired with indecent haste due to fatigue in the wing structure, induced by the Viper turbojets. By the end of September 1971 the last operational flight of a Shackleton MR.3 was carried out and all the Shackleton MR.3s had been retired by

The initial Shackleton MR.3 Phase I could be described as an MR.2 with tricycle undercarriage, but it featured improved crew facilities, better sound-proofing and the ASV.21 radar. This view shows the tip tanks and original outboard engine nacelle profile. *Author's collection*

The Bristol Siddeley Viper was installed in the outer nacelles of the Shackleton MR.3 Phase III. The reprofiled nacelles featured a jet pipe and a door to seal off the intake. The original idea was to boost take-off performance by installing liquid rocket engines such as the DH Super Sprite or solid rocket motors in the inboard nacelles.

the year's end. It wasn't only fatigue that affected the Shackleton force, as one MR.3 of 203 Sqn, WR986, had to be broken up at Luqa on Malta after the wings had been 'damaged beyond repair by rodents.' Perhaps the few surviving old hands from Coastal Command Sunderland days bemoaned the loss of the ship's cat.

The 'Super Shackleton'

The trouble with the Shackleton, aside from being slow, noisy and uncomfortable, was the aircraft's roots in the piston age, with the Griffon engine becoming increasingly difficult to maintain. By 1964 the MR.2s were approaching a point where they needed major refurbishment. In 1963 a total of 147 unscheduled engine changes had been made, equivalent to one engine change for every 170 flying hours. When compared with the Handley Page Hastings, an aircraft of similar vintage, that type's Bristol Hercules had one unscheduled change in 1600 flying hours. Another potential problem with the Shackleton MR.2 was skin cracking, something that threatened to significantly reduce the type's useful life.

As for the MR.3s, the use of the Viper turbojets for take-off took some of the strain off the Griffons and Coastal Command was confident that the MR.3 could last until the mid-1970s. Fitting the MR.2s with auxiliary engines was fleetingly considered, but the MR.2s were deemed unsuitable due to their tail-dragger undercarriage and the £5m development cost this would have incurred. The Air Council in 1962 approved a third phase of modernisation

of the 61 Shackleton MR.2s, to extend their lives into the early 1970s and by reconditioning the 32 MR.3 aircraft; they could soldier on until the late 1970s. However the in-service date of the Shackleton's replacement, initially to OR.350, then OR.357, was slipping continually and was by 1964 looking like being 1975, meaning there would be a 'gap' between the rundown of the Shackleton MR.2 force and the deployment of the OR.357 type. This would mean that 'Coastal Command will cease to be viable and maritime elements overseas will disappear. The Royal Air Force would thus be unable to meet its national commitments.'

Coastal Command's view was that replacing the Shackletons before 1970 would avoid the need to refurbish them, thus saving money and hasten the modernisation of Coastal Command. The year 1970 would soon become a critical date in the acquisition of a new MR aircraft. The MR.2s would be retired first, on the delivery of the new type and the MR.3s taken out of service by 1975. That was the plan, but as ever in UK weapons procurement, things went awry. In fact the search for a Shackleton replacement dated back as far as 1951, just as the aircraft was entering service, which is a fairly standard state of affairs in the military aircraft business. Back in 1951 the flying boat versus landplane debate was still in full swing and, as shown in Chapter Three, would continue until 1953 if not later. Apart from the flying boat question, the other debate, as described in Chapter Four, was whether to split the maritime patrol force into two roles with long-range types to patrol the oceans and the medium-range 'Kipper Fleet' to patrol the North Sea and coastal waters. Some officers in the Air Staff advocated the use of light, short-range types such as the Shorts Seamew in a similar role to the 'Scarecrows' of the war years.

Mr A E Woodward-Nutt, Director, Military Aircraft R&D at the MoS wrote on 31st October 1951 to the Director General of Technical Development and explained that the Air Staff were '...considering the requirement for a Shackleton replacement for maritime reconnaissance duties.' Woodward-Nutt attached a copy of a draft, unnumbered, Air Staff Requirement, which starts off with 'The performance and operational characteristics of the Shackleton have not proved entirely satisfactory. An early replacement of this aircraft is therefore most desirable.' This referred to the MR.2 that had barely entered the fray in late 1951, but appeared to be damned even at this early stage. The document continues to state that 'The aircraft must be highly versatile. Its main function is to kill submarines…'

Killing those submarines was to be carried out using homing torpedoes such as Dealer-B and Pentane, at least two of each per aircraft, and the 250 lb (113kg) 'direct aim anti-submarine bombs'. There is no mention of depth charges in the requirement but that old Coastal Command favourite, the 3in (7.6cm) rocket projectile, was also to be carried, although this was dependent on the assessment of its effectiveness against modern submarines. Gun armament comprised a pair of 'free' forward-firing 30mm ADEN cannon, a tail turret with two 20mm Hispano cannon and a removable mid-upper turret with a pair of Hispanos.

Experience with the Shackleton MR.1 had identified noise within the aircraft as a specific problem that should be addressed in any new aircraft, and the requirement stated that 'Noise is to be reduced so that normal speech is possible without raising the voice in the sound-proofed crew stations. Particular attention is to be paid to sound-proofing of the tactical, the pilots, and the rest stations.' All of which was obviously lifted from R.5/26 Issue 2.

Oddly enough, as discussions progressed, it transpired that the draft requirement was '…intended to provoke thought, and does not reflect general opinion.' and that the entire document was to be discussed at a meeting of Air Staff and MoS personnel in the middle of October 1951. The two main topics of the discussions were to be subsidiary roles for the aircraft and whether any aircraft resulting from the draft requirement would be bigger and heavier than the Shackleton and thus need '…bigger, stronger airfields.' In the event the meeting of the Operational Requirements Committee was rescheduled for 29th November 1951. Meantime, Squadron Leader G J Chandler at the Air Staff R+D Projects OR.2 had received from Bristol Aircraft 'a brochure on a "maritime reconnaissance landplane" developed from the Bristol Britannia.' Chandler was aware that the RCAF had been in discussions with Bristol on a maritime version of the Britannia (ultimately the Canadair CL-28 Argus) and that this new brochure '…would seem to be a "pot-pourri" of the Canadian requirements and of the draft requirement for a Shackleton replacement.'

The meeting went ahead and the first item discussed was the role, with submarine hunting being primary, but all agreed that convoy escort should carry equal weight in the requirement. Subsidiary roles were to include '…shadowing of enemy vessels, search and rescue, mine-laying and attacks on lightly-armed merchant vessels' for which those nose-mounted ADENs would be ideal. Group Captain S Lugg, Director of Radar Engineering (Air) suggested that a further role should be airborne early warning, but this was dismissed as the committee's consensus was '…that a special aircraft would be required for this purpose.' What would the attendees have made of the Shackleton AEW.2 of twenty years hence?

The committee then considered the range and flight profile of the type, and noted that OR.16 were of the opinion that '…a minimum radius of action of 1,250nm (2,315km) with an eight hour patrol was required in the anti-submarine role in the North Atlantic.' Fortunately for the crews, the committee advised that two crews would be required for any sorties in excess of eight hours while a four hour patrol at 1,000nm (1,850km) was the ideal mission profile. Mr A S Crouch, R+D (Projects) at the MoS advised that this would probably require an aircraft with an all-up-weight (AUW) of 135,000 lb (61,235kg) for the eight hour patrol and 120,000 lb (54,431kg) for the four hour patrol, both types being powered by Napier Nomad engines. All-in-all this was looking quite similar to the R.2/48/OR.231 flying boat specification that had been knocking about Whitehall since late 1946 and had by 1951 made very little progress, there being grave doubts about the future of flying boats.

Towards the end of the meeting the committee turned to a general discussion of maritime patrol types and the structure of the Coastal Command force. The future order of battle should comprise a short-range aircraft for coastal defence, a medium to long-range air-

While the Air Staff were considering a Shackleton replacement, the Canadians were looking for a new maritime patrol aircraft. This Canadair proposal based on the Britannia bears a startling resemblance to the Bristol Type 175MR. Note the remote-controlled tail guns and lack of wingtip tanks. *Canadair Photo via Terry Panopalis*

craft that would provide 'the main force of Coastal Command' and 'a very long-range aircraft to close the Mid-Atlantic "gap" and for some overseas commitments such as the Indian Ocean.' In concluding the meeting the chairman, Group Captain H P Broad, DDOR.1 at the Air Ministry, took the view that '…it looked as though any new OR for a Shackleton replacement would come out very much the same as the Shackleton itself.' Broad also said that until Coastal Command had 'fully assessed the Shackleton the question of its replacement could not be profitably decided.' The meeting broke up, having decided that '…no further action be taken on producing a Shackleton replacement requirement until a decision had been taken on the future structure of the MR force.'

So, despite having advised that no further action be taken on replacing the Shackleton, Group Captain Broad's committee had laid out the basic structure of the maritime force, the need for a long-range type and the outline of a sortie. While not a hard and fast requirement, it would give the aircraft companies an idea of what Coastal Command were thinking in the early 1950s. Vickers and Airspeed had the medium-range mission covered with their Viscount, Varsity and AS.69 maritime studies, while Avro had the main force role sewn up with the Shackleton, despite its drawbacks. This left the long-range 'gap filling' type, the 'Super Shackleton' for want of a better term, and its possibility as a Shackleton replacement.

Maritime Heavies

Bristol Aircraft at Filton were first to throw their hat in the 'Super Shackleton' ring with the 'maritime reconnaissance landplane' based on an aircraft that was still to make its first flight in 1951. Bristol Aircraft had plans for the Type 175 Britannia as it had the range, cabin space and payload to form the basis of a heavy maritime patrol aircraft. Developed in response to the Brabazon Committee's Medium Range Empire Type III requirement, specification C.2/47, the Type 175 was to carry 48 passengers on Empire routes. The Mk.1 was to be powered by Bristol's Centaurus 662 and this type had an AUW of 103,000 lb (46,720kg) but ultimately the Britannia Mk.2 was to be powered by the Bristol Proteus turboprop and increased in size to an AUW of 140,000 lb (63,503kg) for transatlantic routes.

To prepare the Type 175 for the maritime role required a number of fundamental changes that would be the hallmarks of the conversion of any airliner to maritime patrol. The first was a change of engines and the second being the fitting of a weapons bay under the cabin floor. Other modifications included a new forward fuselage with bomb-aimer's station, Boulton and Paul Type N turret with two 20mm Hispano cannon, a chin radome for the ASV.13 or the forthcoming ASV.21 radar, a new raised cockpit and a dorsal Bristol B.17 turret with a pair of 20mm Hispano cannon. The wing acquired wingtip fuel tanks, rails for 3in RP and pylons

This drawing of the Bristol Type 175MR shows the major changes from the Britannia airliner, particularly the new forward fuselage with turret and radome. The other major change was the adoption of the R3350 turbo-compound reciprocating engines. *via Duncan Greenman, Bristol Airchive*.

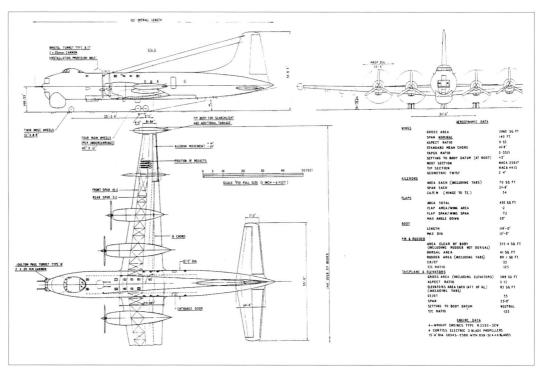

for stores such as air-to-surface missiles. The Proteus turboprops were to be replaced by Wright R3350-32W turbo-compound radials rated at 3,700hp (2,759kW), as fitted to the Lockheed Neptune, to provide better endurance than the turboprops.

The Type 175MR weighed in at 150,000 lb (68,039kg) but the Napier Nomad-powered version, the Type 189, was '…underpowered by four Nomads, each developing 3,050hp (2,274kW) and would need to be operated at a lower AUW of 135,000 lb (61,235kg)'. Apart from the engine installation and using diesel or kerosene, the Type 189 differed little from the Type 175MR. The Bristol Centaurus was also considered for the Type 175/189, possibly as a substitute for the Nomad whose development was protracted and in 1952, looking somewhat precarious.

Avro at Woodford and Chadderton had designed the Anson that until the Shackleton came along, was the only other bespoke MR landplane designed in Britain. Avro would be more than happy to have a crack at replacing its

A Bristol 175MR of Coastal Command launches a Fairey Sea Skimmer missile against a Soviet battlegroup. The Air Staff considered the Bristol type better suited to the role. *Adrian Mann*

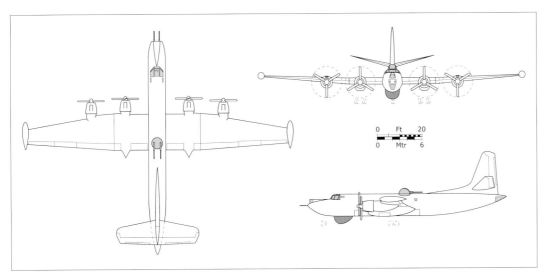

The Avro Type 716, seen here with R3350-85 engines was intended to address the many shortcomings of the Shackleton MR.1. The type was to be powered by the Napier Nomad or Bristol Centaurus, with the R3350 as an option. The Air Staff were unimpressed and sent the Woodford design team back to their drawing boards.

91

One of the changes on the Type 716 and 719 was to the undercarriage with the adoption of an eight-wheel bogie similar to this Vulcan type. The single wheel of the Shackleton MR.1 and MR.2 was replaced for two reasons: it allowed heavier AUW and the flatter profile reduced the depth of the undercarriage bays in the wings.

Avro artwork showing a Type 719, the putative Shackleton MR.4. The Air Staff were unconvinced by this type, and its rival the Bristol Type 175MR. The Shackleton MR.3 was developed instead, applying lessons learned from the MR.2 and the new sensors such as ASV.21 and sonobuoy technology. Ultimately the MR.2 outlived the MR.3 by becoming the RAF's AEW platform.
Avro Heritage

own design and address the criticism that had been levelled at the Shackleton, much of which derived from its wartime roots. The first attempt was the Type 716 that dated from April 1950 and tentatively called the Shackleton Mk.3. This was an entirely new aircraft with a fuselage similar to the existing Shackleton but was soon changed to one with a wider, circular cross-section to carry a pair of Pentane torpedoes side-by-side in the bomb bay. A completely new stressed-skin wing was designed with squared-off tips spanning 120ft (36.6m) and fitted with slotted flaps to improve take-off performance. The Type 716 had a nose like a Shackleton MR.2, but that was about the only similarity as the empennage was a single fin with 15° of dihedral on the tailplanes. The fin was to fold so that the Type 716 could use existing hangars. It was to have an AUW of around 113,000lb (51,256kg) and be powered by four Nomads (Napier E.125) or four Bristol Centaurus. The intention was that the Type 716 would address the crew comfort aspects for which the Shackleton MR.1 had been criticised, with particular attention to noise and vibration. The cockpit was moved forward of the propeller disks by 8ft

(2.4m) to reduce the noise levels for the pilots.

As ever, the cost of this new design was a bit much for the Air Staff and MoS and so Avro 'In view of the absolute necessity for economy in design and manufacturing effort and for the production of the aircraft at an early date...' suggested that the design be modified to use the outer wings of the existing Shackleton MR.1 and 2. The fuselage would be new as would the inner wing and centre section. These inboard wing sections would now span 40ft 6in (12.3m) providing '...greater fuel capacity and aerodynamic efficiency.' and would be fitted with high-lift flaps to improve take-off and landing performance at the higher AUW. This redesigned aircraft would be called the Type 719, would have a wing span of 132ft (40.2m) and an AUW of 115,000lb (52,163kg). A tricycle undercarriage was adopted, something that had become the norm on large aircraft by 1950, and the main undercarriage comprised eight-wheel bogies in the style of the Type 698 Vulcan. This produced a flatter design that allowed the wheels to be stowed within the inboard nacelles without 'breaking into' the wing's structure.

The Type 719 was to be powered by the Napier E.145 Nomad, a somewhat simplified version of the Nomad that used single-rotation propellers rather than the contraprops of the E.125. The Nomad and Centaurus versions were to use the same DH Propellers four-bladed airscrews of 15ft (4.6m) diameter. Both engine types were to use a common mounting, which would allow the Type 719 to be developed and produced with Centaurus engines but be upgraded with fully-developed Nomads at a later stage. The Nomad installation, even with its leading edge radiators, produced slightly less drag than the Centaurus mounting that only had its oil coolers in the leading edge. Comparison of the two engines' performance showed that there was little to choose between them in range and speed, unless the transit was above 15,000ft (4,572m) where the Nomad had an advantage. The fuel system was basically an improved version of the MR.2's, with the outer wings retaining the MR.2's tankage with 1,149 Imperial gallons (5,223 litres) combined capacity, while the new inboard and centre section tanks carried 2,412 Imperial gallons (10,965 litres).

The Type 719, for an anti-submarine mission, was to be equipped with 40 sonobuoys, 40 marine markers and a pair of Pentane torpedoes in the bomb bay plus another 18 sonobuoys and a variety of markers and flares

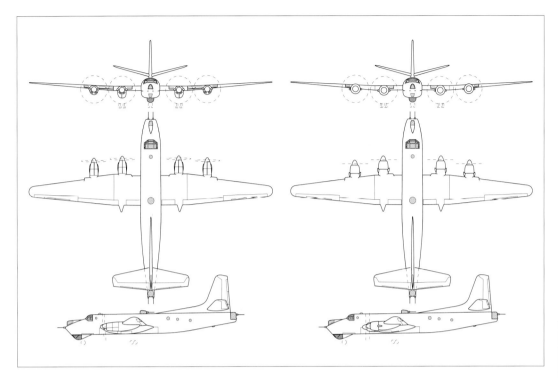

The Avro Type 719 came with two engine options: the Napier E.145 Nomad 2 on the left and the Bristol Centaurus 660 shown on the right. The Centaurus version was developed as insurance, in the light of the Nomad's continuing development delays, but mainly due to uncertainty about its future.

within the cabin. In total an anti-submarine stores load of 9,234 lb (4,188 kg) was to be carried, similar to that of the Shackleton MR.2, but the larger bomb bay allowed more flexibility and higher warloads. Alternative loads for the bomb bay were 65 depth charges, up to twenty-one 1,000 lb HC bombs or up to 55 of the 250 lb anti-submarine bombs. For the SAR role a single Mk.3 Airborne Lifeboat could be carried if modified bomb doors were fitted.

Avro compared the Centaurus and Nomad-powered Type 719 with their Griffon-powered Shackleton MR.2 and given the better fuel consumption of the Centaurus and Nomad, plus the Type 719's increased fuel capacity, the still air range of the new aircraft was more than 1,100 nm (2,037 km) better than the MR.2. In practical terms this meant that the Type 719 could perform an eight hour patrol at 1,000 nm (1,852 km) while the MR.2 was capable of four hours at 800 nm (1,482 km). The Nomad allowed faster transit times, estimated at a saving of four hours on the sortie to 1,000 nm (1,852 km). The Nomad also offered reduced operating costs due to it wide choice of fuel grade and much reduced noise and vibration levels, which might sound surprising for a 1940s-era diesel engine, but the Nomad wasn't a normal diesel. As noted in Chapter Three, the Nomad was to be the future of long-range oceanic operations but needed a lot of development, and money, to prove its worth. To speed its development for the later marks of Shackleton, Avro supplied the second proto-

type Shackleton MR.1 (VW131) to Napiers at Luton for fitting out with Nomads.

Avro also drew up a design study for a flying testbed with Nomad engines in the shape of the Type 717, a cleaned-up Lincoln with its inboard Merlins substituted by Nomads and the outer engines and wing deleted, to be replaced with Tudor outer wings. The resulting aircraft was to be used for long, potentially record-breaking, flights to prove the worth of the Nomad. Unfortunately for Napier, the best efficiency of the Nomad was achieved at high altitude and long flights of more than 13,000 nm (24,076 km) at such heights in the unpressurised Lancastrian were not feasible for crews on oxygen for the entire flight.

Avro's Type 717 was intended to be a demonstrator/testbed for the Napier Nomad E125. It combined elements of the Lincoln and Tudor, with a pair of Nomad E125 mounted on the inboard stations. Avro's hopes of record-breaking flights were dashed due to its lack of pressurisation. This provisional drawing was produced using material from Avro Heritage.

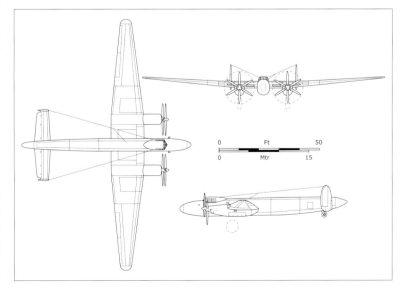

The Winning New Boys Lose Out

In February 1952, at the request of the VCAS, Air Chief Marshal Ralph Cochrane, the Operation Requirements Branch OR.2 Squadron Leader Chandler examined the brochures from Bristol and Avro. OR.2 explained that while no firm requirement for such an aircraft had been issued 'Sales enthusiasm on the part of both Firms has resulted in the preparation of brochures …' and the contents were pretty much based on the companies' guesswork of what a requirement would be. The basis of each was similar: 'a reasonable patrol time' at 1,000nm (1,852km) with an 8,000lb (3,629kg) stores load.

The assessment of the Bristol 175MR and Avro Type 719, as presented in the brochures, gives an insight into the thinking of the Air Staff at the time, and makes an interesting comparison with the proposals of the next decade. In the case of the Bristol design, Chandler saw it as the clear winner, concluding '…the Bristol 175 offers the better solution and is also capable of development to higher weights and correspondingly longer endurances.' All this sounds very odd considering new boys Bristol were up against Avro with their Shackleton experience, so what were the Air Staff's reasons?

Firstly the only advantage they saw in the Type 719 was that it used the Shackleton MR.2's outer wings, while the fact that the rest of the airframe was completely new was seen as a major disadvantage. The 719 was '…too small to give a reasonable patrol time.' Nor did it have, in their opinion, any '…development future in the role.' Lastly the wing loading was too high to operate from Coastal Command's runways.

As for the Bristol 175, it was quite the reverse state of affairs; the only disadvantage was its wing loading being too high, particularly for tropical operations. Advantages were many, including that its higher AUW of 135,000lb gave '…approximately twice the patrol time of the Avro 719'. Being a development of an existing, civil, design '…should considerably reduce the cost and time necessary for development and production.' While being below the design weight of the civil Britannia '…it should have a reasonable development future.' Lastly the Bristol 175 '…would be capable of benefitting from any subsequent increase in engine power which may result from trials on the Nomad engines.'

This, of course, was a paper exercise against an officially non-existent requirement and shows how, when given vague numbers to work with, a company can produce a completely unsuitable project study. In fairness, Squadron Leader Chandler appreciated this and his report concluded with 'It is reasonable to conclude that if A.V. Roe were given the suggested performance figures required they could produce and better aeroplane than the 719 to meet the requirement.

Prior to preparing his detailed report for Cochrane, Chandler had also examined both proposals against R.2/48/OR.231 and had, as mentioned above, found both wanting in take-off performance due to their wing-loading of 70lb/ft² (341.6kg/m²) being deemed too high. Chandler listed four options to address this situation, none of which would be cheap. The first was to proceed with the existing designs and fit rocket-assisted take-off gear, the second involved building new longer, stronger runways while the third was to redesign the wings of the aircraft. The fourth might just betray Chandler's background: 'Build a flying boat', which, as noted in Chapter Three, had been the standard solution to the wing-loading problem since the Great War.

So the search for a Shackleton replacement now encroached upon the ongoing R.2/48/OR.231 Sunderland replacement process, with the argument on the pros and cons of flying boats that had split the Air Staff into landplane/flying boat camps. Neither faction prospered nor did they know it would take almost two decades before the true Shackleton replacement appeared on the scene and that would be based on an airframe of similar vintage to the Britannia. The Shackleton MR.2 evolved, via the MR.2A, into the MR.3 that despite Avro's efforts in 1950-52, was not a larger new design but an improved MR.2 incorporating the lessons learned from the Type 716 and 719 studies and carrying the ASV.21 radar. The Shackleton MR.3s were retired before the MR.2, which on the arrival of the Nimrod was fitted with the AN/APS.20 radar from the Fleet Air Arm's Fairey Gannet AEW.3s to become the Shackleton AEW.2. In this guise, the Growler and its Griffons soldiered on into the Nineties, a unique sound in the sky above north east Scotland in the latter half of the Cold War.

The real winner was the Royal Canadian Air Force whose Canadair CL-28 Argus, basically a Bristol Type 175 Britannia, with a much-modified fuselage and Wright R3350 engines, entered service in 1957 and soldiered on until 1982. Could that have been a viable Shackleton replacement? Perhaps, but as the decade rolled on, the Shackleton would need to be replaced.

Opposite page:

Shackleton MR2 WL754 of 42 Sqn with its dustbin in attack configuration. Interestingly WL754 was later converted to an AEW.2 airborne early warning Shackleton with 8 Sqn. *Author's collection*

The real winner of the Air Staff's search for an improved Shackleton was the Royal Canadian Air Force. Canadair opted to take the Bristol Britannia and by fitting a new fuselage and R3350 Turbo-Compound engines, produced the Canadair CP-107 Argus. Argus 20723 of 404 Sqn is seen here overflying a Royal Canadian Navy *Oberon*-class submarine. *Terry Panopalis Collection*

Having survived the 1957 Defence Review, which focused on air defence and deterrence, the fact was that a Shackleton replacement was still required. As luck would have it, NATO had embarked on a series of programmes in an attempt to standardise its members' military hardware. A maritime reconnaissance type was one such project and this became NBMR.2. However, determined not to become involved, the Air Staff had other ideas.

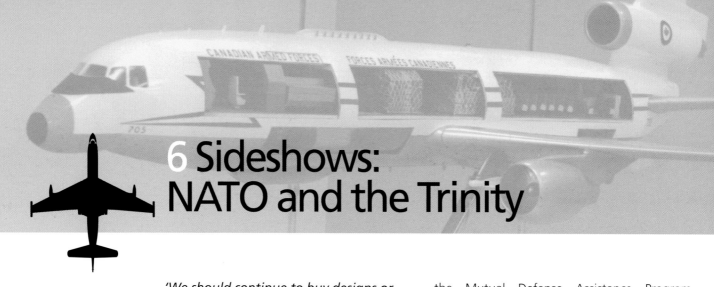

6 Sideshows: NATO and the Trinity

'We should continue to buy designs or finished equipment from the Continent where it pays us to. But we should not yield to pressure to buy equipment that we could make better or more cheaply ourselves, merely to induce the Continent to buy more from us than they otherwise would or to display co-operativeness.' Report of Policy Committee on Interdependence in Research and Production – PCIRP/P(59)6, 1959

'The possibility of obtaining three (and possibly four) derivatives from one basic R+D programme has great attraction from a financial aspect.' Air Commodore I G Esplin, DOR(A), December 1961

The burgeoning Soviet submarine threat had prompted a discussion amongst NATO members about how to go about countering it. Air power was still seen as a key facet of anti-submarine warfare and the United States, as described in Chapter Four, supplied the Lockheed P2V Neptune to European air arms under the Mutual Defense Assistance Program (MDAP) in 1952. Coastal Command received their first Neptune MR.1 in 1952, and these complemented the RAF's Shackletons, allowing these to operate around the Empire, something forbidden for the Neptune under the terms of the MDAP. Canada operated the Neptune as the CP-122 from 1955, the French *Aéronavale* operated 64 P2V6s and P2V7s from 1953 until 1983 and the Dutch *Marine-Luchtvaartdienst (MLD)* from 1953 until 1982. The *Força Aérea Portuguesa* (FAP) also operated the Neptune, but only after 1961 when twelve P2V-5 were transferred from the *Marine-Luchtvaartdienst*.

The search had begun on 14th December 1956 when the NATO Council outlined the need to replace the Neptunes in service with NATO members and it made sense to have a single type across the alliance. One problem encountered by the Western Allies during World War Two was the logistical and tactical difficulties in operating different types of kit. Basic items such as rifles posed a problem with the British Lee-Enfield using a .303in (7.7mm) calibre round while the American M1 Garand

The Lockheed P2V Neptune served with a number of NATO countries. By 1956 the Neptune was established in service and, as ever, its replacement was being planned. This was the origin of NBMR.2 *US Navy*

British Overseas Airways Corporation's first DC-7C Seven Seas, G-AOIA. The English Electric Co. had high hopes of interesting the NBMR.2 Group of Experts with their DC-7 maritime patrol proposal. This would be based on airline-surplus airframes that would be re-engined, fitted with ASV and weapons bays. EECo hoped that their group's expertise in engines, electronics, radar and airframes would prove a winner. *British Airways Speedbird Heritage Collection*

used the 0.3 (7.62mm). Nor was this restricted to rifles, with the Americans generally using US equipment, while the British basically used both British and American kit supplied under Lend-Lease. This of course complicated the logistics chain, made training more difficult as British personnel trained on British kit could also be posted to units equipped with American weapons. If NATO was to be a fighting force under a unified command, its armies should realistically have standard equipment, ideally developed to a common requirement. While this was possible for the Warsaw Pact nations, who used Soviet-supplied weapons, the same could not really be expected of the competitive capitalist West with arms companies to support and many of the NATO countries, particularly the UK and France, with considerable overseas interests to look after, specifically their colonies.

The NATO Council in 1957 issued NATO Basic Military Requirement No.2 (NBMR.2) for a maritime patrol and anti-submarine aircraft to replace the Lockheed P2V Neptune in the NATO air and naval forces. NBMR.2 called for a common airframe and engine configuration, with only the internal equipment differing between the countries. The outline requirement was for a Neptune replacement with a payload of 6,000 lb (2,721 kg) comprising ASV radar, sonobuoys, sonar operators' systems and weapons. The type was to perform a four hour low-altitude patrol at a distance of 1,000 nm (1,852 km) from base, rising to eight hours patrol at 600 nm (1,111 km). Given the distances to the patrol areas, transit speed was important and 300 kt (560 km/h) was the cruising speed for

transit. This speed would require gas turbine propulsion, with the Rolls-Royce RB.109 Tyne turboprop being the preferred engine. The requirement also stated that in the event of an engine failure, the type should be able to complete its mission on one engine! A tall order, one that the UK Air Staff were not at all convinced would be possible, having specified at least three engines for its oceanic patrol aircraft. Oceanic was the key, as the Air Staff considered NMBR.2 to be aimed at operations closer to base; North, Baltic and Mediterranean Seas, rather than their Atlantic and Indian Ocean needs.

In the end, 21 projects were considered; five of these were British and included the last time a British flying boat was proposed for a military requirement and a proposal to boost the market in new airliners by converting the old, but ultimately the British dismissed the entire programme. The competition process was judged not only on meeting the requirement, but by the amount of collaboration between the various countries. This would be the main stumbling block in the NBMR.2 story, particularly for the British and Americans who, as might be expected, followed their own national interests. As ever, the French were keen on industrial collaboration, if they maintained design authority.

One example of this drive for co-operation was the proposal put forward by the English Electric Group in December 1958. English Electric, in the shape of D. Napier and Son, were in collaboration with the Douglas Aircraft Company, Avions Marcel Dassault and Hamburger Flugzeugbau. Their tender involved converting

The frosty relations between The Admiralty and The Air Staff during the Battle of the Atlantic might not have been helped by discussions on the Douglas C-54 Skymaster. Here three Douglas DC-4s are seen in a variety of guises: left to right – Camouflaged USAAF C-54 Skymaster and US Navy R5D-1 freighters with a USAAF C-54 passenger transport in bare metal finish. *USAF*

airline-surplus Douglas DC-7 airliners into maritime patrol aircraft by fitting Napier Eland turboprops and maritime patrol equipment. Interestingly the DC-7's predecessor, the Douglas DC-4 in its USAAF C-54 transport form had been considered as an ideal Very Long Range (VLR) anti-submarine/general reconnaissance (GR) aircraft by the Admiralty as far back as February 1942. This was prompted by a letter from the Admiral J N Edelsten who had discovered that the C-54 was a long-range aircraft and contacted ACAS(TR), Air Vice-Marshal R S Sorley on the matter. Edelsten was actually very realistic in his request, ending it with 'If the Air Force view is that this aircraft is unsuited to the requirement, I shall be satisfied.'

Aside from the fact that the Americans would not provide them, (a few months after the Pearl Harbor attack they needed every aircraft they could get) the Air Staff's view was that the C-54 would need a lot of work to convert it to the GR role. A year later, the Admiralty was again agitating for the C-54; such was the need to cover the mid-Atlantic 'Black Hole'. The Air Staff did not particularly appreciate the Admiralty's interference and advised on what would be required to convert the C-54. The modifications listed initially for the C-54 were '12 x 250 lb depth charges but no defensive armament' and 'centimeter *(sic)* ASV and normal radio equipment for G.R. duties'. The depth charges would need to be carried externally under the wings, although it was not known if this was possible. The lack of any defensive

armament was feasible as air attack was unlikely in the mid-Atlantic as the C-54 GRs were to operate from Bathurst (now called Banjul in The Gambia), the Azores and Gibraltar. However, there was concern that the C-54 could not be fitted with machine guns or cannon for flak suppression on the attack runs. The C-54 GR was pretty much killed by a letter from Group Captain A R Wardle, DDORI, who informed Edelsten that while the C-54 was an aircraft with great range and carrying capacity, 'There is however, no doubt that it would be impossible to produce from the C-54 a reconnaissance aircraft with anything like the range of the Liberator and of course it would be nothing like so efficient as a military aircraft.' Wardle's reply was terse and had been toned down from its original draft, which Sorley had described as '…a bit too fruity for you to send to Edelsten…' and the matter was dropped. The Admiralty and the Air Staff did not have a very good working relationship on the subject of maritime air cover throughout the Battle of the Atlantic. Wardle's 'fruity' letter may just have had a part in that.

As for the re-engined DC-7 of 1958, English Electric Aviation's Managing Director Lord Caldecote in a letter to Sir Thomas Padmore, Second Secretary at the Treasury, described it as having great importance to the British aircraft industry. Caldecote wrote 'It is clear from the broad economic standpoint and from the point of view of selling more civil aircraft that it must be better to use existing surplus aircraft, and so

help to take them off the market rather than to manufacture a new aircraft for this purpose.' Caldecote's view was that sales of British airliners to European and US airlines would increase if English Electric's consortium bought up their DC-7s, encouraging them to buy British aircraft. Caldecote advised Padmore that the Ministry of Supply was aware of EEA's proposal, but thought the Treasury ought to know about the export opportunities that could open by taking surplus DC-7s off the market. Douglas Aircraft had already proposed a DC-7D powered by Rolls-Royce Tynes, the engine preferred for NBMR.2, but English Electric had their own in-house engine builders in D. Napier and Son.

Meanwhile at the Ministry of Supply, the English Electric proposal was being examined. It was claimed that the Eland-engined DC-7 would '…meet or exceed all the requirements of the specification.' The Eland DC-7 could also be used for alternative roles such as trooping, '…carrying over 100 passengers' and that 'As a four-engined aircraft it will be much more satisfactory for long over-water missions than the twin-engined aircraft.' and would '…be economically preferable to any new design.' The Permanent Secretary at the MoS, Cyril Musgrove, who on 23rd December 1958 wrote back to Padmore with 'The proposal has merits. But it was not one of the short list of three designs (out of the 15 submitted) which were recommended for further study.'

Musgrove's letter reveals some interesting facts about the NBMR.2 process, specifically that the '…present favourite is the joint Breguet/Avro project based on two Tyne engines, a smaller and cheaper machine than the DC-7.' Musgrove goes on to state that the process had a long way to go because '…nobody has said they will put up any money for the aeroplane, except the French and there is no United Kingdom requirement – our Shackletons will see us out for some time to come.' Then, Musgrove pointed out that if American pressure were to be applied to the NATO group, the competition might be re-opened '…and in that case, in view of the Douglas backing of the English Electric proposal, it might again figure in the calculations.' Despite this, Musgrove contended that NATO will '…firm up on the Breguet/Avro proposition.' However, even with the obvious carrot of cheapness, always a winner with the Treasury, English Electric's proposal had been dealt a blow by the NATO Group of Experts, who had issued NBMR.2, that '…it is recommended that no further consideration be given to your DC-7 project.'. Realistically, the

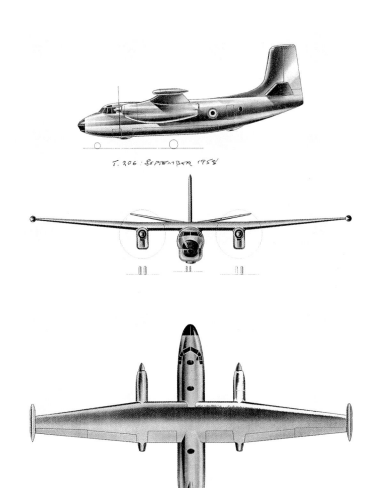

T. 206 : SEPTEMBER 1958

DC-7MR was never a practical proposal, mainly due to its use of old airframes (something that cropped up time and again in postwar UK defence procurement) but since a new airframe was preferred, Britain's aircraft companies drew up design studies for NBMR.2, all of them in the 60-80,000 lb (27-32,000kg) AUW class.

As described in Chapter Five, the Bristol Aeroplane Company at Filton had earlier in the Fifties proposed a variant of their Type 175 Britannia as a Shackleton replacement. This Shackleton replacement was much larger than the type required for NBMR.2, so a clean-sheet design was tendered in the form of the Bristol Type 206. A high-wing monoplane, reminiscent of a Handley Page Herald, the Bristol 206 was powered by a pair of Tynes mounted under the wings whose deep nacelles also housed the main undercarriage. As an alternative, using in-house engines, was the Type 207 that was essentially a Type 206 with two Armstrong Siddeley Mambas

The Bristol Type 206 with its high wing offered an unrestricted operators' cabin and long underfloor weapons bay. A further development, the Type 207, featured a BE.53 turbofan in the tail to boost transit speed. Neither proposal won over the committee of experts.
Duncan Greenman/ Bristol Airchive

Fairey's submission to NBMR2 was Project 83 from June 1958 that followed the standard configuration of mid-wing/twin Tyne of the Avro 745 and the winning Atlantic. The Project 83 aircraft weighed in at 81,000 lb (36,734kg) but was faster than the speed specified in NBMR2.
AgustaWestland via Tony Buttler

Saunders-Roe tendered the P.208 flying boat for NBMR2. It could be described as the pinnacle of anti-submarine flying boats; narrow hull, pylon-mounted wing and hydroskis that would allow fast dipping sonar operations. By 1958 the flying boat had had its day and the P.208 was not given serious consideration.

and a Bristol BE.53 turbofan in the tail. The ASV-21 radar was housed in a retractable dustbin under the forward fuselage, ahead of a capacious weapons bay. The tapered wing was fitted with tip pods that held a searchlight (port) and the guidance system (starboard) for the pair of Martin AGM-12 Bullpup missiles (until 1962 this was known as the ASM-N-7) that were to be carried on pylons under the outer wings. These were to be used against surface targets and were no doubt selected for the design study due to the lack of a British ASM rather than as a nod to international co-operation.

Fairey's Project 83 Maritime Patrol Aircraft (MPA) was proposed to the NATO Expert Group

and shared the configuration, and appearance, of the ultimate winner, the Breguet Br.1150 in being a mid-winged twin-Tyne type. Fairey built on their experience of developing the Gannet carrier-borne ASW type and examined various powerplants before adopting the Tyne as per all the other design studies. The Project 83 had a take-off weight of 81,000 lb (36,741kg), a wingspan of 115ft (35m) and a length of 80ft (24.4m). Fairey had considered mixed turbojet/turboprop propulsion, but considered the increase in transit speed did not warrant '…the inherent maintenance and complexity disadvantages which outweigh any superficial gains in fuel consumption or weight.' Nor were Fairey making a return to the coupled engine such as the ASMD.3 Double Mamba used on the Gannet AS.4 due to the '…time involved in developing the necessary gearing systems…' and were more than happy to use the proven Tyne 11.

As noted above, the NBMR.2 tender process saw the last submission of a flying boat for a military specification. Saunders-Roe, (Saro) arguably the most technically advanced of the smaller aircraft companies in Britain, had interests ranging from lifeboats to the Black Knight research rockets used to test re-entry vehicles for the Blue Streak ballistic missile. As described in Chapter Three, Saro were renowned for their flying boats, but by 1958 had diversified into the high-technology areas of an industry that would soon be called aerospace. The Ministry of Supply had wanted to see Saro go under, but

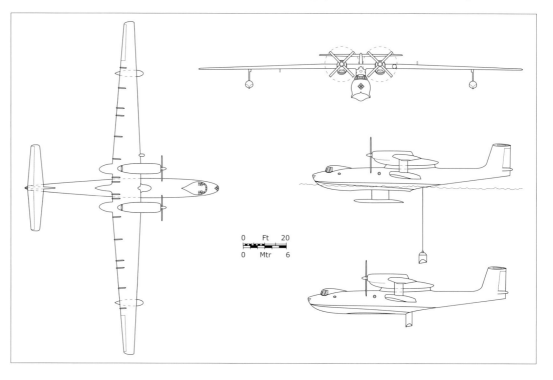

0 Ft 20
0 Mtr 6

Saro's innovative and skilled engineers kept them afloat.

The Royal New Zealand Air Force in 1953 approached Saro for a flying boat to replace their Shorts Sunderlands, and since the RAF's R.2/48 boat (See Chapter Three) was too big, asked if a smaller flying boat could be developed. Saro responded with the P.176, an anti-submarine patrol aircraft with an AUW of 70,000 lb (31,751kg) and powered by a pair of Napier E.151 Eland turboprops, rated at 3,500ehp (2,609kW). This was a handsome parasol-winged design with the engines in nacelles above the leading edge of the tapered, high aspect ratio wings and fitted with a T-tail. This could have conducted a four hour patrol at a distance of 750nm (1,389km) with a transit speed of 240kt (444km/h). When NBMR.2 was issued five years later, Saro looked again at the P.176 and updated it as the P.208 with Tyne 11 turboprops and increased the AUW to 73,000 lb (33,112kg). In addition to meeting the four hours at 1,000nm (1,853km) patrol requirement, the Tyne engines also increased the cruise speed to 300kt (556km/h) at an altitude of 39,000ft (11,887m). Being a flying boat all the sensors, such as the ASV-21 radar, that were installed in ventral locations on maritime patrol aircraft had to be retractable and protected from water ingress by sealed hatches.

Saro claimed that the P.208 was built to operate on the open ocean with the latest in dipping sonars. To alight on the sea in rougher conditions than a conventional flying boat boat, the P.208 was fitted with a hydroski, an innovative low-drag system that allowed easier take-off and landing. Another aspect of the hydroski was that it could allow the flying boat to skim across the surface at low speeds while towing a sonar array. In this mode, the drag of the hydroski was a fraction of that of the flying boat hull. With this towed array, the ever-innovative Saro engineers considered this a better solution to tracking submarines than sonobuoys. Alas, the P.208 was dismissed by the Group of Experts, who were convinced that the landplane would perform better than a flying boat. Saro had learned much about hull design since the war, leading to their adoption of the high beam-to-length ratio hull to reduce drag, both in flight and on the water, but this was not enough to convince the group who, like the Air Staff, were not interested in a flying boat.

While Saro's P.208 was the last British boat to be tendered for a military requirement, on the shores of Lake Constance in southern West

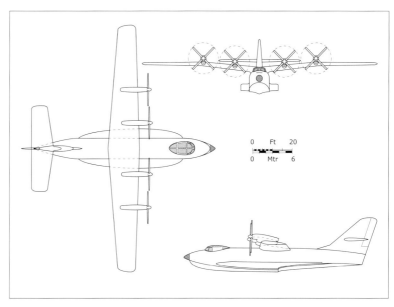

Germany, Dornier Flugzeugwerke was drawing up a design for a flying boat to meet NBMR.2. Dornier had a long history, on a par with Saro, in the seaplane business and having re-established a design and manufacturing operation at Friedrichshafen in 1954 were back in the aircraft business. Dornier proposed the P.340, a high-wing monoplane amphibian with long, fairly narrow sponsons along the hull that would carry the anti-submarine stores. The first design was the P.340/1-01, dating from March 1958 which established the basic arrangement of the P.340 with a pair of Rolls-Royce Tyne turboprops in overwing nacelles driving 15ft 9in (4.8m) propellers. Additional power for take-off provided by a Bristol Siddeley Orpheus turbojet buried in the fin root with its intake closed off in the cruise by a section of the fin fillet that retracted into the fuselage. The 1-01's original hull shape was, in plan view, a biconvex lens shape that saw the fuselage mounted on the extended sponsons that formed the lower hull, but this was soon changed to the hull and sponsons being separate. The wings and tailplanes had slightly swept leading edges and straight trailing edges, but the broad fin was quite highly swept. The flight surfaces had neither anhedral nor dihedral as the hull depth and high-mounted engines kept the propellers clear of spray.

The Dornier P.340-2/01 used the same airframe but was fitted with four Allison T56 turboprops, again mounted above the wings, driving 15ft (4.6m) propellers, with the outboard engines placed at mid span. The wings now had dihedral on the inner section as far as the outboard engines, with the outboard section level, giving a slight gull-wing appearance.

Dornier's design studies for NBMR.2 built on their expertise in flying boats and amphibians. The P340 2-01 was a handsome gull-winged design with Dornier's trademark sponson. The four Allison T56 turboprops were mounted high on the wings clear of any spray. The P340 3-01 was fitted with three T56, one on each wing and one on the tail fin, while the P340 3-02 used three Tyne in a similar configuration.

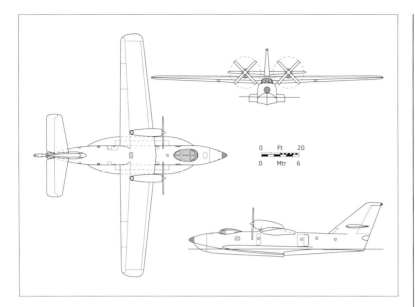

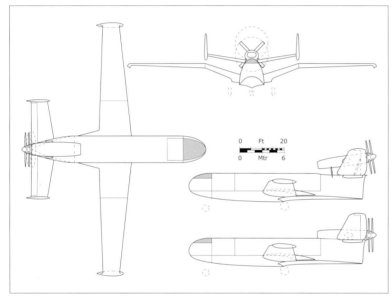

The P340 4-07 was Dornier's submission for the NATO MR requirement. It used the recommended Rolls-Royce Tyne turboprops with a Bristol Orpheus turbojet installed in the rear fuselage, with doors covering the intake and jet pipe. The 4-07 was a rather handsome design and could have prospered had it been selected.

One of the more radical proposals from Dornier was the P340 6-07 that combined some of the advanced developments in flying boats with some of the oldest. The new aspects included the low-mounted gull wing with integral floats and sponsons while the old was the tilting engine nacelle that harked back to the Do 26 whose pusher propellers tilted up to keep them clear of spray on take-off. The P 340 6-07 used a Coupled Tyne driving contraprops.

A model of the Dornier P340 4-07. The use of sponsons on the hull provided good stability and plenty of space for fuel and weapons. The nose-mounted radar could be replaced by bomb-aimer's position and a retractable scanner with 360° coverage under the rear fuselage.
Ingo Weidig/Dornier Museum Friedrichshafen

The P.340/3-01 from March 1958 and the 3-02 from April 1958 were essentially 1-01s with the Orpheus turbojet removed and a new, taller unswept fin and a third engine installed in the fin leading edge above the tailplane, with the 3-01 using the T56 and the 3-02 fitted with a third Tyne.

The P.340/4 series formed the basis of the study submitted to the Expert Group and this was, as expected, Tyne powered with an Orpheus in the rear fuselage, fed by an intake in the fin leading edge. The definitive P.340/4-07 retained the original 1-01 hull with a highly-swept fin, with a pair of doors covering the Orpheus intake rather than a retractable section of the fin fillet. The wing was level, with a swept leading edge and straight trailing edge, with the Tynes mounted in overwing nacelles. The P.340 was a handsome design, but Dornier Flugzeugwerke's seaplanes always did look the part.

As noted in Chapter Three, the late Forties and early Fifties were possibly the period during which seaplanes, particularly flying boats showed most innovation, for least return, being dismissed out of hand by air forces around the world. Dornier was no exception to this and in the last design study for the P.340, the 6-07, pulled out all the stops. Their trademark sponsons were deleted, but replaced by a new hybrid sponson/wing root with wingtip floats. The new tapered gull wings were mounted low on the hull, with the roots extended rearwards as vestigial sponsons. The streamlined fuselage tapered to the rear, with a pair of high-mounted dihedral tailplanes with endplate fins.

Aside from the wing configuration, the most innovative aspect of the 6-07 was its coupled powerplant. This comprised a pair of Rolls-Royce Tynes installed in a dorsal pod, and coupled to drive a contraprop with two 15ft 9in (4.8m) propellers. This would, like the Fairey Gannet, allow one engine/propeller to be shut down in the cruise or patrol to maximise endurance, with the additional benefit of having no asymmetric thrust effects. The 6-07 showed Dornier's ingenuity by using a technique last seen on their Do 26 flying boat from 1938. The Do 26 was powered by four Jumo 205 diesels installed in tandem in tractor/pusher overwing nacelles. The rear engines could be raised through an arc of 10° to lift the propellers clear of spray during take-off and landing. Dornier applied this to the 6-07 by installing the powerplant on jacks that could pivot the pod through 18° to keep the propellers out of the spray plume. Despite this radical design, possibly just a *Bierdeckel* (beermat design) showing how flying boat design could progress, it was the more conventional P.340/4-07 that was submitted to the NBMR.2 Expert Group, but it did not fare well in the competition as it failed to make the final three.

France also had a long tradition with flying boats and therefore a flying boat was an obvious contender for the requirement. Although not considered for the role outside its native France, the Breguet Br.1250 was a handsome twin-Tyne aircraft that shared some of the components of the Br.1150 Atlantic landplane. Configuration-wise, the Br.1250 was similar to the Martin P5M Marlin with its deep hull, gull-wing and weapons bays in the extended engine nacelles. The *Aéronavale* operated the Marlin for five years from 1957, but only in small numbers from Dakar in Senegal.

Nord Aviation, another French company with a long heritage, had examined a maritime patrol variant of their Nord 2500 Noratlas twin-boom transport as the 2505 and 2510, but neither was suitable for NBMR.2 as they lacked development potential and were piston-engined. Nord proposed an aircraft that utilised the twin-engined, mid-wing double-bubble fuselage configuration; the general arrangement preferred by the Group of Experts and shared the look of the Fairey Type 83, Avro 745 and Breguet Br.1150. No designation has been found for the Nord NBMR.2 study and the only published information is derived from a pair of

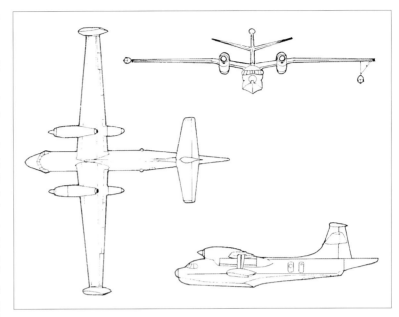

Breguet also proposed their Br.1250 flying boat that used many components from the Br.1150 Atlantic. The Br.1250 shared many aspects of the Martin P5M Marlin: Deep hull, gull-wing, high tail and weapons bay in the engine nacelles. The NBMR2 Expert Panel was not impressed by flying boats and so the Br.1250 remained but a paper study. *via J C Carbonel*

Nord's nameless NBMR.2 proposal was more or less a clone of the BR.1150, Avro 745 and Fairey 83, but with a Nord Aviation look about it. This provisional drawing was prepared from two photographs of a desktop model. Although the requirement recommended the Rolls-Royce Tyne, the locations of the jetpipes suggest a different engine, possibly the ASM Mamba.

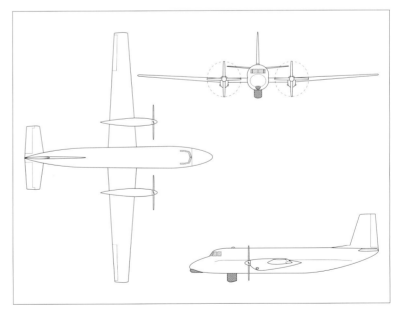

photographs of a model in J. Cuny's *Les Avions de Combat Francais, Tome II*. The general arrangement drawing shown here is provisional and drawn from these photos. The type has the Nord 'look' about it, reminiscent of the *Frégate* transport of similar vintage.

AWA's Maritime Wheelbarrow

Armstrong Whitworth Aircraft described their bid for NBMR.2 as being the application of an economic approach to aircraft manufacturing. While not quite the modular approach that Vickers would later apply to the VC10, AWA opted to use as many existing components from their current designs as possible. AWA's current transport aircraft was the AW.650 'Freightercoach', a twin-boom passenger/cargo type powered by four Rolls-Royce Darts. The

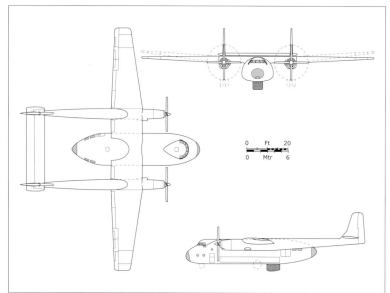

The twin-boomed AW.651 would have made a functional maritime patrol type due to its capacious cabin and stores capacity. The AW.651 married the Argosy wing and booms with two Tynes and a new fuselage with a 'flattened egg' cross section carrying weapons and ASW stores in sides of the lower fuselage. *Avro Heritage*

Armstrong Whitworth drew up the AWP.15, based on the AW.650 transport. As ever, it used the Tyne engine and the new fuselage pod was widened at the base to carry weapons bays along each side. The aircraft was unpressurised as Armstrong Whitworth considered this unnecessary.

AW.651 was similar; but used a pair of Rolls-Royce Tyne 11 with 16ft (4.9m) de Havilland four-bladed propellers. The later AW.660 used the same wings, engines and tail unit, combined with a new fuselage pod to produce what became the Argosy transport for the RAF, known colloquially as the 'Whistling Wheelbarrow', while the AW.670 was a civilian car ferry. Armstrong Whitworth considered the AW.651, with its Tynes, ideal as the basis for an MR type to meet NBMR.2 and, when combined with a new fuselage pod for the maritime task, this study was given the project designation AWP.15.

AWA decided to forego a pressurised cabin, as the type was unlikely to fly above 20,000ft (6,096m) even in transit to the patrol area, and this would hold a number of advantages over a pressurised fuselage, namely structural weight and allowing a versatile layout within the cabin. The weapons bays were to be arranged along each side of the fuselage with the operators' workstations along a '...central crew operating area.' The original AW.650 and 660 utilised a variant of the Avro Shackleton wing, but this was to be redesigned for greater fatigue life. Changes to the wing involved deletion of the outboard engine nacelles and modification of the inboard nacelles to take the larger Tyne 11 engines plus a slight modification to the inner section of the wings to accommodate a wider fuselage pod. The lower fuselage was much wider than the transport, due to the inclusion of a pair of 29ft (8.8m) long bomb bays along each side to carry the 6,000 lb (2,722kg) of stores required. A radome for the ASV radar retracted into a bay under the floor towards the rear of the fuselage pod. A crew rest area and look-out positions took up the 'tail cone' of the fuselage while up front, the flightdeck was raised above the main deck level to improve the pilots' field of view and accommodate a bomb aimer's position in the nose.

Not to be left out, the Canadians submitted a couple of bids for NBMR.2, both four-engined, a result of the concern the RCAF shared with the RAF about operating twin-engined aircraft over the Atlantic. The first was the CL-67A powered by four Armstrong Siddeley Mamba 10 (possibly being developed as the P156 rated at 2,200shp (1,640kW) and the Tyne-powered CL-67B. The CL-67B would have fitted with the NBMR.2 experts' stipulations on powerplant, but with four Tynes, they were less than impressed. Similarly the Italians got involved with the Piaggio Company proposing the P.155 flying boat that was not dissimilar to the Martin Marlin and Breguet Br.1250. The initial P.155 design was a transport fitted with two Pratt & Whitney R.2800 reciprocating engines, but the P.155AS for NBMR.2 would have been fitted with the usual Tyne turboprops and a pair of turbojets could be installed in the tail if required.

Avro's bid for NBMR.2 was the low-wing, twin-engined Type 745 powered by the Tyne 11, although variants with four Rolls-Royce Darts or Armstrong Siddeley Mambas were considered. The Avro 745 had a span 112ft 3in (34.2m), length 82ft 9in (25.2m), height 31ft 9in (9.7m) and an AUW 80,000lb (36,287kg). The Air Staff much preferred more than two engines for their reconnaissance aircraft, preferably four, not only for safety reasons but mainly because the aircraft could continue its mission if an engine failed. A twin-engine aircraft would need to abort and return to base if an engine had to be shut down. However, it was the Expert Group not the Air Staff who were assessing the designs and twin-Tyne aircraft were preferred. The Avro 745, designed using Woodford's experience and expertise from the Shackleton, reached the final three in the NBMR.2 process. The three types were the Avro 745, the Nord Aviation proposal and the Breguet Br.1150. Despite Avro's track record in maritime patrol aircraft in developing the Shackleton, it was the French Breguet Br.1150 that was selected on 28th October 1958. However, selecting the design was just the start, as there would be great deal of bargaining to be done before the first Breguet 1150 took to the air. That bargaining was not just amongst the active participants in the project.

Therefore by 1958 there were four basic configurations to choose from: Large four-engined, flying boats, twin booms and the medium twins. Of the twenty-one proposals only the fourteen most feasible studies have been described above and it was from these fourteen types that the NATO maritime patrol type was to be selected.

Hard Cash and Horse-trading

Possibly the best account of the British view of the Breguet Br.1150 NATO MPA process is to be found in the Treasury documents that relate to NATO weapons procurement. On 30th January 1959, the NATO Armaments Committee met to discuss the findings of the Group of Experts on Maritime Patrol Aircraft (MPA). They endorsed the choice of the Breguet Br.1150. The Armaments Committee's report was distributed to the NATO nations and listed some of the reactions to the choice. Norway considered the Breguet aircraft too large for it to use in its territorial waters and therefore would not contribute funds to the project. France said it would take 50 rather than the 72 they originally planned. West Germany would take 15 and '…make a contribution towards the research and development of the two prototypes.' The Netherlands would take 20 and would also contribute to the development programme. Portugal said that they '…required 24 of the most

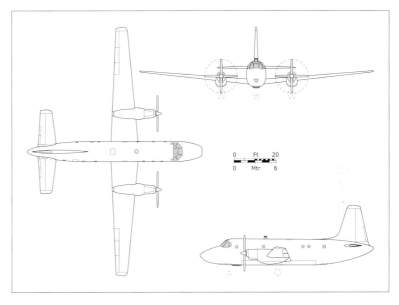

Avro's Type 745 was tailored to the NBMR.2 requirements but with the UK Air Staff uninterested, the prospect of a British aircraft unwanted by the RAF being operated by European air forces weighed heavily against it. The Type 745 would be resurrected for AST.350 as the Type 775.

…And the winner is…the Breguet 1150 Atlantic. Despite the high percentage of British content and using Avro-developed construction techniques, the Air Staff would not entertain the Atlantic, mainly due to it being twin-engined. However, proposals for the Atlantic would land on Air Staff desks on a regular basis.

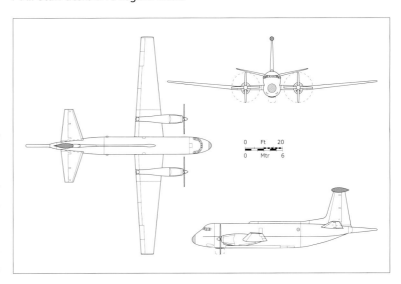

The Breguet Br.1150 Atlantic was adopted by a number of European nations for NBMR.2. The Atlantic would serve for more than five decades. It would also haunt the Air Staff for many years. The Atlantique Nouvelle Generation, renamed Atlantic 2, seen here had replaced the original model in French service by 1996.
© *Graham Wheatley*

modern and effective maritime patrol aircraft available; their present aircraft are obsolete.' But due to 'economic considerations' they would not be able to contribute to the programme. Belgium, oddly enough said that it had no NATO requirement for such an aircraft, but did have a national requirement '…of four to six aircraft for the defence of the Belgian Congo' and did not '…wish to associate itself with either research or development'.

The report now turned to the British view of the matter. Britain's representative on the committee was Vernon Bovenizer, Counsellor of the UK Delegation to NATO, who outlined Her Majesty's Government's view of the matter by saying they had '…no present requirement for aircraft of the Breguet 1150 type and could not therefore enter into any commitment to buy it or assist in the project financially.' However, if such a requirement did arise, '…they would consider the purchase of the Breguet…' and would '…not support the development of any parallel project.' and would happily co-operate on any equipment for the Breguet that might have a British requirement. Bovenizer assured the Committee that the UK would '…continue to place technical knowledge and experience at the service of the Group of Experts…' and would lend three Tyne engines, which cost £4m to develop, for the first prototype.

The Americans assured the Committee that they would contribute funds under the MWDP, but expected the Europeans to contribute the bulk of the money, which prompted the Chairman, Ernest H Meili, (Assistant Secretary General of NATO for Production and Logistics) to express the committee's '…disappointment at the very limited financial contribution to this project…'. Meili observed that the project was a fine example of multi-lateral co-operation but as soon as the matter turned to money, there was '…an almost general drawing back when it came to the point of financial contributions to get the project going.' Despite this, the Committee decided to ask the French to nominate a 'responsible agency' to supervise the research and development of the Breguet Br.1150 programme. This agency was SECBAT: *Société d'Étude et de Construction du Breguet Atlantic,* which oddly enough uses the English spelling of Atlantic! This organisation was tasked with supervising the development and, at a later stage, the construction of the BR.1150 and the first Atlantic prototype lifted off from Toulouse on 21st October 1961.

A second Working Group would oversee the financial aspects and consist of France, Germany, The Netherlands, UK and USA, but the French objected to the UK's inclusion as '…the choice of members should be based on those prepared to participate financially.' Meili was of the opinion that while it would be 'pleasant' to have a financial contribution from the British, '…their participation technically would be welcome.' By August 1959 the participants had agreed on the finance but there was a £300,000 shortfall, which the Belgians offered to contribute in return for some of the manufacturing being done in Belgian factories. The entire business was being viewed with astonishment in Britain, where on 10th August 1959 William Strath, Permanent Secretary at the MoA wrote to Sir Richard Powell, Permanent Secretary at the MoD, on the matter. His view was that there should be a British contribution to the project, mainly because only the British had any experi-

ence in developing such an aircraft and that 'A V Roe are the firm in Western Europe with by far the greatest technical expertise in the subject of maritime reconnaissance aircraft. If they could be included in the consortium, and this I think could only be done if the UK were a contributor to the development costs, we should be assured of a better aircraft.'

In essence, Strath was saying that it might be prudent to contribute to the programme in order to have some influence in it, on the off chance that Britain did opt to buy the NATO MPA. If Britain *did* elect to purchase it at a later date, better to have Avro, with their experience and track record in the field, as a major, if not lead partner in the process. By offering half of the shortfall, £150,000, Strath was sure that this would allow Avro to have a positive influence on the aircraft.

Strath's letter appears to have prompted another look at the NATO MPA as three days later the possibility of a Breguet purchase was being discussed by the Ministries, only to be dismissed on 3rd September 1959 by the Treasury's J E Herbecq who took the view that the £150,000 contribution would be the thin end of the wedge and it would end up costing the UK millions for minimal influence. Herbecq concluded his letter with '…until there is a genuine UK requirement for an aircraft of this type, it would be premature to take any initiative in SHAPE or NATO to secure the adoption of the aircraft…' which was a fairly rational response from the Treasury. Evidently Strath was rather annoyed by Powell and Herbecq's attitude and wrote to Powell on 1st October to reiterate his points and stated that further Shackleton purchases would be very expensive, especially if delayed. Strath signed off with 'If, in your view, the chance of the RAF wanting some maritime reconnaissance aeroplanes is so small that it is not worth £150,000 to obtain control over the project, or so small that we can face an exorbitant price for Shackletons instead, then we can let the matter drop.'

The official line from Whitehall was that the Shackletons would not require replacement until 1970, so there was little point in paying for the development of an aircraft that failed to meet the British requirement. The more or less final word came from Frank Roberts, British Permanent Representative on the North Atlantic Council, who wrote to Powell and made the point that while the politicians stated '…in general declarations how anxious we are to co-operate in European projects and yet to do nothing…whenever a specific project comes up in NATO approved by our allies for actual co-operation in a defined field.' Roberts said that it felt as if he was attending the same old meetings on these projects as he attended two years before. Only the Air Ministry and Ministry of Supply, soon to be merged as the Ministry of Aviation, were in favour of contributing, the MoD was set against it as the Air Staff said they had no requirement for it and the Treasury were not keen to throw money into what they saw as a potential money pit. There was good reason the MoD and Air Staff were against NBMR.2: they had something else up their sleeves.

Trinity

The Air Staff's disinterest in NBMR.2 certainly wasn't down to distaste for collaboration, but merely due to the lack of a requirement, which the NBMR.2 would not have met anyway. The Shackletons would be useful until the early Seventies and the improved performance of the MR.3s would see those through until then, so there was no real urgency for a new MR type. The only other Air Staff requirement for a large aircraft was GOR.347, for a large Patrol Missile Carrier to carry the Douglas GAM-87 (latterly AGM-48) Skybolt air-launched ballistic missile (ALBM) issued in 1960. The aircraft was to be capable of long endurance with a large payload and the internal volume to carry extra fuel tanks. These characteristics were also required for strategic troop transports and maritime patrol aircraft so why not produce a common airframe for all three roles? Add into the mix the possibility of developing a civil airliner from this common airframe and all concerned were very interested in what became known in British aviation circles as the 'Three-in-One'.

Meanwhile, at the Ministry of Aviation (MoA), a debate on how to equip the RAF from 1970 onwards was ongoing. In 1961 the RAF had, or was about to receive into service, over twenty types of aircraft. The Air Staff hoped to slash this variety by introducing multirole types. A replacement for TSR.2 to meet OR.345 and cover the tactical strike and interception roles, a heavy tactical transport to OR.351, medium tactical transport to OR.377 plus the three-role deterrent carrier/strategic troop transport/maritime reconnaissance (MR) aircraft and in that type the MoA saw an opportunity. At a Chief of the Air Staff (CAS) conference on 24th March 1961, the future equipment of the RAF was under discussion, particularly the trooping, deterrent carrier and maritime reconnaissance types, specifically the aspects these tasks had in common. The three roles required long range or endurance, high-speed transit and a large payload or cabin volume for equipment. The Deputy Chief of the Air Staff (DCAS) Air Marshal Sir Ronald Lees opined that if a common airframe could be developed to cater for the three roles, the financial savings would be breathtaking, given that the cost of funding development of the MR type would be £75m, the deterrent carrier £150m and the troop transport £130m. A common basic airframe would ease maintenance, training and other support functions while the possibility of a civil version would be an added bonus. Although a common type was considered to be 15% less efficient in each role than a bespoke aircraft, the savings would be worthwhile and the case for the 'Three-in-One' would be 'over-riding'.

The Vice Chief of the Air Staff (VCAS) was less enthusiastic, with Air Chief Marshal Sir Edmund C Hudleston making the point that of the three, maritime reconnaissance was 'the odd man out' and favoured a 'Two-in-One' approach with a clean-sheet MR type. However, the VCAS also noted that this could be addressed by 'compromising on the Operational Requirement', thus making the Three-in-One feasible. A month later, on 26th April 1961, Sir Henry Hardman, the Permanent Secretary at the MoA, was informing the DCAS that the time was ripe for the MoA 'to make

studies of supersonic and subsonic three-in-one' types. This soon changed, with Hudleston telling the CAS, Marshal of the Royal Air Force Sir Thomas Pike, that '...unless you require supersonic performance, variable geometry is wasteful and inefficient.' Hudleston also took the view that 'I cannot reconcile supersonics with maritime work.'

Origins of Trinity

The initial musings on a 'Three-in-One' type appear in a discussion document, *Future Aircraft and Weapons Systems*, drawn up by Air Commodore W R Brotherhood, DOR(C), in May 1958. Air Commodore Brotherhood proposed that a strategic transport aircraft to replace the Bristol Britannia be developed and that this might also form the basis of a deterrent carrier and maritime patrol type. At this time only Vickers at Weybridge had a type on the stocks that would fit the requirement: the VC10. Designed to operate from the hot-and-high airfields of East Africa such as Embakasi in Kenya and Entebbe in Uganda, the VC10 was ideal for the strategic trooping role, being powerful and capable of carrying a large payload out of restricted airfields. These airfield restrictions – clearing a 50ft (15.2m) obstacle in 6,000ft (1,829m) – were imposed upon designers by the Air Ministry, Air Staff and BOAC, all of whom who wanted to operate out of short or hot-and-high airfields around the Empire. This desire had been the cause of many a problem for Britain's aircraft companies, notably Vickers with the V1000 and latterly the VC10. Quite why the obvious solution of pouring more concrete was not adopted is a moot point, but was the ultimate answer that came too late for Vickers, whose VC10 could leap into the air while Boeing's more successful 707 clawed its way off the newly-extended runways on the East African Plateau.

Vickers grasped the Three-in-One concept and produced a series of design studies to fit the roles and in fact extended the range of roles for the VC10. Scheme A was for the strategic transport role and was almost identical to the Type 1106 that would meet the forthcoming ASR.378/Spec. C.239, with the aircraft fitted with a large cargo door on the port side of the forward fuselage. For the Patrol Missile Carrier (such deterrent carrying aircraft were called Pofflers by the Air Staff and Air Ministry) Vickers outlined a modified Scheme A, with the Type 1106 type fitted with four underwing hard points and large auxiliary fuel tanks in the cabin.

Trinity – The various guises of the Vickers VC10 shown in artwork from a Vickers brochure. Top to bottom: Trooper/Transport, ALBM Carrier, Tanker, Bomber and lastly, Maritime Patrol. By the mid-Sixties, BAC had identified ten roles for the VC10. *Brooklands Museum*

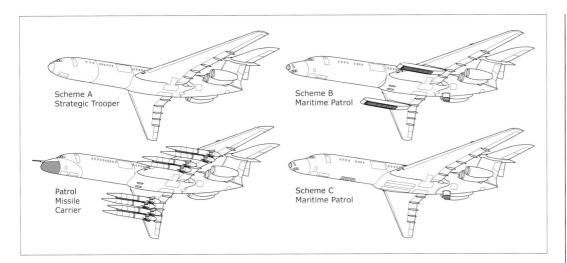

Scheme A
Strategic Trooper

Scheme B
Maritime Patrol

Patrol
Missile
Carrier

Scheme C
Maritime Patrol

Vickers proposed four VC10 variants for the Three-in-One requirement. Scheme A was the strategic trooper, which was modified to become the deterrent carrier. Scheme B was a standard Scheme A transport with a new nose and a ASV.21 under the rear fuselage. The weapons were to be carried in a pair of underwing cocoons. Scheme C took Scheme B and added a weapons bay in the wing/fuselage interface and sonobuoy launchers under the forward fuselage.

The anti-submarine role needed a bit more effort and Vickers drew up two studies, Scheme B and C, as initial proposals for a maritime patrol VC10. Both had a new nose housing a bomb-aimer's station, the forward cabin fitted out with workstations for the tactical operators and a retractable ASV-21 radar in a capsule under the rear fuselage. The aft portion of the cabin was to have six cylindrical fuel tanks holding a total of 76,000 lb (34,473kg) of fuel. For an emergency trooping role, these tanks could be removed through a cargo door fitted on the port side of the cabin, just ahead of the main fuselage construction joint.

The two MR studies differed substantially, but the main change for the Scheme B was '…the addition of BLC by the substitution of leading edge flaps for leading edge slats…' in an effort to 'improve take-off performance and low-level manoeuvrability for use in the maritime role.' BLC was boundary layer control, whereby air was blown out of a slot, or sucked in through a slot, in the wing to modify the flow regime in the layers of air closest to the airframe. This allowed the aircraft to fly more slowly and thus manoeuvre better at low speeds without the need for drag-inducing flaps, both for operations from short fields and while manoeuvring at low altitude on patrols. The boundary layer control system would make the VC10 capable of operating from 6,000 foot (1,829m) runways at the maximum AUW and to make a 180° turn in 26 seconds, an important capability in anti-submarine warfare.

The Scheme B maritime patrol VC10 was described as a modified Scheme A transport carrying its weapons load in a pair of underwing cocoons, similar to those designed for the Vickers Valiant, carried on the inboard pylons intended for the Poffler to carry its Skybolts, plus the boundary layer control systems.

Scheme C was to be modified from the B, but with a sonobuoy bay under the forward fuselage and '…a large semi-faired fuselage bombbay…' in an enlarged wing root fairing under the wing centre section torque box. Scheme C's fairings were lighter and produced less drag than the panniers of the B, but required more modifications and a heavier installed weight than the removable pylon and panniers. Both schemes had an estimated AUW of 380,000 lb (172,365kg). Vickers pointed out that Scheme B or C had additional roles including 'Colonial policing duties [that] can be carried out if alternative weapons are carried in the bomb bay.'

A further development of the VC10 was Vickers multirole/modular VC10 that utilised the VC10 wings, powerplant and rear fuselage, essentially the airframe aft of the fuselage construction joint. This could be mated to a variety of role-specific forward fuselages. Two basic fuselage sections were proposed, with the first based on the Scheme A transport while the second used a new fuselage complete with ventral weapons bay, a reprofiled nose with radar and new cockpit glazing. Vickers' rationale was that the needs of air forces would change as a conflict progressed. At the start, offensive operations would see roles such as bombing, air defence and anti-submarine warfare in most demand while later transports, medevac and other support roles would be needed. Rather than keep a large number of airframes for each role, a series of fuselages could be kept in store and attached to the aft section as circumstances change.

At first glance, the VC10 provided the best solution to the 'Trinity' of roles, certainly better than some of the other types such as the BAC 208 and AW681, both modified tactical transports found lacking in the deterrent carrier and strategic transport roles. The Shorts SC.5/21B

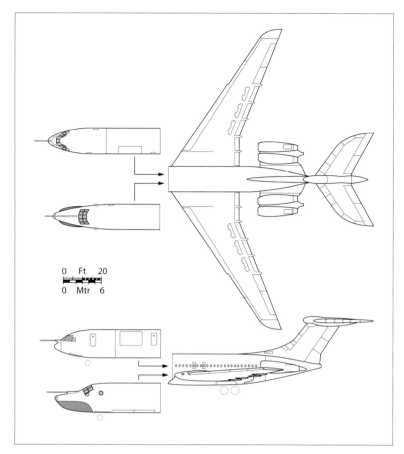

0 Ft 20
0 Mtr 6

Vickers' other proposal was the multirole, modular Vickers VC10, intended to fill a variety of roles, but specifically the 'Trinity'. One proposal from Vickers involved removable forward fuselages tailored to each role. This drawing shows the basic transport variant and the maritime patrol variant with a weapons bay in the fuselage.

demands of a supersonic airliner with the air-field restrictions in force at the time was variable geometry. De Havilland and Vickers were of the opinion that variable geometry (VG) provided more flexibility in civil airline operations than the slender delta being developed by Bristol at Filton. The slender delta, from a Hatfield and Weybridge viewpoint, restricted the supersonic airliner to trans-oceanic operations whereas a VG type allowed high subsonic transit over populated areas without the resultant sonic booms of the delta SST. Vickers believed that '...the advantages of variable sweep all stem, in essence, from the fact that it becomes possible to cater for peak aerodynamic efficiency in off design cases.' What this meant was that VG could allow more efficient landing and take-off speeds, but more importantly make the subsonic portions of a journey more economic '...thus spreading the usefulness of the machine and allowing full use to be made of its enormous potential work output.' By this Vickers intended that the aircraft's operation could be cheaper, but also could be open to use in roles other than point-to-point passenger transport.

De Havilland's studies concluded that an airliner flying at Mach 1.15 would fit the bill as 'This is the highest speed at which, in a standard atmosphere, pressure waves made in the stratosphere will not reach the ground.' Atmospheric conditions would affect this maximum speed, but de Havilland added '...it is highly desirable that an aircraft designed for operation at Mach 1.15 should be capable of economic operation below this speed.' Vickers came to similar conclusions and both agreed that the differing cruise and low-speed phases of flight could be solved by '...variable sweep wings, sized for cruise in the swept back configuration. In swept forward position that same wing can develop very much greater lift.' Swing wings have always been associated with higher structural weight, 'the weight penalty of variable sweep' attributed to the sweep mechanisms and the fact that compared with an 'unbroken' structure of a fixed geometry wing, the structure is heavier as a percentage of the AUW. Vickers pointed out that '...when applied in the right circumstances, (VG) allows a lighter aircraft to be built to do the equivalent job of a fixed wing aircraft. Saving on engine and fuel weight result as supplementary advantages.' Vickers concluded their report by stating that development of such an aircraft was necessary and 'Such an aircraft might be a pure research vehicle, or some military application, but this step must be taken before the civil variable

'Tactical Belfast' was proposed but was too big for the maritime role and could not fly high enough for the deterrent mission. One other possibility for 'Trinity' is the Avro Type 766 which is described as a long-range military transport powered by four turbojets. The Avro 766 dates from 1959, which would place it in the timeframe of 'Trinity', but details are sparse.

When the Three-in-One was under consideration by the Air Staff, their main concern was whether or not the aircraft should be supersonic, which was not particularly useful for the maritime task, apart from during transit, but could be applicable to the strategic trooping and deterrent carrier. Unfortunately, the penalty for supersonic performance was a two-and-a-half times increase in AUW, and as far as the MR role was concerned the Air Staff considered this unacceptable. Not that this prevented the aircraft companies from examining a supersonic aircraft to meet the Air Staff's Three-in-One needs and address the civil supersonic airliner market.

Meanwhile at Hatfield and Weybridge, de Havilland and Vickers designers were investigating the latest fad in British aviation, the supersonic airliner, fitted with that other favourite of the era, the swing wing. Both design offices felt that the only way to meet the disparate

geometry civil transport can become a reality.' Technically speaking, Mach 1.15 is the transonic flight regime and is probably the least attractive speed range for an aircraft to fly within, having the problems of supersonic flight, such as structure and high fuel consumption but without the real advantage – significantly shorter flight times.

Vickers' VG airliner study was 160ft (48.8m) long with an unswept span of 115ft (30m), swept span of 75ft (22.9m) and an AUW in the region of 400,000lb (181,437kg). The wing was set low on the fuselage, with a large glove containing the sweep mechanism and fuel tanks, while the all-moving tailplanes were set high on the rear fuselage. The wings were tapered, with rounded tips, and could be moved from a 20° leading edge sweep to a fully spread position of 65°. Propulsion was provided by four reheated turbojets installed in a block of four across the rear fuselage, fed by a pair of wedge intakes either side of the fuselage. The Vickers airliner could carry 135 passengers for 1,500nm (2,778km) at Mach 1.15, slightly further if the cruise speed was kept to Mach 0.95.

Meanwhile at Hatfield, de Havilland were working on their DH.130 VG airliner and had come to the same conclusions, and by the time de Havilland had become part of the Hawker Siddeley Aviation Group, the entire DH.130 had been transferred to the Hawker Siddeley Advanced Projects Group (APG) at Kingston and became the APG.1011. The APG.1011, also known as the HS.1011, was a handsome needle-nosed design with an area-ruled fuselage 172ft 6in (52.6m) long with a broad, swept

fin surmounted by a triangular tailplane.

The initial design for what became the HS.1011 was powered by four reheated Rolls-Royce RB.80 Conway bypass turbojets, rated at 9,250lbf (41.1kN) without reheat and 16,500lbf (73.4kN) with reheat. Two of these were installed, side-by-side and fed by a single pitot intake, on top of the rear fuselage at the root of the tail fin, with the rear of the nacelles and the jet pipes faired into the aircraft's tail cone. Another Conway was installed in a fairing under the wing glove of each of the mid-mounted wings. These fairings carried the intake trunking and wing sweep mechanism, forming a flat-bottomed nacelle that also held the main undercarriage units. The area ruled fuselage narrowed above this wide fairing to maintain aerodynamic efficiency in the transonic regime, which also impacted the cabin's internal layout, becoming much narrower over the wing, changing the seat layout from six abreast to four.

A model of the HS.1011 airliner that shows the type's clean lines. The HS.1011 had the internal space, speed and range to fit the bill for the Trinity roles. Unfortunately it would have been too expensive and the UK government had backed the Anglo-French project that would become Concorde. *Author's collection.*

Left: Supersonic Trinity – besides Concord(e) British companies had designs under consideration for the high-speed transport aspects of the three-in-one requirement. Vickers' variable-geometry design could carry up to 135 passengers at Mach 1.15. Handley Page's laminar flow delta was studied under the Hawker Siddeley funding umbrella and could carry 130 passengers at Mach 2.2. The HS.1011, née DH.130, could also carry 130 passengers and was more suited to the Air Staff's needs.

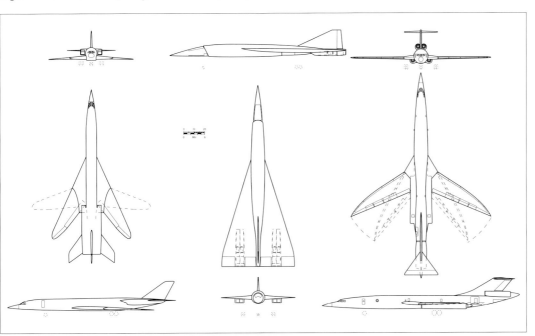

The wings pivoted about a point much closer to the fuselage than on the Vickers design, meaning that there was no space for fuel in the wing glove, so the fuel was carried in the lower fuselage and wings. The tapered wings had straight trailing edges with the straight leading edge curving backwards into an elegant blade shape from a point about 50% along the leading edge. Fully-swept with 58° leading edge sweep, the wingspan was 119ft 9in (36.5m) and with the wings with 31° sweep the span increased to 131ft 6in (40m). In fact the HS.1011 was an elegant aircraft indeed.

Rolls-Royce had high hopes for its RB.80 Conway engine and intended developing its potential by increasing the bypass ratio to improve efficiency. This led to the RB.178, the Super Conway, which when combined with a new reheat system produced a turbofan rated at 27,500lbf (122kN). This increased thrust allowed Hawker Siddeley to redesign the HS.1011 with a trio of RB.178s by deleting one of the Conways in the tail. The drawing of this aircraft is labelled 'APD.1023', which was a follow-on project to the HS.1011 investigating a transonic airliner carrying 117 passengers as a comparison with the HS.121 Trident. The intakes were also modified to be variable geometry, modifications that would allow the aircraft a dash speed of Mach 1.9. This made the aircraft more suitable for the deterrent carrier role, carrying four Skybolts internally, two side-by-side in the forward cabin, one over the wing box and a fourth carried on a rotary launcher in an unpressurised section of the rear cabin. Skybolt release involved the missile's navigation system being updated by an astro-navigation system before the weapons bay rotated and the missile ejected. A conveyor system within the cabin moved the next missile to the launch bay before updating the launch position and releasing it. Missile launch could take place at any point in the flight regime from the Mach 0.75 patrol speed to the dash 'strike' speed of Mach 1.9.

Hawker Siddeley advised that if given the go-ahead in 1962 they could develop the HS.1011, first as a military transport to enter service in late 1970, with its civil airliner variant available from early 1971. The deterrent carrier could be in service six months later, while the most complex variant, the maritime patrol type would enter service in early 1972. There was no perceived need to replace the Shackletons in the short term, but this timescale would allow the maximum time to develop the new 'breakthrough' technologies in anti-submarine systems that the Air Staff believed would revolutionise ASW.

Based on these studies, the companies considered a VG, transonic transport aircraft most suitable for the 'Three-in-One' roles. Vickers' design study fell by the wayside as BAC geared up for the Filton-led Concord (the 'e' was added later) project. Hawker Siddeley on the other hand, viewed the VG transonic airliner as an alternative to the Mach 2.2 SST being developed by BAC and foresaw the potential problems with sonic booms and their impact on routings: oceanic routes, which given the range of Concorde meant it would be restricted to Transatlantic operations.

By the time the Air Staff were examining the proposals in late 1961, Vickers had been absorbed into the British Aircraft Corporation and de Havilland into the Hawker Siddeley Aviation group. Vickers' VG supersonic transport work became subsumed in the Concorde project and their VG SST came to nought, which left the HS.1011 as a possible Three-in-One aircraft. It had the speed for the strategic trooping role, could carry up to four Skybolts and had the high transit/low patrol speed needed for the maritime and deterrent carrying roles. The HS.1011 was well-suited to the strategic trooping role, being based on an airliner design. As a deterrent carrier the HS.1011/1023 could transit to a patrol area at transonic speeds, enter a racetrack patrol pattern at 30,000ft (9,144m) and cruise at Mach 0.75. On being given the launch instruction, the HS.1011 was to accelerate and climb to its launch point at 50,000ft (15,240m), release its Skybolts at speeds up to Mach 1.9, before shedding speed and altitude for a subsonic return to base.

In the maritime role, the HS.1011/1023 could beat any transit speed requirements, taking an hour and fifty minutes to reach its patrol area at 1,000nm (1,852km), a distance the Shackleton MR.3 could take six hours to complete. The HS.1011 did leave a lot to be desired in its patrol speed as, even with the wings spread and high-lift devices deployed, the HS.1011/1023 struggled to meet the patrol speed of 150kt (278km/h), 200kt (370km/h) being the minimum drag speed in search configuration.

So, faced with the disparate requirements of these roles, a 'Goldilocks' solution was not available and ultimately the Air Staff's hopes of killing three, if not four, birds with one stone were far too ambitious. Each role was compromised by the needs of the others, with maritime patrol requiring so many changes that the resultant aircraft was effectively a new version. The Three-in-One type was quietly dropped, a measure made easier by the cancellation of the Douglas Skybolt by the US Government, thus

In the 1970s the RCAF was intent on replacing its ageing Canadair Argus fleet with a new, more flexible type. McDonnell Douglas produced the multi-role tanker, transport, maritime DC-10. Like the RAF a decade earlier, the RCAF discovered that each role was compromised and ultimately acquired the Lockheed CP-140 Aurora. *Terry Panopalis Collection*

removing that requirement. This left the strategic trooper, where high speed and payload over a long distance was the key aspect, while the maritime patrol aircraft needed rapid transit followed by a low-speed patrol with quite nimble manoeuvrability to prosecute its targets.

Interestingly, the Royal Canadian Air Force in the early 1970s became interested in the multi-role type along the lines of the Trinity to replace the CP-107 Argus, long-range transports and provide tanker support. The RCAF examined the Boeing 707 and the McDonnell Douglas DC-10 as the basis for their Long Range Patrol Aircraft (LRPA) requirement. The Boeing 707 was already in service with the RCAF but the wings of the DC-10 were built in Canada and was a more modern design, making the DC-10 LRPA attractive. Ultimately the DC-10 LRPA did not prosper and the Argus was replaced by the Lockheed Orion-derived CP-140 Aurora.

The strategic trooper to Specification C.239/OR.378 was filled by the Vickers VC10, specifically the Type 1106 that combined the basic airframe of the Type 1103 'Combi' with the engines and fin fuel tank of the Type 1151 Super VC10. Although the Air Staff were still hopeful for a breakthrough in submarine detection, they soon realised that one airframe to fill many roles was not feasible and so it was back to the drafting panel for the Operational Requirements Branch. A bespoke maritime aircraft was what was required and a draft requirement for this was soon in circulation. This would become OR.350 and would see the resurrection of some familiar proposals, dusted off, enlarged and proposed more or less because there was no other large aircraft contract for which the myriad of British aircraft companies could tender for. However, the Air Staff had set the bar high.

The final outcome of the Three-in-One studies was the Vickers Type 1106 VC10 which served for 43 years. Initially a troop transport, the VC10 later served as a freighter and tanker. Not quite the Three-in-One, the Air Staff rebuffed Vickers' overtures for a maritime patrol version, but it performed well in any role it took on. *Graham Wheatley*

An Avro Vulcan MR.3 of Coastal Command makes a low-altitude pass over a *Yankee*-class SSBN running on the surface. The Vulcan's sprightly performance at low-level would have made it ideal for the role. Unfortunately the type would need a completely new 'fat' fuselage complete with operators' workstations, sonobuoy dispensers and computer systems to meet OR.350. *Adrian Mann*.

'The requirements of AST OR.350 are fairly critical and only by taking full advantage of the latest developments in aircraft and engine techniques is it possible to meet the target with a moderate sized aircraft.' Summary of second MoA report on AST OR.350, February 1962

'...the chosen aircraft must be able to carry the new detection and weapons systems which we expect to emerge from around 1968 onwards and it must have considerable development potential...' Air Commodore R H C Burwell, D of Ops (M, Nav & ATC) January 1962

As noted above, the RAF had been party to the NBMR.2 maritime patrol competition, but since the UK's prerequisites for the role differed so much from the NATO requirement, particularly in the endurance category, the Air Staff had opted out. They considered the role to be of such importance that a national answer was the best solution but the first attempt at this failed as it became clear that the multi-role 'Three-in-One' type was compromised in all three of its intended roles. The dissatisfaction with these earlier attempts to replace the Shackleton and the uncertainty about how anti-submarine equipment might develop, prompted the Air Staff into two lines of action. The first was to modify the Shackleton, under the long-extant OR.320, to improve its performance in the form of the MR.3 Phase III with ASV.21, an improved electronic warfare suite and Viper turbojets, while the second was to draw up a completely new requirement. This requirement, OR.350, was to include all the latest developments in engines, airframe, avionics and still have potential for further development after its intended service entry in 1968.

The first task of the OR Branch was to identify the future threats against which the aircraft was to operate, particularly Soviet nuclear submarines carrying ballistic missiles, and drew up a few characteristics that could be expected in the period 1970 to 1975. The main development in submarines during the late 1950s was nuclear propulsion's provision of increased shaft horsepower, which when combined with

pumpjet technology (a form of ducted fan for ships) would lead to submerged speeds up to 38kt (71km/h). This was as fast, if not faster, than many surface warships and, when combined with the ability to dive deeper than 2,000ft (607m), made the submarine a difficult target for the current ASW weapons such as the Mk.30 torpedo.

Another aspect of the threat was the improved capability of Soviet submarines and their weapons. Long-range SLBMs such as the SS-N-8 *Sawfly* allowed the Soviets to conduct their patrols in areas closer to their home ports and, as noted in Chapter One, avoid the need to pass through Greenland/Iceland/UK gap. This would require maritime patrol aircraft that could operate in the Norwegian and Barents Seas and, if operating from the UK, reach these areas quickly if the Soviet submarine fleet sortied unexpectedly.

Having identified the threat, the Air Staff turned to laying out the aims for this Staff Target: AST.350, issued on 18th July 1960 and subsequently a requirement, ASR/OR.350, issued in July 1961. Faced with a new generation of enemy submarines, the initial thinking was to have a complete reassessment of Coastal Command's tasks post-1970. These tasks included:

a) Rapid surveillance of sea approaches – daily radar map.

b) Early warning and defence against nuclear Soviet missile submarines – infrared detection of the boost phase of a missile launch.

c) Defence of Allied Polaris submarines.

d) A requirement for the use of air-to-surface guided weapons against single ships and smaller targets.

e) Anti-submarine operations against deep-diving submarines (1,000-2,000ft) with speeds of 30-40kts.

In short, the putative Shackleton replacement was to conduct wide-area surveillance to provide early warning of attack by SLBMs, defend the forthcoming *Resolution*-class submarines carrying the Polaris deterrent, attack surface ships and, lastly, find and destroy nuclear submarines. That the anti-submarine role comes last in this list is surprising, but possibly reflects the difficulties involved in finding nuclear submarines in the first place. As described previously, nuclear propulsion robbed the maritime patrol aircraft of its primary sensor; radar detection of *schnörkels* or the submarine itself cruising on the surface under diesel power. This left sonar as the main technique for detecting and tracking submarines and in 1960 the comput-

erised methods for sonar signal processing were very much in their infancy. Luckily, the early Soviet nuclear submarines were somewhat noisy creatures and could be detected and tracked by sonar.

Next item under consideration was endurance, with the primary factor being the need to cover a new 'Black Hole' in the seas north and west of Norway and this required a radius of action of 1,000nm (1,852km) with an eight hour patrol once on location. Eight hours was deemed the maximum patrol time as it '…represents the maximum period of optimum efficiency with which a crew can operate on patrol; this is supported by medical evidence.'

This crew efficiency and the need to increase the on-station utilisation of the aircraft saw the OR Branch examine flight times to and from the patrol areas, which were governed by the cruise speed of the new type. Faster was better, as transit time was non-productive time but 'Supersonic transit speed would be astronomically expensive' and so a transit speed of 450-500kt (833-926km/hr) was considered best. Typically, a turboprop type could cruise efficiently at 350kt (648km/hr), taking two hours and forty five minutes to fly the required 1,000nm (1,852km), whereas a jet aircraft at 500kt (926km/hr) would make the same flight in two hours. This higher cruise speed would also be useful in the ocean surveillance role, allowing a greater area to be covered, while in the deterrent support role the aircraft could '…follow-up (or in war, attack) intelligence reports of any missile-firing submarines.'

What became apparent was that the choice lay between a jet aircraft with an AUW of 200,000 lb (90,718kg) that could transit at 450-500kt (833-926km/hr) and a turboprop (or mixed jet/turboprop) type with an AUW of 140,000-160,000 lb (63,503-72,575kg) whose transit speed was 350-400kt (648-741km/hr). So, from the transit point of view, the jet-powered aircraft was an obvious choice but for the patrol and particularly the attack, the turboprop-powered would be more suitable. The jet aircraft was considered a more flexible choice as, despite being more expensive, it could be used for troop transportation and 'Internal Security' and had more development potential. Jet propulsion allowed the engines to be placed on the rear fuselage, which created less noise and vibration than wing-mounted jets and particularly, turboprops. The noise and vibration affected crew and the equipment, with good soundproofing necessary for sonar operators, and so rear-mounted turbofans would make the cabin quieter and more comfortable for the crew. It should be recalled that noise and vibration was a constant, and justified, gripe of Shackleton crews.

By March 1961 the Air Staff OR Branch had been advised that a turboprop aircraft capable of 400kt (741km/hr) was possible, although such speeds would place the propellers at the limit of their efficiency. However, there was another reason for selecting a jet: the turboprop as an engine class was in 1961 seen as having a limited future in the face of developments in turbofans 'The development of turboprop engines is likely to have ceased by 1970 and if obsolete engines are used, it will not be possible to develop the aircraft in service…' Nor were the OR branch tempted to re-engine the type in the future with jet engines as that …'would be impracticable or costly and would almost certainly cause a repetition of the sort of situation currently facing the Shackleton.' By this they meant the fitting of Viper turbojets and subsequent airframe fatigue that would see the Shackleton MR.3s withdrawn before the MR.2s.

This produced a quandary for the Air Staff as the combination of high-speed transit with low-speed manoeuvre and patrol 'are not easily combined in one aircraft and can be achieved only by an aircraft designed specifically around them.' So, having dismissed the multirole Three-in-One type on the grounds of compromise, the Air Staff was now faced with melding two disparate flight regimes in a single airframe to carry out a critical task. Having outlined what they wanted out of the new type, the Air Staff and the Operational Requirements Branch had to conduct discussions and draw up a requirement for issue to the aircraft companies. This would be Operational Requirement OR.350.

OR.350 to meet Specification MR.218 was issued on 18th July 1960 and called for a maritime patrol aircraft to replace the Avro Shackleton from 1968. The chosen aircraft '…must be able to carry the new detection and weapons systems which we expect to emerge from 1968 onwards…' and must have 'considerable development potential'. The basic requirement was for an aircraft to replace the Shackleton MR.2 and MR.3 by 1968. It was to be capable of conducting a low-level patrol for eight hours at a distance of 1,000nm (1,852km) from a base equipped with a 6,000 foot (1,829m) runway. A transit speed of 400kt (741km/hr) was required with a patrol speed of 225kt (417km/hr) with 180kt (333km/hr) for short periods while attacking a target.

Requirement

The type's main role would be anti-submarine warfare (ASW), to locate and destroy Soviet submarines with subsidiary roles including 'Internal Security' around the shrinking British Empire, plus transport, specifically an emergency trooping role. Another task that the OR.350 type should carry out was search and rescue (SAR) with equipment to home in on search aids. These included the Search And Rescue And Homing Beacon (SARAH) and Search And Rescue Beacon Equipment (SARBE) which were radio location devices. Lindholme Gear was also to be carried in the bomb bay as the airborne lifeboat had finally left the inventory. A crew of twelve was specified, comprising the Captain (in overall command of the aircraft and not necessarily a pilot), two pilots, two navigators (tactical and routine), an engineer/crew chief and six air electronics officers.

For ASW operations the specified warload of 17,000lb (7,711kg) was to comprise six torpedoes to OR.1186 at 1,000lb (454kg) apiece, two nuclear depth bombs weighing in at 1,200lb (544kg), twelve Size C sonobuoys at 80lb (36kg) each, 120 of the 22lb (10kg) Size A Sonobuoys, 840 PDC (practice depth charges) each weighing 5lb (2.3kg), 36 Marine Markers at 20lb (9kg) and 24 pyrotechnic signals. For use against surface ships the type was to be fitted to launch and control guided missiles such as the AJ.168 Martel, but the Nord AS.30 or Martin Bullpup (already in UK service) were considered more likely candidates. The inclusion of ASGWs was prompted by the intention to fit the Atlantic and Argus with missiles, rather than any pressing requirement from Coastal Command. When the Canadians were approached about the OR.350 type as a replacement for the Argus, the possibility of a stand-off ASW weapon was discussed. The Canadians were concerned that submerged submarines might have a SAM capability. The RAF was sceptical, but Vickers in the 1970s developed such a system as the submarine-launched air-flight missile (SLAM) while in the 1980s, the Soviet Navy fitted the Kilo-class with the SA-N-8 Gremlin and SA-N-10 Gimlet SAMs.

Detection

The requirement called for a suite of sensors including a sonobuoy receiver to meet OR.3548, ASV.21 radar, SLAR, MAD, Autolycus Mk.3, Clinker wake detection system and equipment to detect 'firing of Polaris-type missiles'. This last item was probably based on infrared detectors to pick up the hot exhaust plume of the missile in its boost phase. On the small matter of radar kit, the ASV.21 was to be used in low-level patrols against submarines and small vessels while the sensor for the high-level surveillance role was to be a new side-ways-looking airborne radar (SLAR) to generate a 'surface picture' and allow the aircraft to cover a much wider area than the ASV.21 would allow. Clinker was an infrared linescan system, essentially an updated Yellow Duckling (see Chapter Two) that the Admiralty had been working on since the mid-Fifties. Another item listed in OR.350 is 'Improvements to visual search (Television)' which possibly alludes to a new low-light TV system being developed to meet AST.610 and ASR.1006.

In selecting a maritime aircraft, endurance was the most important factor and OR.350 stated eight hours patrol at 1,000nm from base. The patrol was to be conducted at low level with the speed specified at 225kt (417km/h) with 180kt (333km/h) while manoeuvring on attacks and capable of a 180° turn within 40 seconds within a radius of 4,500ft (1,372m). The type was required to use existing Coastal Command airfields with their 6,000ft (1,829m) runways. As discussions continued, it was apparent that the vaguest aspects of OR.350 were the patrol and transit speeds, which ultimately depended on the type's propulsion system. The Air Staff was of the opinion that the transit phase was 'non-productive' that is, wasted time and its reduction was a necessary part of the requirement. To this end the requirement laid out what was required for a propeller-driven aircraft – 350kt (648km/h or Mach=0.6) and for jet-propelled aircraft – 450kt (833km/h or Mach=0.8). Higher transit speed was sought but a 50kt increase would raise the AUW from 140,000lb (63,503kg) to 160,000lb (72,575kg) for a turboprop and 200,000lb (90,718kg) to 250,000lb (113,398kg) for a jet and of course, that would put costs up.

In effect there were two Staff Targets; one for a turboprop, another for a pure jet and this was quite probably where OR.350 fell apart as it was more or less the NMBR.2 and Three-in-One processes but with longer endurance and single role aspects respectively. This was a recipe for indecision, lack of clarity and delay. Add into this mix the Air Staff's belief that new ASW techniques would be developed by the time the Shackletons needed replacement and the result was a lack of focus and no real sense of urgency. The Air Staff ran parametric studies to find their ideal type, and the results pointed towards a turboprop aircraft with a transit speed of 350kt (648km/hr) weighing in at 131,000lb (59,421kg) and a jet, transiting at 450kt (833km/hr) with an AUW of 220,000lb (99,790kg).

The surprise runner in the OR.350 stakes was the Breguet Br.1150 Atlantic, yet again from Avro (or rather, Hawker Siddeley) at Woodford, who in December 1961 had resurrected the 1958 agreement with Breguet. The French appeared keen to see a resumption of Anglo-French co-operation, with Jean Chamant, Vice-President of the French National Assembly writing to Julian Amery, British Minister of Aviation on 28th December 1961 to state that Amery's Parliamentary Secretary, Monty Wood-house, had witnessed a demonstration of the Atlantic. Woodhouse '…reported favourably on the Breguet project.' and was reported to have '…laid great stress on the advantages of re-establishing Franco-British co-operation in this sphere.'

Chamant and Woodhouse's work appeared to have been somewhat futile since the Atlantic, despite being the winner of the NATO NBMR.2 competition and being assessed for OR.350 as the Shackleton replacement, fell far short of the requirement on endurance and particularly, transit speed. The MoA asked the DCAS, Air Marshal Sir Ronald Beresford Lees, who explained that the existing Atlantic could not meet OR.350 but understood that '…Hawker Siddeley are undertaking a study of the development of the Breguet Atlantic, but no details of this project have been formally issued.' Lees also pointed out that even the developed Atlantic would be unsuitable as it lacked development potential once it entered service.

Amery thought Lees' answer still gave the French hope of selling the Atlantic to the UK, albeit as a joint project, and wrote to Chamant explaining that 'I can give you no firm assurance that the Breguet Atlantic will prove an acceptable aircraft to replace the Shackleton.'

Replacing the Shackleton

From the ashes of the Three-in-One studies the Air Staff drew one conclusion: a maritime reconnaissance aircraft should be bespoke. Not that this deterred the aircraft companies when it came to tendering for OR.350. Unsurprisingly in the climate of the Post-Sandys White Paper era, just about any aircraft that came close to meeting the endurance, payload and transit speed was dusted off, fitted out with maritime kit and proposed as a design study for OR.350.

Transport aircraft were thought highly suitable for the role, but the capacious fuselage, built to carry bulky items, suffered from high parasitic drag. The large payload met the weapons load requirements, but deploying these weapons was problematic and the freight holds needed sound insulation for the operators' stations. Converted strategic bombers fared little better and being optimised for high speeds at high altitudes, invariably in a straight line (although the Avro Vulcan was quite nimble and had potential) were unsuitable for the tight low-altitude manoeuvres required for ASW. Both these classes of aircraft had the range and the payload, but neither possessed that other must-have that had emerged from Shackleton operations, crew comfort. As noted earlier, Coastal Command had conducted studies on crew efficiency and, assuming that the workstations and environment were conducive to efficient working, considered eight hours the maximum that an operator could be expected to work. Aircrews had complained about the conditions within the Shackleton MR.1 when it underwent trials at the A&AEE and some of the MR.1's problems were addressed in the MR.2 and particularly the MR.3. The converted transports and bombers also lacked crew comforts and it had been dissatisfaction with the maritime versions of the Lancaster and Lincoln bombers that had led to the Shackleton in the first place. There was one class of aircraft that was designed from scratch for comfort, speed and endurance, all combined with a pressurised, heated and sound-insulated cabin, rest facilities plus that all-important galley – the jet airliner.

On 19th March 1962 the British and Canadian Air Staffs held discussions in Ottawa on replacing the Canadair Argus and Avro Shackleton around 1970 under a Canadian/UK Combined Operational Requirement (COR). The British outlined OR.350, particularly its Issue 2 form that was still under consideration in London while their Royal Canadian Air Force (RCAF) host, Group Captain Roberts, pointed out that '…the RCAF were not as advanced as the RAF in the consideration of a new MR aircraft…' and that the Canadians had '…just begun consideration of an Argus replacement.' The gist of what the Canadian officers said was that 'it would not be possible to agree a detailed OR during the visit of the Air Ministry Officers to Ottawa.' The Canadians agreed with the need for crew comfort, but then the discussions took a somewhat odd turn. The RCAF's Wing Commander Ingrams, '…raised the desirability of the new aircraft being able to hover, and said that this capability might be obtained if the aircraft had a nuclear powerplant.' All present agreed that this was indeed 'desirable' but '…out of the question…in a 1970 aircraft.'

Interestingly when the Canadians queried parts of OR.350 that appeared to be definitive and therefore not up for discussion, the RAF DDOR.4, Group Captain Dutton, pointed out that 'The present document was, however, a target not a requirement.' This was March 1962 and the definitive OR was yet to be issued! A further, separate, list of questions on the OR was issued on 5th April 1962 by OR.13, Wing Commander Bentley, with one of the questions relating to whether 'In view of the latest pronouncements on causes of cancer…the aircraft will require to be cleared for smoking.'

By 11th May 1962 the Royal New Zealand Air Force (RNZAF) Air Staff was becoming interested in OR.350 and were sent a copy of the OR and added to the distribution list for informa-

The 'too big and too expensive to operate' VC10. British United VC10 Type 1103, G-ASIW cleans up as it departs on a long-haul flight. Note the flaps and slats on the wing that conferred excellent low-speed handling, ideal for hot/high airfields, but for the maritime role boundary layer control would have been required for low/slow ASW operations. Author's collection

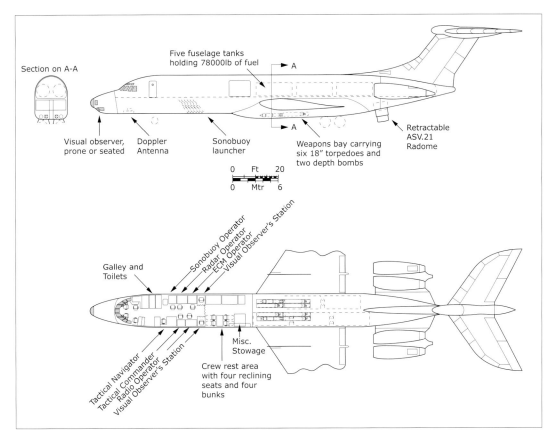

Section on A-A

Five fuselage tanks
holding 78000lb of fuel

A

A

Visual observer,
prone or seated

Doppler
Antenna

Sonobuoy
launcher

Weapons bay carrying
six 18" torpedoes and
two depth bombs

Retractable
ASV.21
Radome

0   Ft   20
0   Mtr   6

Sonobuoy Operator
Radar Operator
ECM Operator
Visual Observer's Station

Galley and
Toilets

Tactical Navigator
Tactical Commander
Radio Operator
Visual Observer's Station

Misc.
Stowage

Crew rest area
with four reclining
seats and four
bunks

The ink was hardly dry on OR.350 when newspapers announced that the VC10 was the winner. They were wrong and although the VC10 Scheme C was proposed for OR.350, it was not selected. Note how much of the fuselage is effectively unused and fitted out with fuel tanks. Vickers would persevere with the Maritime VC10 but it was always too big and too expensive to procure and operate.

tion. Oddly enough, just a few days previous to this, it appeared that the cat was out of the bag. Newspaper reports stated that a replacement for the Shackleton had been selected and that there was 'Commonwealth interest in the type.' Then on the 22nd May the Japanese air attaché, wrote to Wing Commander R Whittan at the Air Ministry asking if the VC10 was to be used as 'the future anti-submarine aircraft.' This caused a bit of a flap, since the VC10 had pretty much been ruled out as the OR.350 type for various technical reasons, but also because modifying Coastal Command's hangars around the country to take the VC10 would be very expensive at £40,000 for each hangar. Squadron Leader R S Brand advised that the Japanese Air Attaché should be informed that the VC10 was too costly to operate and could not manoeuvre at the low speeds and altitudes required.

On 4th June 1962, two French staff officers turned up at the MoA, ostensibly to brief the Director of OR(B), Air Commodore P C Fetcher. As it turned out, the French were interested in a replacement for the Atlantic and wanted to be 'kept in the loop' on OR.350. This related to the undertaking that the UK had entered into on leaving the NATO NBMR.2 programme, which stated that the UK would consider the Atlantic for any future MR requirement. The

next morning, a Breguet representative called Air Commodore Fletcher and advised that if they did not receive the information the officers had requested, 'Hawker Siddeley will not get the information in time for them to complete their contract.' This might explain how the Breguet Atlantic, a seemingly hopeless outsider for OR.350, entered the running.

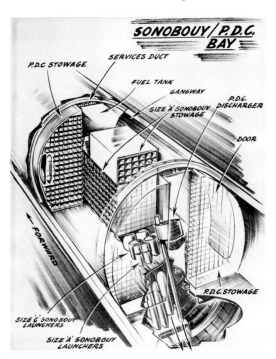

SONOBOUY/P.D.C.
BAY

SERVICES DUCT

P.D.C. STOWAGE

FUEL TANK

GANGWAY

SIZE 'A' SONOBUOY
STOWAGE

P.D.C.
DISCHARGER

DOOR

FORWARD

P.D.C. STOWAGE

SIZE 'C' SONOBUOY
LAUNCHERS

SIZE 'A' SONOBUOY
LAUNCHERS

The VC10MR was to be fitted with a sizeable stock of sonobuoys and practice depth charges (PDC). These were to be launched via an array of launchers in the forward underfloor cargo bay. The launchers were loaded from a storage area in the forward cabin. *Brooklands Museum*

The Runners

As noted above, OR.350 attracted tenders from all the builders of large aircraft in Britain, as well as some overseas types and basically, if the aircraft could carry a payload for long enough, it was tendered. To deal with the overseas bids first, the main players were the Canadair Argus and the Lockheed SC-130 Hercules. The Argus was dismissed almost out of hand as being little better than the Shackleton and despite only recently entering service with the RCAF, was probably due for replacement in the early 1970s. Lockheed's tender, the SC-130, lacked endurance as it could only patrol for three and a half hours once on location. Oddly enough, fifty years later the C-130J Hercules was the primary maritime patrol type in the RAF.

Hawker Siddeley's design office and factories at Woodford and Chadderton had the edge in developing a maritime patrol aircraft, having provided the original purpose-designed maritime reconnaissance landplane in the shape of the Avro Type 652A Anson through the general reconnaissance Lancasters and Lincolns to the Shackleton MR.3. Hawker Siddeley intended developing the Shackleton replacement and when it came to OR.350, certainly sought to give the Air Staff more than enough choice. Avro had tendered the Type 745 for NBMR.2 and had made the final three, only to be pipped at the post by the Breguet 1150. How that competition would have panned out had the UK remained in the process is anyone's guess, but even the French acknowledged Avro's expertise in the field.

An Embarrassment of Riches

As has been pointed out previously, the only company that had experience of a modern maritime patrol aircraft was A V Roe, which had become part of the Hawker Siddeley Aviation (HSA) in 1960. Woodford was the company's large aircraft division and had produced the Shackleton and the Vulcan, experience that would be valuable in the OR.350 process. Due to the size of the group and the companies that had merged into HSA, a number of designs were suitable for conversion to the maritime patrol role.

Avro dusted off the Type 745 that had been tendered for NBMR.2 back in June 1958. Given that the 745 could not meet the OR.350 transit speed or the endurance demanded, Hawker Siddeley opted to enlarge the airframe by just over 10% producing an aircraft of similar size to the Breguet Atlantic, with a wingspan of 125ft 6in (38.2m). A Rolls-Royce RB.168 turbofan was installed in the tail and it was renamed the Type 775/400, the 400 alluding to the transit speed in knots. The main external change made to produce the 775/400 was the reworking of the tail to install the turbofan, fed by a dorsal intake at the base of a new swept fin. Unfortunately, given the vagueness of the Air Staff's requirement and the fact that the 775/400 was powered by Tyne turboprops and a Spey turbofan, Hawker Siddeley felt that they should also address the higher transit speed regime of 450 knots. This involved a major change to the wing, with 30° sweep applied to the leading edge giving a wingspan of 118ft 9in

A comparison of the Breguet Atlantic and Avro's Type 775/400. The aircraft are of similar size, but the 775 is fitted with a Spey turbofan in the tail to boost the transit speed to the 450kt specified in OR.350. Unfortunately a swept wing was better suited to the transit speed. *Avro Heritage*

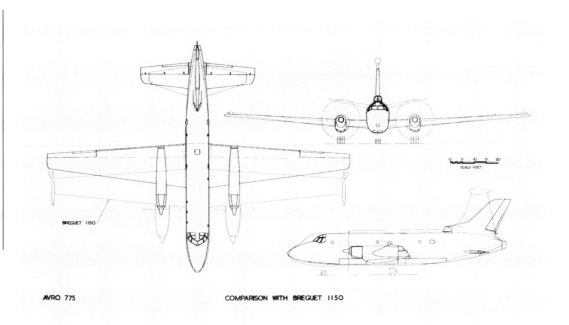

AVRO 775   COMPARISON WITH BREGUET 1150

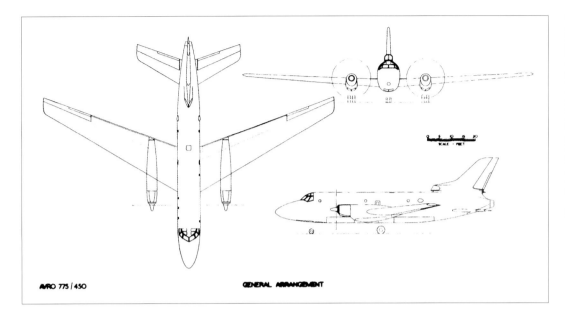

AVRO 775/450

GENERAL ARRANGEMENT

The Type 775/450 modified the 775/400 by fitting wings and tailplanes with 30° sweep to fully utilise the extra thrust of the Spey. This raised transit speed to the 450kt specified and produced a rather attractive 'Mini-Bear'.
Avro Heritage

(36.2m) and similarly swept tailplanes to produce the Type 775/450. This aircraft would be capable of transiting the 1,000nm (1,852km) to its patrol area at 420kt (778km/h) at 30,000ft (9,144m), carry out its eight hour patrol at 200kt (370km/h) flying at 1,500ft (457m) before returning to base at 450kt (833km/h) and 30,000ft (9,144m) to land after a sortie of thirteen hours and twenty-five minutes. As the Type 775 came under closer scrutiny from the MoA and Air Staff, it was considered too small to meet OR.350, smaller than all the other proposals apart from the Atlantic. In fact the MoA's Final Report on OR.350 pointed out that to carry out a similar sortie, the Vickers Vanguard required almost double '…the weight of equipment and power services' available on the

Avro Type 775 and if this was the norm for the other types, the 775 might fall short on capability.

Woodford also examined the Breguet Br.1150 as the basis for an OR.350 type and apart from anglicising the name to Atlantic, Hawker Siddeley intended a number of major modifications to the aircraft. Under the same Type 775 designation, Hawker Siddeley gave the Atlantic the same treatment, Spey turbofan in the tail cone and 30° swept wing, to produce a rather fetching version with swept wings, but retaining the same tailplanes. This significant change to the planform would allow the Atlantic to make the required 450kt (833km/h) transit speed. Avro had previously looked at modifying the Atlantic as a Skybolt carrier by

The mission profile of the Type 775/450 is effectively a snapshot of the OR.350 performance requirement. Transit at over 420kt and an 8 hour patrol at 200kt, followed by a return transit at 450kt.
Avro Heritage

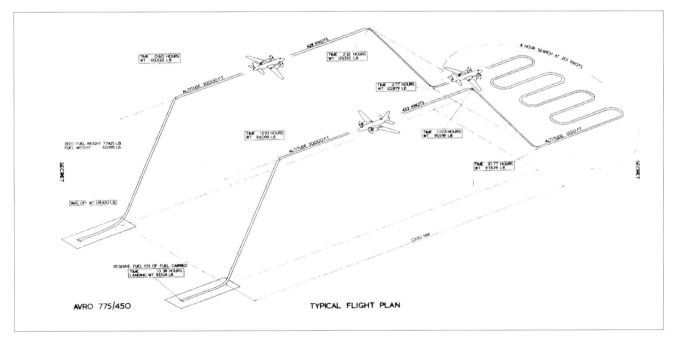

AVRO 775/450

TYPICAL FLIGHT PLAN

Avro's Type 775 was considered too small for the RAF's needs but, as this sectional drawing shows, it had absolutely no wasted space. Avro also appear to have learned from their Shackleton experience and installed rather more comfortable crew rest facilities! *Avro Heritage*

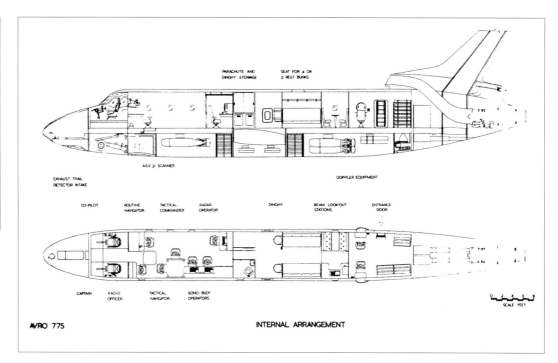

AVRO 775   INTERNAL ARRANGEMENT

stripping out the maritime kit, adding extra fuel tanks and a Rolls-Royce RB.153 under each wing, increasing the aircraft's ceiling to ensure the Skybolt attained its maximum range. The additional engines were also applied to a standard maritime Atlantic to produce the Atlantic 2A, but two underwing RB.153s were less efficient than a single Spey in the tail of the swept-wing Atlantic nor could the Atlantic 2A make the required transit speed. Since OR.350 also included a trooping role, the Atlantic was capable of carrying 27 fully-equipped troops with minimal changes, or, if the maritime kit was removed and the radar and weapons bays converted for cargo, could carry 59 troops in rear-facing seats. The Air Staff would become increasingly sceptical of mixed powerplants as experience of the Shackleton MR.3 Phase 3 would soon show that the Viper turbojets put additional stress on the airframe, leading to fatigue and ultimately prompted a push for the Shackleton replacement to be available sooner than was originally specified. Overall, the Air Staff were sceptical of turboprops, believing engines like the Tyne would lead to a 'Griffon scenario' in the 1970s, as the Griffons were by 1960 proving unreliable and costly to maintain. Whatever would they have made of the Griffons finally leaving RAF service in 1990 fourteen years after the Tyne had left the RAF?

Avro, having been a major partner in the Breguet 1150 for NBMR.2, took that type and tailored it to OR.350. A Spey turbofan was installed in the tail-cone and the wings swept at 30°, all in an attempt to meet the 450kt transit speed and two-plus engine specified in OR.350. *Avro Heritage*

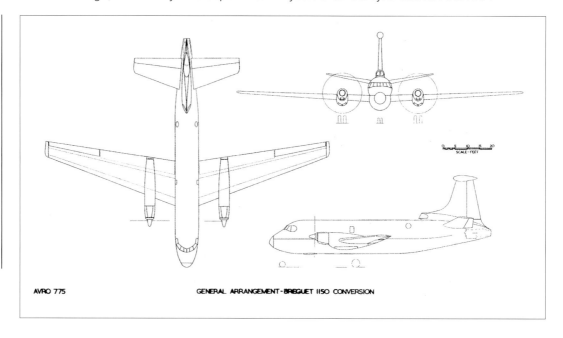

AVRO 775   GENERAL ARRANGEMENT - BREGUET 1150 CONVERSION

Vulcan MR

Avro had long nurtured the notion of using the Type 698 Vulcan strategic bomber as the basis of an airliner, the Avro Type 722 Atlantic (not to be confused with the Breguet aircraft!), but by the late 1950s this had fallen from favour as bespoke jet airliners came into service. Concurrent with the Atlantic studies, Woodford had drawn up the Type 718 as a trooping aircraft that combined the Vulcan prototype's wing with a deeper double-deck fuselage. The Type 718 had seating for 24 troops in what had been the bomb bay plus an upper deck cabin with seating for a further 66 troops. The soldiers' kit was to be carried in freight holds in the nose and rear fuselage. By 1962 the Vulcan B.2 was in service and it formed the basis of a 62-seat troop transport design study by marrying the wings and rear fuselage of the B.2 with a new double-bubble forward fuselage comprising a cabin for 34 troops in rear-facing seats aft of the flight deck. The lower lobe of the forward fuselage held fuel and the nose undercarriage, with the bay under the flight deck available for freight. The other 28 troops were carried in a pressurised capsule within the bomb bay, accessed by a hatch in the pod floor, with emergency exits in the roof through which access was gained to the upper surface of the aircraft. Like every other large aircraft of the time, the 'fat' Vulcan was also proposed as a Skybolt carrier, one AGM-48 under each wing, with the bomb bay taken up by two tanks carrying a total of 3,800 Imperial gallons (17,275 litres) of additional fuel.

Avro proposed using the 'fat' Vulcan as the basis for an MR type to meet OR.350. The upper cabin behind the flight deck was to be fitted with workstations for a crew of nine operators, with the bay under the flight deck used for a H2S Mk.9 radar or later, the ASV.21 radar antenna. Since this scanner suffered from a restricted field of view, a second, smaller ASV set was installed in the tail cone to provide 360° coverage. The bomb bay was split horizontally with the lower section carrying weapons, the upper section housing sonobuoy dispensers and additional stores. This area could be accessed in flight to reload the dispensers and since the Vulcan MR would be operating at low level while this was happening, the area wasn't pressurised.

Hawker Siddeley's designers certainly had a wide range of airframes to work from and could be accused of chucking any type remotely capable of conversion to the MR role at the requirement. That would be unfair, but two types in

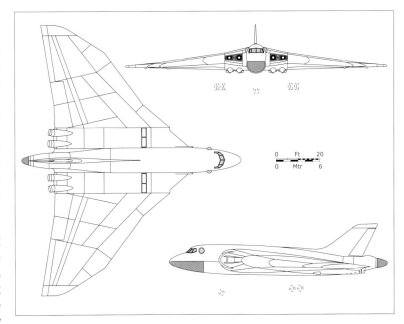

Avro proposed modifying the Vulcan B.2 airframe by fitting the wings, engines and tail assemblies to a new enlarged fuselage with operators' cabin and split weapons bay. Also of note is the fore-and-aft scanner system (FASS) for the radar to allow 360° coverage in azimuth.

particular spring to mind. The first was the HS.1011/1023, last seen being proposed for the Three-in-One studies and going nowhere fast, and the second was an aircraft that would make progress, and being proposed for OR.350 while still very much a paper study – the Armstrong Whitworth 681.

The full, official, list of Hawker Siddeley studies for OR.350 was submitted to OR.13 on 31st August 1962 and comprised eight types:
1) Avro Shackleton Mk.3 Phase 3 Updated
2) Breguet Atlantic Mk.1
3) Breguet Atlantic Mk.1 development (Mk.1A)
4) Breguet Atlantic conversion to meet OR.350 (Mk.2A)
5) Avro 775
6) APG/1010A – a new turboprop aircraft to meet OR.350
7) APG/1011F – Variable Geometry Transonic MR aircraft
8) AW.681 Maritime Reconnaissance version

The first three on the list failed to meet the endurance, payload and transit speed criteria, but were included for comparison with the other five. There was another aircraft from the Hawker Siddeley Group that was considered but not submitted for OR.350. The de Havilland design office at Hatfield had drawn up a maritime patrol study based on the de Havilland DH.106 Comet IV airliner. The Comet, a design that predated the Shackleton that was to be replaced, was deemed too old and lacking in development potential and so the maritime Comet disappeared from sight.

Hawker Siddeley's Advanced Projects Group at Kingston specialised in design studies for

Avro's Vulcan MR to meet OR.350 used an enlarged fuselage with a cabin fitted out for seven ASW operators and a rest area, with access to the sonobuoy dispenser in the rear fuselage. The Vulcan's capacious bomb bay was much reduced in size with the upper section used for sonobuoy and PDC storage. Note also the fore and aft scanner system for the radar, a modified H2S Mk.9 in the nose and ASV.19 in the tail cone. *Avro Heritage*

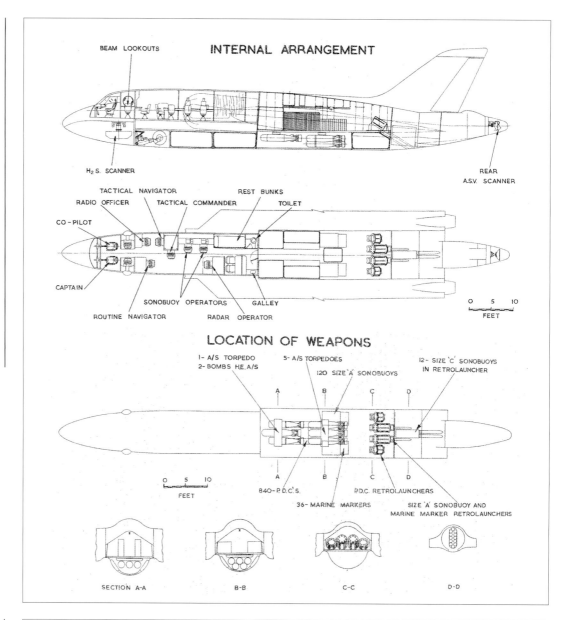

The HS.1010A originated with the Hawker Siddeley Special Projects Group at Kingston and like the later Avro Type 784, was powered by Tynes. Originally tendered for OR.350, it could also carry four Skybolts under the wings! *Avro Heritage*

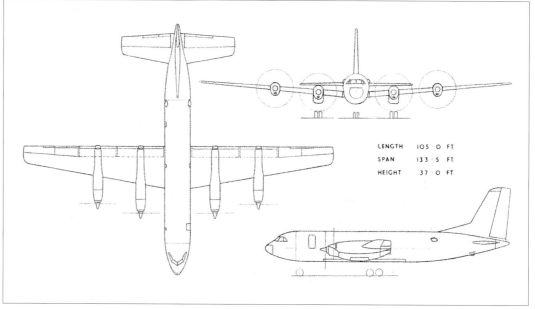

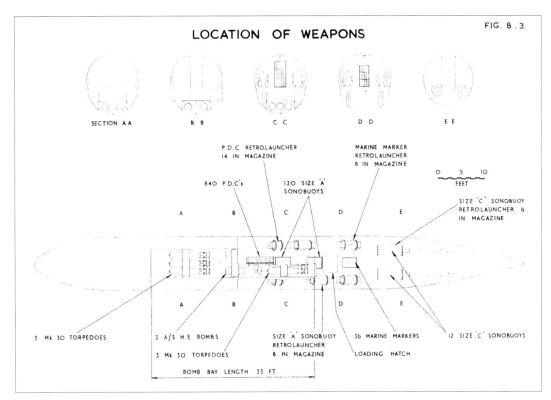

LOCATION OF WEAPONS FIG. 8.3.

Despite its size, much of the HS.1010's cabin was taken up by sonobuoy and marine marker dispensers. This freed-up the weapons bay to accommodate a large payload and at 35ft (10.7m) long, it could carry the full suite of ASW weaponry. *Avro Heritage*

high-technology aircraft and were involved in many of HSA's projects in the 1960s. They drew up two designs for OR.350, the APG/HS.1010 turboprop MR type and the APG/HS.1011, variable geometry transonic airliner. The APG.1010 was a fairly conventional type with double-bubble fuselage, a low-mounted straight wing spanning 133ft 6in (40.7m) and a fuselage length of 105ft (32m), powered by four Tyne 20 engines driving de Havilland 16ft (4.9m) propellers. The 1010's AUW was 156,100lb (70,806kg) with a 17,000lb (7,711kg) warload and could transit at 380kt (704km/h). The main drawback with the APG.1010 was its transit speed, which did not meet the requirement. One method that APG proposed to improve the type's speed and endurance was to dispense with the standard ASV.21 radar and its retractable scanner by fitting two antennae, one in the nose, the other in the tail cone to provide unrestricted 360° coverage with minimum drag. The APG/1010 could also carry 42 fully-equipped troops with minimum changes, or 90 with the ASW kit removed. Four Skybolts could also be carried under the wings, but this role would require a beefed-up structure and undercarriage to handle the higher AUW of 193,120lb (87,598kg)

Conventional is not a charge that could be levelled at the APG/HS.1011 and HS.1023 whose chief advantage in the role envisaged by OR.350 was its ability to sweep its wings, light

the reheat on the RB.178-1B engines and move rapidly to its patrol area or prosecute a target in another sector. This was how the area surveillance role would have worked, particularly when dealing with Soviet submarines launching their SLBMs. As noted in Chapter One, at this point in the early Sixties, Soviet ballistic missile submarines could only launch their weapons from the surface and this would take time. Since OR.350 called for the aircraft to conduct area surveillance with a SLAR and be capable of detecting missile launches, it made sense to have an aircraft that could patrol at high altitude, detect a target, either on the SLAR or via the IR launch detection systems, and accelerate to attack the target while it was still vulnerable. This was another reason for having an aircraft capable of high, if not transonic, speed.

The HS.1011 was a large, 180ft (54.9m) long, variable-geometry type with four RB.178-1B Super Conway turbofan engines, one under each wing root and two installed side-by-side atop the rear fuselage. The wingspan was 131ft 6in (40.1m) fully spread and 89ft 9in (27.3m) fully swept. The HS.1023 was based on a design study for a transonic airliner to replace or complement the DH.121 Trident. The HS.1023 was fitted with three RB.178-1B engines and was, by having better fuel consumption, more tailored to the patrol missile carrier role, but the HS.1023 would re-appear later in the Shackleton replacement story.

Woodford examined the HS.1011 for the Three-in-One OR.347 requirement. It was for OR.350 that most of the maritime studies were conducted. As time progressed, by late1962 these became more focussed on the three-engined HS.1023, shown here on the right.

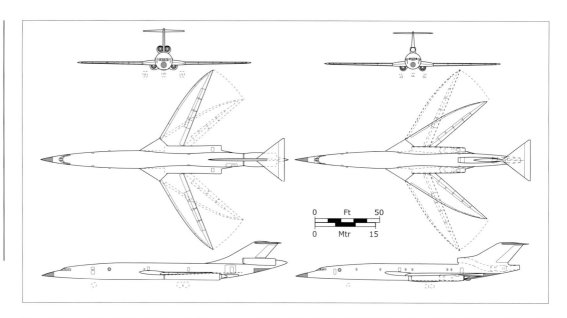

The HS.1023 was intended as a transonic transport in the same passenger capacity class as the Trident. The HS.1023 is seen here fitted out for the maritime patrol role. Note the weapons bays fore and aft of the wing box, and the size of the Sonobuoy storage and dispenser compartment. *Avro Heritage/Bill Rose*

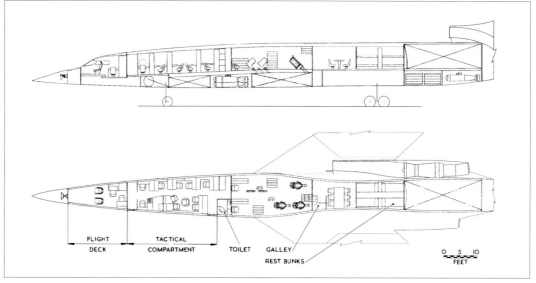

Opposite, top:
To the uninitiated the Avro 776 looks like a military DH.121 Trident. Although it shared some components with that type, the wings and fuselage were bespoke and designed by Avro at Woodford and Chadderton. The Type 776 would play a greater role in the later OR.357 process.
Avro Heritage

The Air Staff's principal objections to the HS.1011 and 1023 were their size and therefore cost and the development time-frame. Apart from buying the aircraft, operating a supersonic type was expensive, even when the number of engines was reduced to three on the HS.1023. This was summed up in the 'First Report by MoA on A.S.T.350' which stated 'Operating costs for supersonic flight rise steeply and would not be justified in the maritime role'. Nor were the Air Staff or MoA convinced by HSA's assurance that the aircraft could be developed by the late 1960s, stating that '…it is unlikely to be available in the required time scale.' Further examination concluded that the HS.1011 could conduct its entire mission, meeting OR.350 in full with its wings fixed at 25° to 30° 'This being adequate for speeds up to 0.8 Mach…' and therefore variable geometry would not be required. The MoA report summed up the VG type with 'Variable sweep aircraft are only justifiable by use of the aircraft in some alternative role in addition to the AST.' Since this had already been addressed in the Three-in-One study, that was the end of the HS.1011 and HS.1023 for OR.350.

A New Avro Type Appears

Despite the Avro 775 meeting OR.350, the design team at Woodford took the view that they could produce a much improved aircraft, with more growth potential. Avro had concluded that the pure jet was the best solution to OR.350 as '…propeller turbine engines have reached the end of their development potential with the Tyne.' At a meeting in the Woodford works on 8th June 1961, Avro's Chief Aerodynamicist, Mr Scott-Hall, described what would be called the Avro Type 776. Subsequently Squadron Leader R

S Brand of OR.13a outlined the 776 in a letter to the DDOR.1. This, Brand related, had started out as a four-engined jet, with the engines in pairs on the rear fuselage *à la* Vickers VC10, but soon resorted to a tri-jet configuration using RB.178 turbofans. One engine was mounted on each side of the rear fuselage with the third buried in the tailcone, fed by an intake at the tailfin root. The aircraft looked like a DH.121 Trident and 'Certain components of the fuselage are borrowed from the DH Trident…but the swept wing is an Avro design.'

Originally intended to carry a FASS radar system, the ultimate Type 776 carried an ASV.21 radar in a deep chin fairing with a large radome. The MAD was in the form of a towed 'bird' deployed from a pod on the port wing, with a similar pod on the starboard wing carrying a searchlight. With 16,920lb (7,675kg) of stores, the Type 776 would have an AUW of 170,000lb (77,111kg), could cruise at 30,000ft (9,144m) with a transit speed of 460kt (852km/h) and when flying on two engines during the patrol, the Type 776's speed would be around 210kt (389km/h). If required, on conversion to a transport, the 776 could carry 100 passengers or if used as a Skybolt carrier, two missiles for a thirteen hour patrol. Brand concluded his letter with 'I feel that it represents the most promising solution so far by any firm for a replacement for the Shackleton in 1970.' The Type 776 was well received, but its timing, late 1962 would see it play a greater role in the follow-on to OR.350.

The STOL Transports Drop In

Armstrong Whitworth Aircraft (AWA) at Baginton near Coventry had become part of the Hawker Siddeley Group back in 1935 but with the consolidation of the UK aviation industry into two main groups in 1961, Armstrong Whitworth Aircraft merged with Gloster Aircraft to become Whitworth Gloster Aircraft. AWA had been unsuccessful with their AWP.15 bid for NBMR.2 and since OR.350 called for a pressurised aircraft with a fairly high transit speed, decided the AWP.15 would be unsuitable. Fortunately for AWA, the Air Staff had also just issued the requirement for their next major aircraft project: OR.351 for a tactical transport aircraft. A number of bids had been submitted for this requirement, the most successful being the BAC.208 (née Bristol Aircraft Type 208) and the HS.681 (née AW.681).

Both combined what is now considered the classic military airlifter configuration of high wing, underwing podded jets and a capacious

cabin with a large tail door-cum-ramp. Common to each type was the use of the latest engine technology for short take-off and landing (STOL). The HS.681 was to use a quartet of Bristol Siddeley BS.53/5 turbofans, rated at 18,100lbf (80.5kN), which ultimately became famous as the Pegasus on the Harrier. An alternative engine was the Rolls-Royce RB.177/11 (a major redevelopment of the RB.142 Medway), rated at 18,600lbf (82.7kN) with these engines to be installed in individual pods, carried on underwing pylons, with two vectoring nozzles per side. The BAC.208 was to be fitted with four BS.53s mounted in pairs on a single underwing pylon, with a vectoring nozzle on one side of each engine, outboard on No.1 and No.4, inboard on No.2 and No.3. Both types were to be capable of Vertical Take-Off and Landing (VTOL) if fitted with banks of lift engines, either BS.59 or RB.175, in wingtip pods.

In response to requests for tenders to meet OR.350, Whitworth Gloster's design team drew up a study for a maritime patrol version of the Whitworth Gloster AW.681. The company

Armstrong Whitworth proposed their AW.681 tactical transport as a maritime patrol type. Unfortunately, to meet the requisite endurance, a pair of large external tanks was added, increasing the drag. This study makes an interesting comparison with AWA's later submissions to OR.357.
Avro Heritage

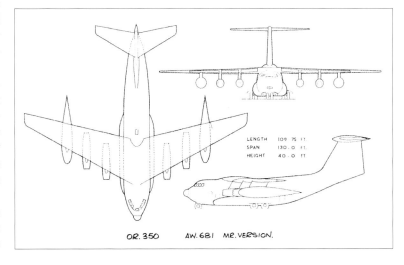

OR.350   AW.681 MR.VERSION.

made it clear that the same engines and airframe would be used, with the RB.177/11A Medway the preferred type, modified for 'straight through' exhaust rather than via thrust-vectoring nozzles, and fitted with a reheat system. Alternative engines were the original BS.53 in its 'straight through' BS.53/6 guise or the BS.100/9, which was to be developed from the vectored-thrust engine destined for the Hawker P.1154. Other changes were the addition of large fuel tanks under the outer wings, installation of maritime equipment in the freight hold and 'provision to close engine intakes during the search.' The weapons were to be carried within bays in the large undercarriage fairings either side of the fuselage. A bomb aimer's station was to be installed in the lower nose below the forward ASV radar, housed in a nose radome, while a second, aft-facing, ASV radar was installed in the underside of the rear fuselage to provide 360° coverage.

With an AUW of 277,300 lb (125,781kg) the AW.681MR met the OR.350 requirement and its patrol time at 1,000nm (1,852km) was maximised by the ability to shut down two engines, hence the provision to close the intakes to minimise the drag from a windmilling turbofan. The type could carry up to 70 troops if the crew rest area furnishings and the sonobuoys and practice depth charge dispensers were removed from the rear of the cabin. However, it was too slow, too heavy and suffered from being a compromise design, rather than a bespoke MR type. Further problems were foreseen with the conversion of tactical transports to the MR role; sonobuoy and weapons carriage. As noted, without major structural modifications, the bays to carry these had to be installed in extended undercarriage fairings or on a converted tail ramp. The OR.351 tactical transports had the floor of the freight hold set at truck-bed height for easy loading and unloading. This was a problem for weapons carriage as the undercarriage was by necessity, short and that left little clearance under the belly, so even if ventral bays were installed, stores could not be easily loaded.

While on the subject of converted transports, Shorts' proposal to convert the Belfast has to be mentioned. Shorts had taken a look at OR.351 and thought that the SC.5 Belfast, what was for a long time the largest aircraft to enter service with the RAF, would be suitable to turn into a tactical transport aircraft. As had been pointed out in the Three-in-One study, the Belfast was good at one thing, albeit slowly, and that was lugging outsized heavy loads over long distances. Converting that capability into a tactical transport that could operate into short fields near the front line would take a bit of effort on Shorts' part, never mind convincing the Air Staff that it would work. Shorts proposed the 'Tactical Belfast' the SC5/21B that was to be fitted with Tyne 32 engines rated at 8,400shp (6,264kW) and fitted with boundary layer con-

This drawing shows why the main problem with converted tactical transport aircraft such as the AW.681 was wasted space in the cabin, particularly in the upper cabin and on the tail ramp. The ASW stores and weapons would be carried in the extended sponsons.
Avro Heritage

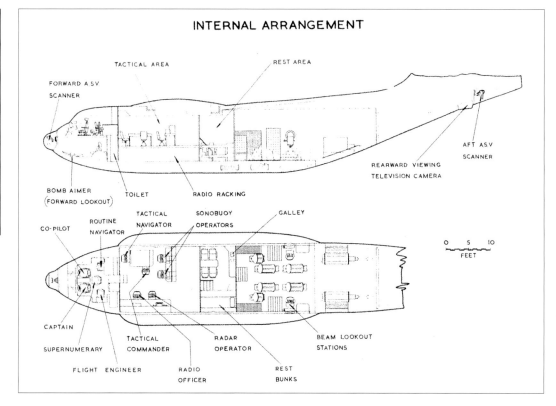

INTERNAL ARRANGEMENT

TACTICAL AREA

REST AREA

FORWARD ASV SCANNER

AFT ASV SCANNER

REARWARD VIEWING TELEVISION CAMERA

BOMB AIMER (FORWARD LOOKOUT)

TOILET

RADIO RACKING

CO-PILOT

ROUTINE NAVIGATOR

TACTICAL NAVIGATOR

SONOBUOY OPERATORS

GALLEY

CAPTAIN

SUPERNUMERARY

TACTICAL COMMANDER

RADAR OPERATOR

BEAM LOOKOUT STATIONS

FLIGHT ENGINEER

RADIO OFFICER

REST BUNKS

0   5   10
FEET

trol system comprising 'blown' flaps and control surfaces. The high-pressure air for this was to come from three Rolls-Royce RB.176 gas generators (intended for the Fairey Rotodyne) installed in a removable dorsal pack. Taking the SC5/21 and turning that into a maritime patrol aircraft was a feat worthy of a magician and the trick failed to impress. The Air Staff summed it up thus: 'The long-range transport version is too slow and, like similar large capacity freighters including STOL transports, would be uneconomic to operate.'

This was also the Air Staff's opinion on the Vickers VC10, which had been tendered for OR.350 in the same guises as for the Three-in-One studies. The VC10 was 'An excessively large and expensive aircraft rejected even in the multi-role concept though meeting O.R.350 requirements.' Further to that was the cost of operating the VC10 which was discussed at length in an October 1961 paper on the Shackleton replacement. The VC10 Scheme B and Scheme C from the earlier Three-in-One studies were examined and the question of cruise efficiency and fuel costs arose. The VC10 had to operate on four engines until the AUW had reduced to 280,000 lb (127,006kg) through fuel burn, then it could operate on two engines and could maintain height on one engine in the event of a failure until the other engines were restarted. So, in the pursuit of efficiency, the VC10 could patrol on two engines and meet

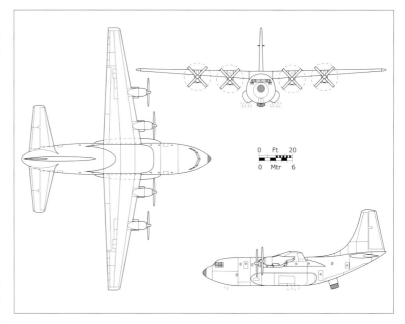

the OR.350 requirement. However, that required an AUW at launch of 380,000 lb (172,365kg) of which 214,000 lb (97,069kg) was fuel! The point was that if the Shackleton force was replaced by VC10s on a one-for-one basis, '...the fuel bill alone for Coastal Command would increase tenfold.' This subject re-emerged during later discussions on OR.357 when an officer observed that the fuel load of the VC10 exceeded the AUW of most of the other contenders for the requirement, including the Avro Type 776!

This provisional drawing shows the Shorts submission for OR.350 was based on the shortened 'tactical' Belfast, the SC.5/21B. Weapons and stores would be installed in bays on the tail ramp.

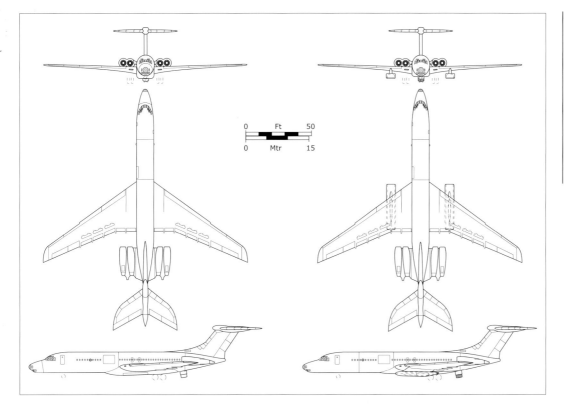

Vickers VC10MR for OR.350 came in two styles. Scheme C was modified with a weapons bay in the wing box while Scheme B was fitted with panniers. The panniers shown here are fitted with Tambour doors.

129

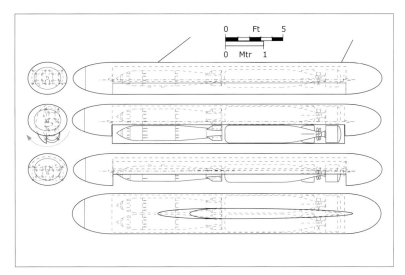

One of a number of VC10 panniers for Scheme B, seen here carrying two WE.177A and two Mk.31 homing torpedoes. This example uses the rotary bomb door system used on Blackburn Buccaneers and Martin B-57 bombers. The weapons are mounted on the bomb door itself and this rotates about the pannier's long axis to expose them for delivery.

Realistically the VC10 would have been expensive to buy, costly to operate and not particularly suited to the role. Not to be outdone, Vickers had another iron in the OR.350 fire in the shape of the Type 952 Vanguard, a four-engined turboprop airliner in the same class as the Lockheed Electra that was being transformed into the P-3 Orion for the US Navy. Vickers had by January 1961 conducted a design study on a Maritime Vanguard for the Indian Government, as a replacement for their Consolidated B-24 Liberators.

Powered by four Tyne 11s, the Vanguard airliner was mid-winged, with the wing box between the lobes of a double bubble fuselage with the upper lobe forming the passenger cabin while the lower lobe carried baggage and

freight. Conversion of the upper cabin into an ASW operators' compartment ahead of the wing and a bay for sonobuoys and their dispensers aft of the wing would not be difficult, while turning the lower lobe into a pair of weapons bays would be feasible. The fuselage was too long for the MR role, but could be easily shortened by removing a couple of frames fore and aft of the wing. A new nose housing a bomb-aimer's station and ASV radar could be added, or the ASV.21 could be installed in a retractable installation in the lower fuselage lobe ahead of the wing.

Vickers at Weybridge's first Vanguard MR study 'Vanguard 350' was submitted to the MoA in August 1961 and was capable of a seven hour patrol at 1,000nm (1,852km) with a transit speed of 320kt (593km/h) at 20,000ft (6,096m). Vickers estimated a 1968 in-service date if an order was placed in 1962. The initial Vanguard MR was based on the Mk.2 airliner, but with the fuselage shortened by 14ft 7in (4.45m) forward of the wing and 8ft 7in (2.6m) aft of the wing, a redesigned flight deck and nose (with bomb aimer's position) and an operators' cabin. Propulsion was to be Tyne 21s rated at 6,000shp (4,413kW) driving 15ft (4.6m) diameter propellers, rather than the Tyne 11s driving 14ft 6in (4.4m) propellers. The AUW was increased to 196,000lb (88,904kg) compared with the airliner's AUW of 146,000lb (66,224kg). A weapons bay was to be installed under the wing centre section, with the radar installed just aft of the wing. Vickers did not

Vickers' Vanguard was a sound design and served as a passenger and freight transport for many years. As a possible MR type it was a viable option, apart from its transit speed, but Vickers did address this for OR.350 by fitting underwing bypass engines.
Author's collection

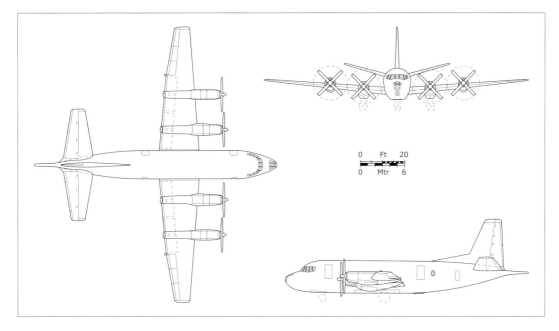

Maritime Vanguard was proposed in August 1961 based on an earlier design for the Indian Air Force. The fuselage was shortened by 25ft (7.6m) and carried its ASV.21 in a retractable dustbin under the rear fuselage. The increased AUW of 202,000 lb (91,626kg) may have required a new four-wheel bogie undercarriage.

intend fitting the ASV.21 but '…fully submerged split-scanner radar system will be available in the timescale quoted', which meant a FASS system with nose and tail radomes. Vickers stated that fitting a retractable ASV.21 would cut the patrol time by 40 minutes. The wings were to be strengthened and the leading edge flaps were to be blown to keep the take-off distance within 6,000ft (1,829m). Launcher bays for Type A and C sonobuoys were in the lower rear fuselage between the radar and tailplanes with a magnetic anomaly detector (MAD) in a tail sting.

By December 1961 Vickers had produced another brochure outlining a further two versions: Vanguard Maritime Minimum Change (MMC) and Vanguard General Maritime Reconnaissance (GMR). The MMC retained the standard airliner fuselage, Tyne 11 engines (with 15ft propellers) and all the windows but with the bomb aimer's position relocated in a ventral fairing between the weapons bay and the nose undercarriage bay. A retractable ASV.21 scanner was fitted under the rear fuselage with a MAD sting at the tail. A large freight door was fitted in the aft fuselage to allow the installation of six fuel tanks or fitting out for the trooping role. The AUW of the MMC was to be 172,500 lb (78,245kg) and could be a cheaper option, available at the end of 1965, than the Vanguard 350. The MMC could patrol for six hours and fifteen minutes at 1,000nm (1,852km) with a 6,000 lb (2,721kg) stores load, which Vickers said '…compares more than favourably with the Shackleton MR.3…' but with twice the transit speed.

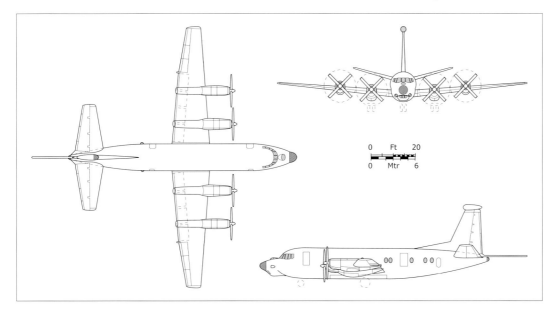

Vanguard 350 of December 1961 took the Maritime Vanguard and added fore and aft radar antennae, boundary layer control and a new nose, plus ESM in the tail fin fairing. The engines were to be Tyne 21s, fitted with 15ft (4.6m) diameter propellers. The fuselage was to be shortened by 14ft 7in (4.4m) ahead and 8ft 7in (2.6m) aft of the wing.

Vanguard Maritime, Minimum Change (MMC) was more or less a Vanguard airliner with a weapons bay, retractable ASV.21 and a pannier for a bomb aimer. This was the cheapest of the Vanguard variants, and lightest at 172,500 lb (78,245kg).

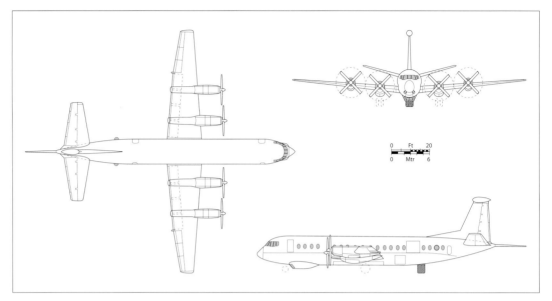

Vanguard General Maritime Reconnaissance (GMR) saw the MMC fitted with a new nose complete with radar and bomb aimer's position. Still essentially a standard Vanguard, its higher AUW of 200,000 lb (90,718kg) would have required a strengthened undercarriage.

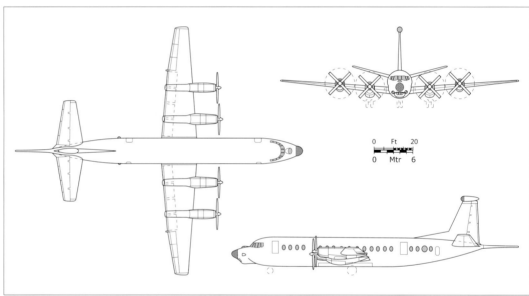

The Vanguard 350 was to be fitted with jet engines, possibly Conway or Spey turbofans, to meet the OR.350 transit speed. In this variant the engines are fitted at the inboard stations, but the fairings for the undercarriage would have increased drag.

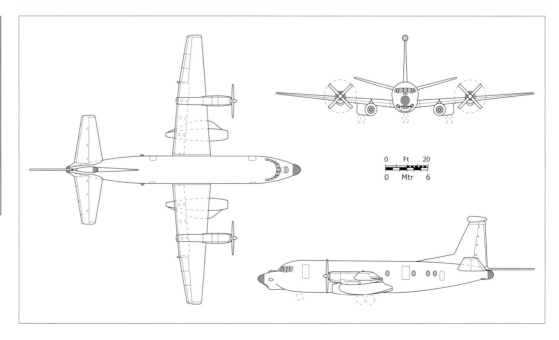

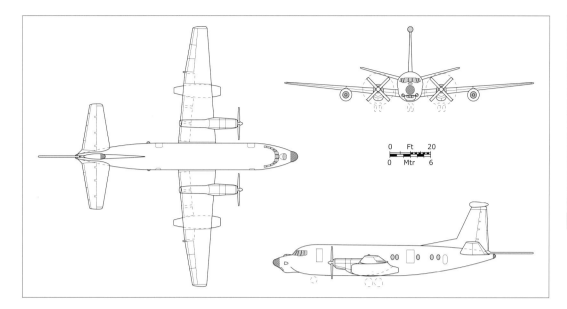

0 Ft 20
0 Mtr 6

Vanguard 350 with outboard engines allowed the retention of the standard undercarriage and therefore less drag than the version with inboard turbofan engines. The Tyne 21s, with increased power may have allowed patrol on two engines in the later stages of a sortie.

The Vanguard GMR was '…a compromise between the performance requirements of OR.350 and a reduced first cost…' of the MMC. The GMR could carry 17,000 lb (7,711 kg) of stores, but with an AUW of 199,750 lb (90,605 kg), needed a 9,000 foot (2,743m) runway to take off, but could patrol for the required eight hours. The GMR was to be fitted with a 'Maritime Nose' with bomb aimer's station and the forward scanner of a FASS radar system, if fitted. The alternative ASV.21 scanner was housed in a retractable fairing aft of the wing. Both the MMC and GMR could carry 103 fully-equipped soldiers in the trooping role or use the cabin for freight. Vickers advised that four-wheel main undercarriage bogies were available for all three proposals to reduce the aircraft's LCN by 50%. All three versions were to be fitted with an ECM pod on the fin.

Perhaps the oddest Vanguard variants to appear in the Weybridge OR.350 studies were the mixed powerplant Vanguards based on the Vanguard 350. No details are available and the studies only survive as drawings in the Brooklands Museum archive and show the Vanguard 350 with two of the Tynes substituted for unidentified jet engines in underwing fairings. These could be Rolls-Royce Conways, given that engine's use on the Vickers VC10, but that would be speculation. The first study has the inboard (No.2 and No.3) turboprops replaced by large engine pods projecting fore and aft of the wing, with large fairings for the undercarriage scabbed on the outboard side of the engine pods. The second version has the inboard Tynes retained, along with the undercarriage bays, and the jets in similar pods fitted at the outboard (No.1 and No.4) positions.

Summing up the Vanguard MR (and all the converted transports) Air Vice-Marshal K V Garside, Senior Air Staff Officer, Coastal Command, wrote 'It is considered that the Vanguard project study well illustrates the objections to this possible solution. The resulting aircraft falls unacceptably short of the requirement.' Garside continues to list the drawbacks of such a conversion including lack of development potential, little or no saving and available only a '…little ahead of an entirely new type.'

Laminar Flow MR

Handley Page Ltd (HP) had invested a lot of time and effort in developing what they considered to be the technology that would revolutionise the aviation industry. As far as HP were concerned the future of air travel did not lie in whisking the elite, the Jet Set, across the Atlantic at Mach 2, nor did they share de Havilland and Vickers' views on a transonic transport. Handley Page saw mass travel at low cost in fuel-efficient airliners as the future and compared with a supersonic transport (SST), HP's airliners would carry more than twice the number of passengers over three times the distance.

The technology HP intended to apply was laminar flow aerodynamics, through boundary layer control. The aim of aerodynamicists is to reduce frictional drag by smoothing the passage of air across the surface of the aircraft, and for simplicity the air flow can be visualised as a series of layers, laminae, moving across the structure. Due to friction, the laminae closest to the airframe move more slowly than those farther away and these slower layers can be turbulent, which increases friction and therefore

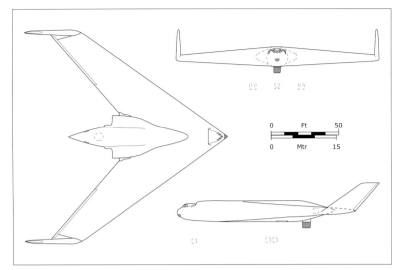

Handley Page's HP.117 was designed around the laminar flow control (LFC) system and as such was an idealised LFC type. This of course compromised it for other roles, but in the MR role, the HP.117 had the endurance, cabin space and weapons load required. Unfortunately it was big, expensive and the Air Staff were unconvinced of the benefits of LFC.

The HP.117 Maritime was based on the airliner and so had a great deal of unused space, formerly used for passengers, within the wings. To maintain laminar flow, the HP.117s weapons were deployed through small ventral hatches on specialised dispensers, thus reducing the flexibility of the weapons load. Note the visual search station in the tail cone.
via RAF Museum

drag in what is called the boundary layer. If the turbulent boundary layer can be controlled either by managing it (as in the case of the Blackburn Buccaneer) or removing it, the reduction in drag can be significant. HP sought to remove as much of the boundary layer as possible over an airframe by using apertures in the airframe skin through which the air in the boundary layer was drawn. This required a great deal of ducting within the wings and fuselage, with the intention being that the engines would suck in the air for use in combustion. The ideal configuration for laminar designs was the flying wing and so HP embarked on designing a flying wing to utilise the technology. A fuller account of Handley Page's laminar flow studies can be found in *Vulcan's Hammer*.

Suffice to say HP tried to gain funding for its laminar flow work, but given the Government's focus on the SST and HP's resistance to the Ministry of Aviation's drive for a rationalised indus-

try, very little funding was available. They did, however, carry on funding the work in-house, in the hope of attracting Ministry money through showing faith in their research work. One design study in particular, a flying wing airliner, had been suggested for the Three-in-One studies and this, the HP.117, was also proposed for OR.350. The HP.117 design was innovative because almost its entire surface was 'laminarised', including the fins at the wingtips. Only the cockpit glazing, control surfaces, access hatches and undercarriage doors lacked the boundary layer suction system. As a result many of these had dual purpose, such as the undercarriage bays providing the access for passengers, so only one set of doors was required.

With an AUW of 309,000 lb (140,136kg) the maritime HP.117 was to be powered by four Rolls-Royce RB.141, rated at 11,800 lbf (52kN) whereas the civil airliner was to be powered by RB.163, rated at 9,850 lbf (43.8kN). Interestingly, in the maritime role the HP.117, although almost 10,000 lb (4,535kg) lighter than the VC10, could carry almost twice the weapons load. Undoubtedly too big for OR.350 and although it depended on laminarisation to meet the endurance criteria, it met the range and speed requirements. This was what concerned the Air Staff as HP's own research had shown that the boundary layer suction system was susceptible to blockage by dust and insects at low altitude so should not be operated below 10,000ft (3,048m).

Despite that, the Air Staff were interested in the HP.117 and in a letter to AOC-in-C Coastal Command dated April 1961, Air Vice-Marshal R N Bateman, ACAS, stated that 'The performance figures of the HP.117 suggests a scaled

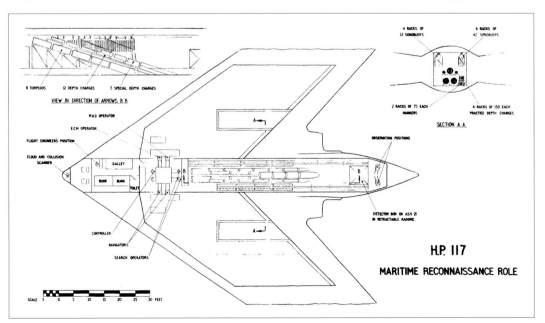

down version could be the most efficient to meet the AST but such a solution could not be recommended until the system has been proved in flight and until more is known about the problems of maintaining the surfaces of the aircraft in low drag condition and its operation under maritime conditions. Interestingly the Final Report on OR.350 suggests that such an aircraft, fitted with pusher propellers '…might show great advantages for a further generation of Maritime Aircraft.'

Outcomes

The Avro 775 met the requirement, but as noted above, the 776 was a much better bet for development, had the transit speed, the payload and met the requirement, with potential for growth. By May 1962 the vagueness that belied OR.350 was being addressed by the MoA's OR.13 branch, no doubt a result of the tsunami of designs that had swept into the MoA from the companies. The obvious solution to this lack of focus was to re-issue the requirement and this was intended for later in 1962. The MoA's Project Design Engineer J H Quick laid out his views in a letter to E E Marshall and stated that '…the ideal aircraft should be about 200,000 lb MTOW, have three or four pure jet engines and will cost about £2m per aircraft.'

The MoA felt Vickers took too long to respond to requests, whereas HSA were less tardy and expressed this view in memos '…departments concerned feel they receive exceptionally rapid Projecting service from Hawker Siddeley, and that Vickers' submissions take too long.' Quick made further comment to the effect that '…VA have regarded the M/R as just one role of the Multi-role concept and have not expended sufficient design effort on the maritime aircraft as a study in its own right.' Criticism of Vickers' 'too big and expensive to operate' VC10 and the 'just passé' Vanguard was summed by OR.13 in the statement '…they had just been thrown at the back of the bottom shelf'. Nor were they impressed after a flight in a Vanguard, which they felt was '…far removed in comfort from what they desire.' Interestingly the Americans had the same opinion of the Orion 'American disappointment in the Orion (ex-Electra) has also militated against the Vanguard. Apparently the Orion is down not only on endurance and payload, but gives an extremely uncomfortable "ride" at low (gusty) altitudes.'

Of all the types submitted, only one had impressed OR.13 – Hawker Siddeley's Type 776. The impression gained was that '…the H/S group have now an instinctive maritime "feel".', which was mainly the result of a 'top-line presentation to Coastal Command.' OR.13 advised that 'VA would be wise to make a similar presentation as soon as convenient.' Whether Vickers made such a presentation is not documented, but the Woodford team had the attention of the MoA and Air Staff, with what could be viewed as the definitive statement on OR.350: 'Hence it is envisaged that the OR will specify a jet (by-passed) engine conventional aircraft of 200/250,000 lb all-up-weight, similar in design to the Avro 776.' With that the way forward was a much more focussed requirement, particularly in the specification of the weapons and detection equipment. Rather than issue yet another amended OR.350 the MoA and Air Staff cancelled OR.350 on 16th May 1963 and issued a revised requirement – OR.357.

With the Air Staff failing to select a maritime patrol type from the multitude of studies that emerged from the UK's aircraft companies, it was left to the Shackletons to soldier on. As described in Chapter Five, the Shackleton MR.3, as seen here in WR970 the MR.3 prototype, would require replacement sooner than its MR.2 predecessor. *Author's collection*

'…but there is at present no indication that a break-through will occur in the next few years that will give maritime aircraft a random search capability against nuclear submarines.' – The Shackleton Replacement – Note by the Air Ministry, 7th May 1963

'PUS (Permanent Under-Secretary) said that the cost of the project was forbidding and had probably not been fully realised. He hoped that no opportunity would be lost to trim the cost.'
Note by DCAS, Air Marshal Sir Ronald Lees, Air Council meeting, 23rd May 1963

What became obvious from the OR.350 tender process was that there was a plethora of types from which a maritime reconnaissance aircraft could be derived. When that had been combined with the OR.347/Three-in-One/NBMR.2 studies, it became clear that only a bespoke MR type would suffice and this should be jet-powered. The other major factor in selecting an MR aircraft was the much-anticipated advance in anti-submarine detection techniques. By the end of 1961 the Avro Type 776 was the preferred type, but as ever in British weapons procurement programmes, the most obvious choice wasn't the one the powers that be opted for. As per standard procedure in UK defence procurement, when in doubt, issue a new requirement and this one would benefit from its predecessors. The Air Ministry did agree on one aspect: 'The general conclusion is that it will be necessary for a new design to replace the Shackleton and AST.357 has been passed to the Ministry of Aviation for feasibility studies on this basis.' There was also the small matter of the Canadian involvement and a possible replacement for the Argus, but that would require consultation with the Canadians.

That new requirement was OR.357, circulated around Whitehall from April 1963 but issued to industry in June 1963. The requirement was intended to clarify many of the vague statements in OR.350: jet engines would be preferred and although turboprops would be marginally cheaper to acquire, the speed of a jet

Soldiering on against the clock. As the OR.350 process dragged on and made little progress apart from leading to the issue of OR.357, the Air Staff became more alarmed by the need to replace the Shackleton. Shackleton MR.2s, such as WB833 (the MR.2 prototype) seen here, would need to be replaced by 1970. *via Tony Buttler*

aircraft would have other benefits. The new type should be in service from 1970, have a minimum of three engines, an ideal transit speed of 500kt (926km/h) with 450kt (833km/h) as the minimum. The take-off distance was relaxed to 7,500ft (2,286m) at 15°C (59°F) or 9,000ft (2,743m) at 20°C (68°F), rather than the 6,000ft (1,829m) of OR.350. This latter runway limit had been set to allow the type to operate out of airfields such as Gibraltar, where an extension to the runway would be rather expensive. Another aspect of OR.350 clarified in OR.357 was stores load, with a breakdown of the stores set out, including 8,400lb (3,810kg) for weapons plus 1,500lb (680kg) for externally mounted ASGW, rather than a vague 17,000lb (7,711kg) total payload. OR.357 also stated that a 20% increase in weapons load and size must be catered for in the resulting aircraft. Specific weapons were detailed including the Improved Air-Launched Anti-Submarine Homing Weapon to meet ASR.1186, '...the conventional stores that will be in the armoury...' and the Advanced Tactical Air-to-Surface Guided Weapon to meet AST.1168, which was to be externally mounted. Nuclear weapons were to include the NATO (that is, US) nuclear depth bombs and the British nuclear depth bomb to meet OR.1177, with the AST.1168 ASGW to be nuclear-capable as well, with their yield unlikely to exceed 10KT.

The vague requirement for ballistic missile launch warning in OR.350 was also clarified, with the system to detect a launch within 100nm (185km) of the aircraft and provide range and

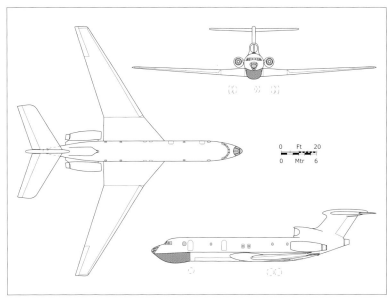

bearing of the launch point. However, this was a nice-to-have if the kit was available rather than a pretext to start a new development project. The 'wide area' radar system was also better defined, as being either a radar, an IR linescan system or combination of both to meet OR.3629, with the possibility of merging the search radar with a SLAR. Other systems were better defined in OR.357 including the Anti-Submarine Warfare Environmental Prediction System (ASWEPS) to provide sea temperature, salinity and state. All-in-all, OR.357 provided a comprehensive outline of what the Air Staff and Air Ministry wanted the aircraft to do, how it should do it and with what equipment, to a degree of detail that made its predecessor look positively amateur. However, that might just have been its undoing.

The Avro 776 was a handsome design, whose Trident heritage is evident. The large chin radome housed the ASV radar, in this case an H2S Mk.9 modified for maritime use. Later versions would revert to the ASV.21.

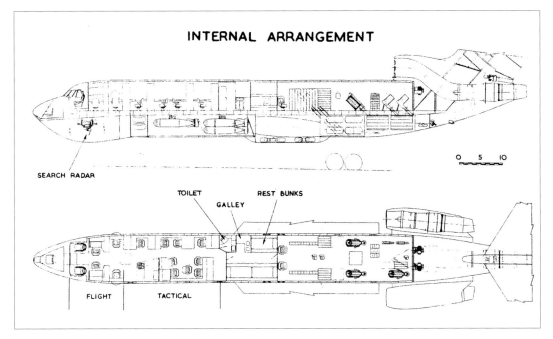

INTERNAL ARRANGEMENT

SEARCH RADAR

TOILET
GALLEY
REST BUNKS

FLIGHT
TACTICAL

The Avro 776 internal arrangement was probably just about the right balance of equipment bays, operators' workstations and weapons stowage. As such it had little superfluous space and this 'ideal' size of the airframe made it the preferred choice for ASR.357. *Avro Heritage*

That harmonious situation lasted precisely six weeks until May 1963 when DCAS, Air Marshal Sir Ronald Lees, wrote to the Minister of Aviation advising that the Chief of the Air Staff had '…decided to widen the speed range specified in OR.357 so as not to exclude a turbo-propeller aircraft.' Lees was following orders here as he was most definitely against turboprops and had laid out his case quite clearly during the OR.350 process. What had concerned Lees back then was that the Tyne, effectively the only turboprop that could be used, might end up like the Griffon – no development potential and the last of the line. The OR.357 tender process was increasingly becoming a re-run of OR.350, even down to Canadian interest and by widening the speed range to a point where the Breguet Atlantic could be considered, even the French might become involved.

In the run up to the issuing of OR.357 in May 1963, the MoA and Air Commodore Fletcher had a number of conversations with the French, who apart from attempting, unsuccessfully, to convince the British that the Atlantic was what they needed, let slip that they thought the Atlantic would need to be replaced before 1980. Fletcher was of the opinion that a joint MR project would be ideal '…in the best of all worlds a UK/Canada/France project might evolve, but this is perhaps too much to hope for.' Fletcher finished his briefing to the Air Staff by stating '…'I believe that our best chance of success lies with the Canadians and that we should not make sacrifices to appease the French until we are certain that the Canadians also require them.'

And so, with little real change from OR.350, the Air Staff in early September 1963 drew up a list of possible types, or rather, a list of categories from which an OR.357 aircraft could be drawn:

- A maximum state-of-the-art aircraft – A subsonic VG type, but this needed a research aircraft such as the ER.204 type
- Conventional jet aircraft – Avro 776
- Conventional turbo-propeller aircraft – BAC Britannia derivative or Avro 784
- Derivatives of planned aircraft – A Maritime AW.681 or the Breguet Atlantic 2A
- Derivatives of existing aircraft – Maritime variants of the VC10 and Trident
- Derivatives of Obsolete aircraft – Maritime Vanguard or Britannia
- Foreign aircraft – Licence building of the Atlantic or Lockheed P-3 Orion

The OR.357 proposals

Ostensibly OR.350 had eliminated some of the rank outsiders from the Air Staff's search for an MR type, with their preference being the Avro Type 776. OR.357 more or less opened up the contest afresh with some types, such as the Belfast, HP.117 and HS.1011 disappearing, but new proposals popped up to replace them. Two of these came from Lockheed, in the form of the P-3A Orion and the HC-130E Hercules. Lockheed Marietta had proposed their SC-130B search and rescue variant (originally designated thus for the US Coast Guard, later renamed HC-130B) for the maritime patrol requirement of the Royal New Zealand Air Force, who had

HC-130H Hercules 1714 of the United States Coast Guard on final approach after a sortie. The HC-130H replaced an earlier model that was originally designated R8V-1G, then SC-130B before becoming the HC-130B. The RNZAF had been offered the SC-130 and the Air Staff briefly considered it. They also considered the BAC.222, a Tyne-powered C-130. *USCG*

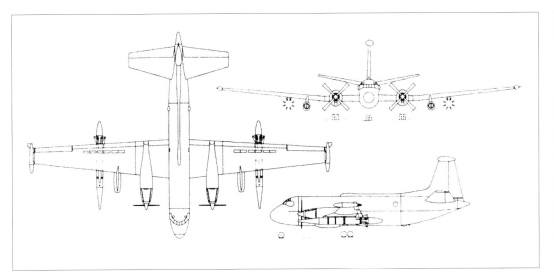

Avro's Atlantic 2A, seen here as a deterrent carrier with a pair of Skybolts. Previously tendered for OR.350, the Atlantic 2A was to be fitted with a pair of RB.153 turbofans on pylons outboard of the Tynes to allow the type to meet the transit speed requirement. *Avro Heritage*

Lockheed P-3A Orion of VP-49 seen here at low level in 1964. The P-3 had encountered a number of problems during development that had put the Air Staff off the acquiring the type. These had been overcome by Lockheed and the US Navy and the Orion went on to have a very successful career. Any Orions for the RAF would have been built in Northern Ireland by Shorts. *US Navy*

found it unsuitable for its needs. The SC-130B was also unsuitable for the RAF's OR.350, but Lockheed then suggested 'the maritime version of the C-130E' or HC-130H as the SC-130 had now become, for OR.357. According to Air Commodore P C Fletcher, DOR(B) in a letter to the DCAS Air Marshal Sir Ronald Beresford Lees, the HC-130E '…would come nowhere near meeting our requirements and would be obsolete before it entered service…'. Certainly the C-130's time on patrol at 1,000nm (1,852km) was around half that demanded by OR.357.

The Hercules, in an Anglicised form, was also proposed for OR.357 by BAC as the BAC.222 to fill OR.351. The C-130's Allison T56 turboprops were to be replaced by Rolls-Royce Tyne 20s and the wings fitted with boundary layer control on the control surfaces. The maritime version of the BAC.222, like all the propeller-driven types, failed to meet the transit speed criteria with 320kt (592km/h) and it remained as a tactical STOL transport. Another (ostensibly) foreign type suggested for OR.357 was the Breguet Atlantic 2A, which as noted in the previous chapter, was a Hawker Siddeley conversion of the Atlantic fitted with a pair of RB.153 turbofans in underwing pods. As also noted previously, the Air Staff viewed mixed powerplant aircraft with suspicion due to their ongoing experience with the Shackleton MR.3. Added to this was Fletcher's observation that '…the French themselves have hitherto shown no interest.' Then there was the MoA's view that the Atlantic was 'on the small side' and it only just met the requirement, although its transit speed was too slow.

Lockheed also proposed the most obvious candidate in the entire process, the P-3A Orion.

Fletcher took the view that the Orion was '…a good in-service aircraft; but it had very little "stretch" left in it and would be 11 years old by 1973.' In fact the MoA stated that the US Navy '…expected to need a P-3 replacement by about 1970. Fletcher also pointed out that there was a warning to be heeded in the development programme for the Orion. A maritime type derived from an existing aircraft looks like a bargain but 'this is not necessarily so'. Fletcher described how the Orion cost 'more than double the cost of the Electra on which it was based.' However, Fletcher also advised that the apparent cost savings of converting an existing type over developing a new aircraft would look attractive and '…we should have the greatest difficulty in securing government approval for a new aircraft.'

Oddly enough, the Orion's biggest fan was Marshal of the Royal Air Force Sir Thomas G. Pike, the CAS, and Fletcher was determined that his mind would be changed. In a letter to the DCAS dated 17th May 1963, and copied to the ACAS(OR), Fletcher says 'Should you find that the CAS is really serious in his belief that we should buy the P-3A, I could prepare a more fully developed case against it for you to use in discussion with him.' A major factor against the Orion was its American origin and purchasing an American aircraft could prove problematic. There was a plan afoot to have the Orion built by Shorts in Belfast if it was accepted for OR.357, but it is difficult to see HSA and BAC swallowing that, given the time and effort they had expended on the various MR requirements. As with the flying boat contract for R.2/48, the Government was keen to place contracts with Shorts in Belfast to maintain employment in Northern Ireland, something that continued into the Nineties with the Shorts Tucano trainer for the RAF.

Maritime VC10s

'They came on in the same old way, and we sent them back in the same old way,' The Duke of Wellington, on the Battle of Waterloo, 1815
The Duke's views on Napoleon's Old Guard are apposite to Vickers and the VC10 – they kept pushing it, in the face of Air Staff resistance, as the solution to the MR requirement. As ever, AUW and cost were the VC10's problem and the MoA took the view that '…the all-up weight would have to be limited to 300,000 lb to permit its use from Coastal Command airfields, the aircraft would have less than 1 hour

on patrol.' Fletcher continued to state that this, plus the high unit cost, were the reasons '…it was a waste of time to put cost figures forward.' The problem in this case was field length, although it was pointed out that if it could operate from Bomber Command airfields, it would have the required endurance. On enquiring about airfield lengths, only Gibraltar and Changi (Singapore) were less than 7,500ft (2,286m) in length. Even so, the VC10 would be 'fuel limited' when operating from many of the airfields Coastal Command used.

Further to this was the observation that to operate the VC10 from Coastal Command airfields the VC10 would need to be re-engined with Medways to match the four and a half hours endurance of the Shackleton never mind the eight hours for OR.357. Within the Air Staff and MoA, the VC10 was a rank outsider, but the process had to be left open and continued apace. Vickers at Weybridge were organising a demonstration flight for Coastal Command and Ministry personnel.

Then on 15th July 1963, a rather interesting memo was sent by Air Vice-Marshal R H E Emson, the ACAS (OR) to the DCAS, Air Marshal Sir Ronald Beresford Lees, describing a situation that had arisen with the VC10. Back in June 1962, BOAC had proposed that the RAF should take over ten of the Standard VC10s that BOAC had ordered as part of the fourteen Type 1106 VC10s the RAF had on order. Emson wrote '…we successfully evaded…' that proposition but that another plot was '…now developing.' This plot involved the same ten VC10s, but they should be taken by the RAF as part of an order for Maritime VC10s. What worried Emson was that the Secretary of State for Aviation might get wind of this and decide that he could put these towards a Shackleton replacement in the '…erroneous belief that it represents a cheap and acceptable solution to the replacement problem.'

Meantime, on 11th September 1963, Fletcher was yet again briefing the MoA on the Shackleton replacement, in particular a '…dangerously attractive…' proposal from Sir George Edwards, Executive Director of BAC. This promised to take the VC10, re-engine it with Medways to meet OR.357, take on the ten airframes unwanted by BOAC and at the same time, by installing new engines, re-invigorate the aircraft for the civil market. Fletcher is sceptical of all Edwards' claims and opines that the proposal was merely launched to sway the results of the feasibility studies that were due in October.

Alternatives from Weybridge

True to their Director's word, BAC produced a pair of documents in October 1963. Entitled 'Maritime Reconnaissance Aircraft to AST.357' it comprised a number of proposals, including a few based on a technical evaluation of swing-wing and fixed-wing types conducted in June 1963. The companies of the recently formed British Aircraft Corporation, despite the diversity of its aircraft products had no type that fitted the AUW range required for OR.357 and so continued to push the VC10 for that requirement. The VC10's main problem, as perceived by the Air Staff and the Ministries was its size. In the Jet Age, particularly with transports, the standard procedure when faced with a requirement that required a larger or smaller airframe was to either stretch the fuselage by adding frames, as in the 36ft 10in (11.2m) stretch to produce the Douglas DC-8 *Super Sixties* or removing frames and reducing the Boeing 707-120 fuselage length by 9ft (2.7m) to create the Boeing 720. Stretching the VC10 posed little problem, as with the 13ft (3.9m) stretch for the Super VC10. Unfortunately the VC10 could not be shortened sufficiently to adapt it for OR.357 and bring it close to the weight category of the Avro 776 that by mid-1963 was the benchmark for OR.357. The reason was fairly simple; being rear-engined the VC10 relied on the long fuselage forward of the wing to balance the weight of the four Rolls-Royce Conways.

Having no direct equivalent of the de Havilland Trident, upon which the Avro 776 was based, BAC at Weybridge, design authority for the VC10, sought a solution to this quandary and opted to design an aircraft to meet OR.357. Weybridge roped in the various members of the BAC consortium, particularly the Bristol design team at Filton and Hunting Aircraft at Luton, but also sought the help of Handley Page Limited at Radlett, with Shorts and Harland in Belfast to be involved in construction. Canadair would also be part of any development work on the Bristol Britannia for ASR.357, having used that type as the basis for the Canadair CL-28 Argus.

Weybridge's studies were wide-ranging and innovative, but all used a standard 12ft (3.7m) diameter fuselage with a common internal layout. The studies were intended to examine the airframes and propulsion systems rather than the weapons and systems associated with the requirement, although BAC did point out that the group had wide experience of complex systems and their integration. The requirement

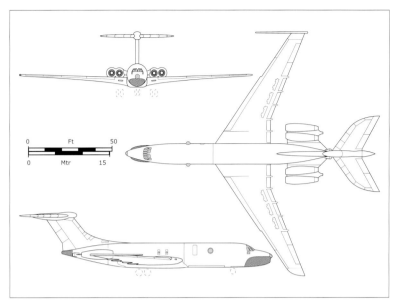

had a very wide speed band, 180kt to 500kt (333-926km/hr) covering the search and transit speed respectively and therefore BAC sought to address this range with a variety of engines powering aircraft with fixed and variable-geometry wings. High transit speed was desirable as it maximised the patrol time as a percentage of the sortie, reduced crew fatigue and allowed '…greater tactical concentration'. High transit speed coupled with a high 'dash' speed would serve to improve the utilisation and ability to react to a changing tactical situation. The high dash speed was seen as valuable in a war situation whereby the aircraft could react to a vessel being sunk or SLBMs being launched. By reaching the area faster, the aircraft could commence the search for an enemy submarine sooner, thus reducing the search area to a minimum.

Initially, three projects were to be examined for a report issued in June 1963. The first study was based on the VC10 Scheme B with four Rolls-Royce RB.178 Super Conway turbofans, essentially a modified version of the existing Transport Command Type 1106. The second was an 'Optimum Maritime Aircraft' with a 25° swept wing and the third a variable-geometry type. All three were to perform the required eight hour patrol at 1,000nm (1,852km) from base as per OR.357. As expected, the VC10 was cheapest to develop, but at 352,000lb (159,665kg) AUW, was too heavy and expensive to operate. With their AUW in the 284,000-300,000lb (128,820-136,078kg) the Optimum Maritime Aircraft and the variable-geometry type showed promise and so a more in-depth study was commissioned, with its scope widened to include a range of engines, a

This MR variant of the VC10 as about as short as the VC10 fuselage could be: a minimum length of fuselage ahead of the wing was required to balance the rear-mounted engines. This resulted in a great deal of empty space within the cabin, which to meet the range requirement had to be used for fuel tanks. Filling these was seen as an extravagance that Coastal Command could not sanction.

141

The VC10 Scheme B was to be fitted with underwing panniers to carry stores and weapons. These would be installed on the same hardpoints that were used to carry the spare engine pods on civil VC10s. This BOAC example, G-ARVH, is carrying such a pod under its starboard wing. *Author's collection.*

turboprop type, fixed and variable-geometry wings and a laminar flow control aircraft. The main thrust of this study was to show the effect of transit speed, over a range of 350-1,000kt (648-1,852km/hr) on airframe size and cost.

In a long-range aircraft such as a maritime patrol type, structural weight is the key, particularly unnecessary weight in the fuselage structure. A reduction in fuselage diameter of 6in (15cm) could reduce the airframe weight by as much as 2,000lb (907kg) for a fuselage 100ft (30.5m) long. Fuselage length was governed by the various compartments under the cabin floor: radar, nose undercarriage, weapons bay, wing box, main undercarriage and sonobuoy bay. The cabin itself would ideally provide a 6ft 6in (2m) headroom in the operations compartment and therefore an optimum fuselage inter-

Research had shown that the ideal fuselage diameter was 11ft 6in (3.5m) which allowed headroom of at least 6ft 6in (2m) for the crew compartment with a decent depth for an underfloor weapons bay. A smaller diameter compromised equipment capacity while larger diameters resulted in much empty space and increased drag. Although the VC10's diameter was ideal, the length of its fuselage resulted in too much empty space.

nal diameter would be 11ft 6in (3.5m) with greater diameter merely adding superfluous weight to that required to carry out the role.

Transit speed was another major influence on all-up-weight, due to larger engines, increased fuel load and consequently, an increase in the airframe size and structural weight to carry these. For example, BAC's studies showed that for an OR.357 aircraft with an AUW of 270,000lb (122,470kg), increasing the transit speed from 440kt to 470kt (815 to 870km/hr) would see AUW increase by 24,000lb (10,884kg) while increasing transit speed by a further 30kt (56km/hr) would incur an increase of 101,000lb (45,805kg). Into this mix was the intention to have a wing loading suitable for the role, with 95lb/ft^2 (463kg/m^2) being considered as the optimum wing loading in this case.

Transit speed also governed the wing sweep, and the wide speed range of OR.357 made the variable-geometry configuration attractive. The VC10's 32.5° sweep was tailored for a transit speed of 470kt (870km/hr) but less than ideal for the patrol portion of the mission, although it did possess good gust response. The optimum aircraft sported a reduced sweep of 22° for a transit speed of 440kt (815km/hr). Increasing the transit speed to 500kt (926km/hr) was best met by applying a wing sweep of 41° and the result of these disparate sweep angles and speed regimes led to the swing-wing configuration being considered. The variable-geometry type was by its nature most suitable for the wide speed range required of a maritime type to OR.357 as the wings would have a range of 10° to 45° sweep. The sweep mechanisms and associated structures of variable-geometry aircraft have long been considered to be adding extra weight to an aircraft and the weight penalty might not always be justified. Certainly in the OR.357 case, if the required transit speed is 500kt (926km/hr) the AUW for a variable-geometry aircraft will increase by 20,000 lb (9,070kg) with 6,500 lb (2,766kg) being structure and the balance the fuel for the engines to propel the aircraft at 500kt. A fixed-wing type on the other hand would produce a 40,000 lb (18,144kg) increase in AUW, due to the airframe being less optimised to higher speeds and therefore needing more fuel to feed engines operating at higher power to overcome additional drag.

With the transit speed and configuration having been addressed, attention turned to the propulsion system. By the late 1950s the turbo-jet was being supplanted by what would now be described as a low bypass ratio turbofan, such as the Roll-Royce RB.80 Conway, rated at 16,500 lbf (73.4kN). As the pace of gas turbine engine development increased in the early 1960s, British designers of large aircraft had a number of engines to choose from. For the OR.357 studies BAC selected the Rolls-Royce RB.177-22, rated at 23,000 lbf (101.3kN) as the datum engine, with the Bristol Siddeley BS.106, rated at 25,000 lbf (102.3kN) as an alternative. The beauty of the gas turbine as an aircraft propulsion system is scalability. As noted in Chapter Three, reciprocating engines were fixed in size and restricted in their power range, and as in the case of the R.2/48 flying boat, the airframe had to be sized to match the engine. Gas turbines' scalability presented the aircraft designer with no such redesign problems, in that an engine's size could be enlarged or reduced to maximise the efficiency of the airframe/engine combination.

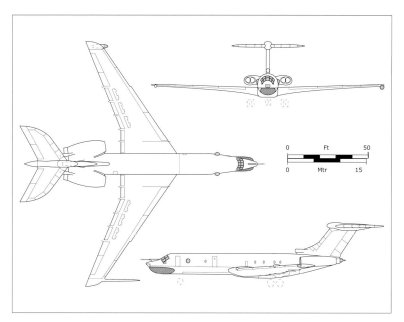

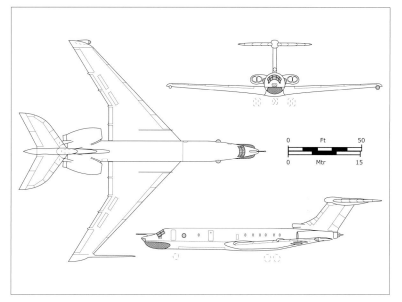

This VC10 variant has its RB.177 engines faired-in for a much cleaner design. This configuration would be married to two other VC10 derivatives, the so-called 'Optimum' types with more highly swept or variable geometry wings or the less-swept, higher aspect ratio wing of the hybrid designs that led to the later BAC Ten-Eleven.

To meet the higher transit speed of 500kt, more than required by ASR.357, Vickers proposed a more highly-swept wing, 41°, on this type powered by four RB.177 or Bristol Siddeley BS.106 engines. However, this produced a very heavy aircraft due to the revised wing structure.

BAC's variable geometry type to achieve a 500kt transit speed was almost 90,000 lb lighter than the fixed-wing equivalent. It could also achieve a dash speed of 450kt from a patrol speed of 180kt.

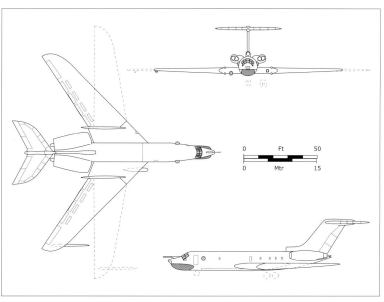

The aft-fan versions of the Spey (left) and Medway (right) were married to the standard BAC maritime cabin. The variant with the aft-fan Medways is fitted with a variable geometry wing, whereas on the aft-fan Spey version the central engine sucks the rear fuselage boundary layer into its fan duct.

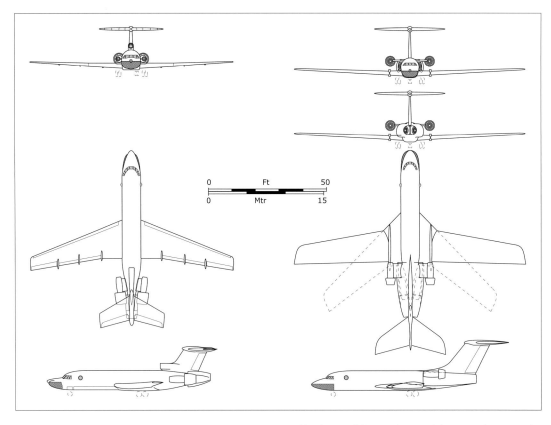

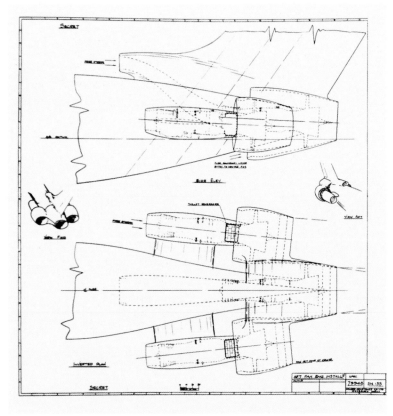

This aft-fan configuration from BAC at Weybridge was designed to suck in the boundary layer on the rear fuselage into the fan duct of the middle engine. Note how the intake in the fin root is located away from the fuselage skin so that it feeds the core engine separately with 'clean' air. *BAE Systems, Warton/North West Heritage Group*

Having addressed matching engine to airframe, another innovation in engine technology was becoming practicable in the early 1960s – the high bypass ratio turbofan. In 1963 the 'big fan' engine so familiar on the large transport and passenger aircraft of today was still in development and its future ubiquity was not obvious. What was being drawn on the large aircraft design studies of the early 1960s was the aft fan. Aft fans use a separate rear-mounted turbine section attached to the engine jet pipe as a free turbine. The turbine disk is surrounded by a larger concentric fan, effectively extensions of the turbine blades that turn with the free turbine within a short cowling. The attractive aspect of the aft fan was that it could be 'bolted on' to the back of pretty much any jet engine, although a low bypass turbofan such as the Conway or the Spey with bypass ratios of 0.25:1 and 0.64:1 respectively, was preferred.

By lowering the jet velocity through increasing bypass ratio, the propulsive efficiency of a jet engine can be improved. The highest bypass ratios could be achieved by using an aft, rather than front fan, mainly due to drag generated by the cowling of the front fan. One other, rather peculiar, characteristic of the aft fan allowed it to be installed in clusters in the rear fuselage, something that was not possible with front fan engines. The aft fan could ingest the boundary layer of the airflow over the rear fuselage and if one of the engines is buried in the rear fuselage and fitted with long ducts, the boundary layer from the fuselage can also be ingested. If a further two engines are fitted, these can integrated into the rear fuselage in such a manner that they ingest the boundary layer of the cowl and ducts of the central engine.

In the early 1960s, higher bypass ratios could only be achieved through the use of an aft fan, but as the decade progressed, the aero-engine companies, particularly General Electric with the TF39 and Rolls-Royce with their RB.211, developed the front fan for large transport aircraft such as the Lockheed C-5A Galaxy and the Lockheed L-1011 TriStar. The front-fan high bypass ratio turbofan shrank the world by bringing low-cost, long-range travel to the masses while the aft fan disappeared.

The US Navy's Lockheed P-3 Orion and the Breguet Atlantic used turboprop power, but the Air Staff's transit speed requirement ruled out the turboprop for OR.357. However, the October 1963 BAC study also examined the prospects for using a turboprop engine on the OR.357 type, mainly as a comparison with the jet-powered types. Vickers had previously submitted a modified Vanguard for OR.350 and OR.357, even fitting turbofans to boost the transit speed for OR.350, but rather than examine a further maritime Vanguard, BAC at Weybridge looked to Bristol Aircraft Ltd at Filton for the airframe. A development of Bristol's Britannia had been proposed as a Shackleton replacement back in the 1950s as the Type 175MR (see Chapter Five), but re-engined with Wright R-3350 turbo-compound piston or as the Type 189 with Napier Nomad compound engines. In another development, Canadair, in conjunction with Bristol, had developed the CL-28 Argus using the wings and tail surfaces of the Britannia mated to a new unpressurised fuselage, incorporating two weapons bays, for the maritime patrol role.

A decade after these developments, Filton returned to the maritime patrol Britannia with a study that benefited from the ongoing development programme and experience gained with the Britannic and subsequent Shorts SC.5 Belfast. Twenty of the RAF's Britannia C.1 and C.2 strategic transports had been built at Shorts in Belfast and the Britannia wing structure was used for the Shorts SC.5 Belfast, but unlike the Britannia with four Bristol Proteus, rated at 4,500shp (3,352.5kW), the Belfast was powered by four Rolls-Royce Tyne 11 turboprops, rated at 4,816shp (3,588kW). As noted in pre-

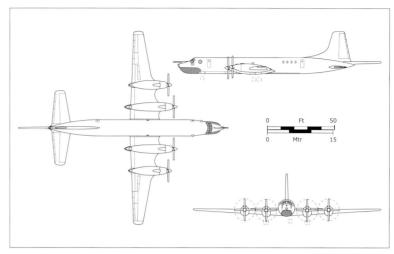

vious chapters, the Tyne was the preferred turboprop engine for European maritime patrol aircraft, mainly because it was the only engine in its class under continuing development in Europe at the time. The Britannia was the only Proteus-powered type to enter production, with its two Canadian derivatives, the CL-28 Argus and CL-44 Yukon being powered by R-3350 turbo-compound engines and Tynes respectively.

Described by BAC as a 'hybrid aircraft' the OR.357 turboprop study was a Britannia with the beefed-up wing centre-section structure from the Belfast mated with the forward fuselage in the 'ideal' configuration used for all these BAC studies. The aircraft was to be fitted with four Tyne 22, rated at 5,540shp (4,131kW) installed in modified Proteus nacelles (unlike the CL-44 Yukon that used bespoke nacelles). As the Canadians had discovered while developing their Argus, the Proteus turboprop was quite unsuitable for the long patrols at low altitude that anti-submarine missions required and so fitted the Argus with Wright R-3350 turbo-compound reciprocating engines. This was not a criticism of the Proteus, but relevant to all gas turbines, whose efficiency increases with altitude whereas the reciprocating engine, particularly in its turbo-compound form, is much more suited to the low-level role. In 1954 when the Argus was being developed, gas turbines could not match the piston engine in fuel efficiency and consequently, endurance.

However, by 1962, the situation had changed. The gas turbine had seen off the piston engine, but for maritime patrol aircraft an efficient engine was still required. The Napier Nomad – that most efficient of aero-engines – had been cancelled but the effort continued to produce an efficient turboprop. Oddly enough

The Bristol Britannia was back for ASR.357, but with the BAC 'standard' MR forward fuselage, akin to that of the VC10MR, and Bristol regenerative turboprops. It was described as a hybrid of the Britannia and Belfast and like all the propeller-driven types, struggled to meet the transit speed.

Bristol Engines had produced the archetype of the efficient turboprop in their first gas turbine, the Theseus of 1945 that was intended to possess fuel consumption comparable with Bristol's big radials. British industry has long been accused as the failing to capitalise on technology it had pioneered, be it the fast breeder reactor or the tilting train, there was an aspect of the Theseus in its Th.21 guise that fell into that category. In the interests of thermal efficiency, Bristol's designers had incorporated a heat exchanger in the flowpath of the Theseus to capture a portion of the exhaust heat and use it to preheat the incoming air before it entered the combustion chamber. Roy Hawkins, a senior research engineer in Bristol Siddeley's ramjet department, summed up the operation of a jet engine (of whatever stripe) as a means of adding heat to a working fluid, thus making it expand and produce thrust, be it out of a jet pipe or on a turbine to turn a propeller. By including the heat exchanger, Bristol Engines aimed to increase the efficiency of the engine by reducing the amount of fuel required to add the same amount of heat.

Unfortunately this failed to inspire the aircraft companies and the Theseus did not prosper. The heat exchanger wasn't without penalty as it added weight, not to mention the increased drag of the larger diameter nacelle, and the Theseus was described as being a failure, superseded by that company's more powerful Proteus. Oddly enough the Theseus was originally proposed as an engine in the 4,000hp (2,983kW) class but was down-rated to 2,800hp (2,087kW) at the request of the Min-

istry of Aircraft Production. So, in 1945 Bristol Engines had a turboprop with a heat exchanger to improve thermal efficiency but it was simpler engines like the Tyne and the Dart that powered Britain's aircraft. The main reason was that the Theseus did not live up to its specification, its specific fuel consumption being higher than expected. One of the reasons for this was that the pressure drop through the heat exchanger – 1700 stainless steel tubes of 0.625in (16mm) diameter – and associated ducting was higher than expected. The Theseus lost its heat exchanger and became the Th.11 and the entire programme was cancelled in 1951 but how a Theseus at its intended full-power would have affected the R.2/48 flying boat tender had it been available is an interesting, but unanswerable, question.

Meanwhile in America, the Allison Engine Company was one of a number of American engine companies that had attempted, but struggled, to produce a successful turboprop. Allisons had by 1954 developed the T56 that powered the Lockheed C-130 Hercules and P-3 Orion. Allisons were by 1961 working on an engine suitable for re-engining the Orion and other long-endurance aircraft such as AEW platforms with an efficient turboprop and in 1963 revealed the T78. This engine was based on the T56 but rated at 5,000hp (3,728kW) and incorporated a heat exchanger, as per the Theseus, and Allison called this a regenerative engine (the Bristol engine is more properly described as having a recuperator). Allison's initial T78 regenerative engine, revealed in 1963, used a pair of short cylindrical heat exchangers arranged normal on either side of the engine's long axis producing a drum-like arrangement with an engine poking out the side. Such drum type regenenerators were more compact that the recuperators used on the Theseus for a similar level of efficiency.

A year or so later, Allison showed off a much-modified engine, with annular mass-flow ducts and heat exchanger arranged along the engine's long axis, lending the bare engine a bottle-like appearance. The T56 was reconfigured to include ducting that carried the mass flow from the last stage of the compressor and, bypassing the combustion chamber, passed it rearwards through an annular heat-exchanger before it returned to the combustion chambers. After adding fuel and combustion, the exhaust gasses passed through the turbine stages and the free turbine driving the propeller before a portion of the exhaust flow entered the 'hot' section of the heat exchanger.

BAC's Britannia-derived proposal for ASR.357 was to be powered by four Bristol Siddeley regenerative turboprops. The concept dated back to the earliest Bristol turboprop, the Theseus. The regenerative turboprop passed its exhaust through a heat-exchanger prior to entering the combustion chamber. This increased the heat addition in the engine, thus making it more efficient.

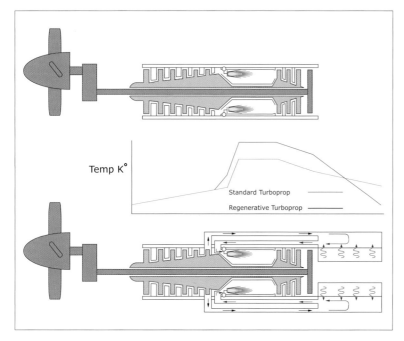

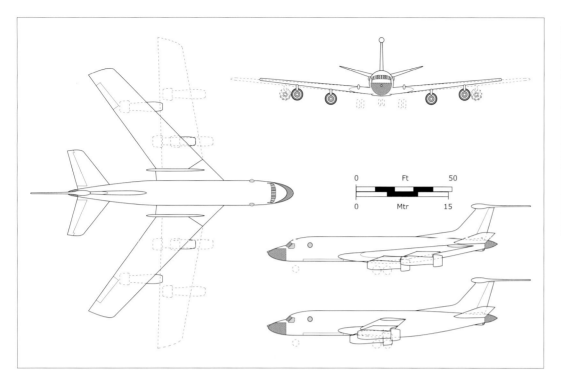

This handsome design mounted its aft-fan Medways or Speys on swivelling pylons on the moving portion of the wings. As with all pylon-mounted engines, this allowed easier upgrade and/or replacement by more efficient engines during the aircraft's service life.

The regenerative engine sounds simple, but in practice it was very difficult, with the heat-exchanger utilising 15,000 stainless-steel tubes with a diameter of 0.25in (6mm) and a wall thickness of 0.005in (0.127mm). In the drum-like version, the internals rotated in a manner similar to an earlier Whirlfire heat exchanger used in an automotive engine developed by Allison's parent company, General Motors. Such a configuration was impracticable for an aero engine due to weight and drag, so the later, axial, version was preferred.

So what was this new turboprop that Bristol Siddeley was proposing for the 'hybrid' Britannia MR? Although there is no mention of regenerators or recuperators, the only engine in the correct power range and which fits the timescale of the early Sixties is the BS.98. Rated at 5,200shp (3,878kW) the BS.98 was a turboprop based on the BS.75 turbofan, as selected for the Fokker F.28 airliner, with reduction gear based on the Bristol BE.25 Orion.

Variable Sweep from Weybridge

As noted above, Weybridge had been examining the application of variable-geometry wings to project studies for OR.357. Vickers, with variable-geometry expertise stretching back to the war, carried out a wider range of studies than Hawker Siddeley, whose studies were restricted to examining the HS.1011 and HS.1023 in the maritime patrol role. The numerous Vickers studies ranged from a straight conversion to variable-geometry of the Optimum Aircraft to what could be described as the ultimate development of the VC10.

This last type, the 1,000kt Aircraft Scheme A, was to be capable of a transit speed of 1,000kt (1,852km/hr), but in reality, to attain the eight hour patrol time a maximum transit speed of 500kt (926km/hr) was recommended, with any periods of supersonic flight eating into that patrol time. The type retained the 'ideal' fuselage but this was to be in Vickers' words, 'slenderised' to match the speed regime by applying a needle nose and tailcone more suitable to high speed. The wings comprised a wide chord fixed glove with the pivot point for the movable wing panels set at about 20% span. The outer

BAC also examined a VG aircraft capable of transits at up to 1,000kt. Power was supplied by four RB.177-22, scaled to 1.17 times, and fitted with reheat. The type could carry out the eight hour patrol, but its fuel supply would have required careful husbanding.

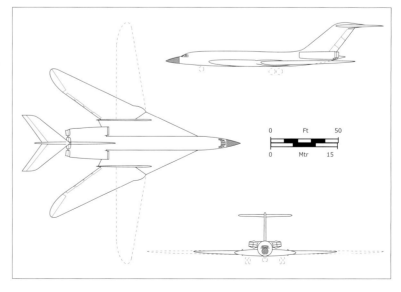

panels shared the planform of the Vickers Type 532 with gently-rounded wingtips. The four reheated RB.177-22 (scaled to 1.17) were mounted, VC10-style, in pairs on the rear fuselage, with a highly swept T-tail completing the aircraft. The 500kt aircraft was to be 170ft (52m) long, with a fully-spread wingspan of 175ft (53.3m). The sweep range of the wings was to be 10° to 65°, allowing the aircraft to meet the required 180kt (333km/hr) patrol speed. Despite the aircraft's variable geometry and supersonic capability, this type still weighed in at 355,000lb (161,025kg) which was less than the original VC10 Scheme A to OR.350. This aircraft was the most radical and spectacular-looking proposal for OR.357, the high-water mark of the OR.357 technological tide. Suffice to say it remained but a few lines and some weight and drag calculations, which is a pity.

The last type examined in the Weybridge report from October 1963 might come as a bit of a surprise as it proposed collaboration with Handley Page Ltd, a company who, by refusing to be part of the merged aircraft corporations, was shorn of government contracts and funding. Having been denied access to the OR.357 tender process, HP had to co-operate with one of the other companies. Handley Page's HP.117 laminar flow control aircraft had been dismissed from the OR.350 bidding back in 1960 due to a perception that the slots and orifices used to suck the boundary layer from the wings would be prone to clogging by salt spray and even insects at the low altitude required for maritime patrol. Handley Page's part in the OR.357 process was to supply the laminar flow control (LFC) technology, a field in which HP had invested heavily in the belief that LFC would herald a new era in efficient long-range

aircraft. It could be argued that HP were travelling the well-trodden path followed by the Napier Nomad before, although HP saw passenger transport rather than freight as the future. They were, of course correct, but it was the large high bypass ratio turbofan rather than LFC that made this possible by the end of the 1960s.

The Vickers study took the 440kt Transit Speed three-engined aircraft and fitted new wings complete with LFC equipment. Each wing carried an additional RB.177 turbofan in a faired-in pod under the trailing edge, bringing the total number of engines on the aircraft to five. These extra engines lacked conventional intakes but were fed through a series of ducts connected to slots and orifices on the wing surface. While HP's aim had always been a fully-laminarised design, such as the HP.117, the installation described here was viewed as a retrofit that could bring partial LFC to conventional aircraft, providing some degree of efficiency improvement.

When operating, these engines sucked the turbulent boundary layer through slots or permeable panels in the wing's skin, thus reducing drag. Reduced drag meant lower fuel consumption, which in turn increased endurance and lowered fuel costs, something that had been a factor in the Air Staff rejection of the larger, high-speed proposals such as the VC10. The intention was that the LFC system would operate during the transit to and from the patrol area, with two or even all three of the fuselage-mounted engines shut down. Vickers hoped to use the LFC system whilst on patrol, but was concerned about the build-up of salt in the slots or permeable panels affecting the engine performance and forming a rough surface on the wings. This roughness affected laminarisation by causing a transition from laminar to turbulent flow sooner than a smooth surface. To investigate this, and the means to prevent it happening, Handley Page had been conducting a series of trials at Weymouth on the Dorset coast.

HP had discovered through trials with test rigs that the higher the flying speed, the greater the number of suction slots required and these had to be closer together, leading to increasing structural weight and complexity. The alternative to slots was permeable panels set into the wing's surfaces. Handley Page tested a wing section fitted with 'Porosint' a sintered bronze material that could be machined and shaped to fit into the wing's LFC structure. Despite its name, it was the permeability of Porosint that HP were interested in, porosity being the meas-

Criticism of the fuel burn for a fleet of VC10MRs probably prompted the adoption of three RB.178 Super Conway or RB.186 Super Medway engines for this variant. The fuel cost was a major stumbling block for the VC10 and, seeking ways to make their aircraft more efficient, BAC looked to Handley Page for expertise on laminar flow control.

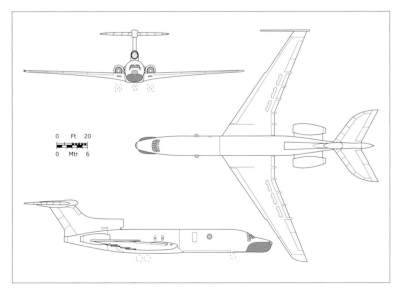

0 Ft 20

0 Mtr 6

ure of how much space is between the grains of a material, while permeability is the degree of communication between these spaces which allow fluid, in this case air, to move through the material. Put simply, porosity is the space in a material, permeability is how that space is connected and HP needed high permeability to suck in the boundary layer. Unfortunately the small size of the pore throats within the Porosint matrix that allowed air to through the Porosint made the panels susceptible to salt blockage and reduced permeability.

Salt wasn't the only concern for HP as it was thought that insects being smashed into the airframe could clog the LFC slots. Consider the build-up of insects and debris on a car windscreen and amplify this by the suction effect of a gas turbine compressor. HP's concerns were not only about the slots becoming clogged but the roughness mentioned above. Handley Page, prior to involvement in maritime patrol studies had been more interested in the insect and dust build-up, which they called 'fly contamination', than salt spray. HP's studies had shown that the problem lay at low altitudes as that was where the insects were most prevalent, but as altitude increased, the dehydration and deep freezing of the insects, combined with higher flight speed tended to remove the debris, so the easy answer was to climb to high altitude as quickly as possible.

Oddly enough, the original tests on fly contamination and surface roughness had been carried out at Blackburn's Brough plant by Dr W S Coleman as far back as 1951. Coleman had built a test section in Blackburn's wind tunnel and injected science's favourite insect, fruit flies, into the airstream. To ease the operation, the flies were anaesthetised with carbon dioxide before injection. Dr Coleman then meas-

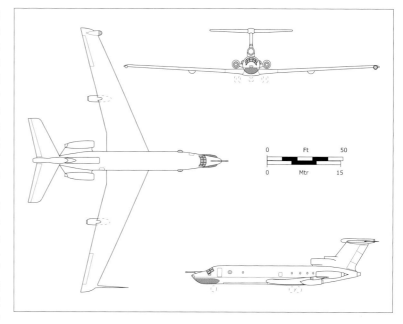

ured the 'roughness' of the wing skin. A single fly debris gave rise to a lump 0.01in (0.25mm) high which was reduced by erosion to 0.005in (0.12mm), with wing and legs being eroded almost immediately after impact and so only the fly bodies contribute to roughness and thus affect transition from laminar to turbulent flow.

The RAE was also interested in this and W E Gray had concluded that fly contamination affected 7% of the chord on the upper surface and 20% of the lower surface of a wing, measured from the leading edge, the higher value on the underside being due to incidence. Meanwhile at Armstrong Whitworth Aviation, whose AW.52 used laminar flow techniques, D Johnson had investigated the point at which the transition from laminar to turbulent flow occurred and how fly contamination affected this. Johnson concluded that transition occurred at 5% chord on the upper surface and

One of the more interesting proposals in the BAC file is this variant configured with moderately swept, 24°, wings and five engines with laminar flow control. The LFC system would be used in the transit at high altitude, sucking in the wings' boundary layer to feed the wing-mounted engines.

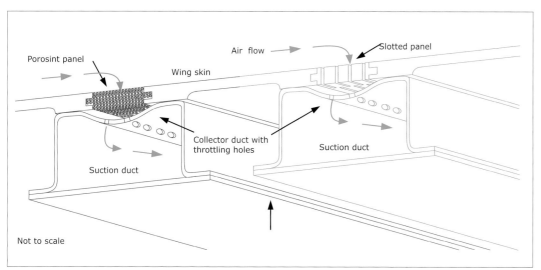

Two BLC systems were under development by Handley Page. The first used Porosint, a permeable metal sheet while the second involved narrow slots in the wing skins. The main concern about BLC was clogging of the pores and slots by salt or insects.

12% chord on the lower, both within the zones of maximum contamination.

Of course, none of these wind tunnel experiments could replicate the freezing and dehydration effects of high altitude flight and thus it was concluded that fly contamination posed little problem for aircraft flying above 40,000ft (12,192m). This raised the question of how to reduce roughness from flies in the climb and descent, with the early stage of the flight being more critical than at the end of a sortie. As before Blackburn and Armstrong Whitworth had investigated this and a number of techniques had been examined. Dr Coleman at Brough favoured a retractable deflector plate in the leading edge, Armstrong Whitworth engineers G Beech and W M Nicholas championed a mechanical scraper that could be pulled back and forth the along the leading edge. Another technique was to expel liquid through permeable panels in the leading edge, to provide a continual flushing that would prevent debris adhesion. Possibly the most unusual suggestion was to expel liquid that formed a layer of ice to provide a surface for the debris to accumulate, then once above the insect contaminated zones, apply the de-icing system to remove the ice and the debris.

The simplest, and ultimate, solution was to keep the LFC system switched off below 10,000ft (3,048m), which would prevent the clogging of the LFC system, but not help with surface roughness and transition. Apart from a test section fitted above the fuselage of Lancaster PA474, Handley Page's LFC technology never reached the flying hardware stage, although HP had detailed proposals to fit a Hawker Siddeley HS.125 with a LFC system as the HP.113 and had designs for a fully-laminarised technology demonstrator, the HP.119. Whether HP's reluctance to be merged, lack of government funding or the usual British trait of not pursuing new ideas to fruition, did for their LFC work is moot. In an era when reduction of fuel costs is of paramount importance, engine efficiency has been pushed to the limit with the big fans and airframe weight reduced with composite structures, conceivably the next step is laminar flow control. Perhaps in a decade or so, a fully-laminarised, composite aircraft powered by super-efficient turbofans will take to the skies, landing (as the old joke goes) with more fuel than it took off with!

So what benefits did laminar flow control provide to Vickers' OR.357 aircraft? HP's assumptions were based on the LFC system being in use for 30% of the sortie duration, with the remainder of the sortie being compromised by carrying the additional engines and equipment for LFC. Firstly there was a 5% reduction in AUW, secondly an increase in the ferry range of some 35% for the same AUW and thirdly, based on 150 sorties per year, a fuel saving of 920 tons (835 tonnes), a matter dear to the Air Staff's heart as this was a major criticism of the VC10. Handley Page's final assertion was that the aircraft could also form the basis of a very efficient long-range transport.

While all this activity was going on in Whitehall, Vickers had organised a test flight to allow the RAF '…to evaluate the characteristics of a jet aircraft at low level over the sea.' A similar trial with the DH Trident was scheduled for the end of October 1963, dependent on the availability of de Havilland's Chief Test Pilot, John Cunningham. Throughout all of the correspondence with Vickers, Air Commodore Fletcher was at pains to point out that the VC10 '…finds little favour with us…' as a Shackleton replacement. Vickers' dogged determination to have the VC10 adopted as such has to be admired; however, their doggedness may have had some substance as in early November it transpired that the Ministry of Aviation was 'solidly behind the VC10 as the Shackleton replacement'. This information had come to the DCAS' attention via the Permanent Secretary at the MoA and the DCAS was incredulous that '…there can be a solid MoA view before the feasibility studies

Handley Page's boundary layer control research would have been applied to the BAC maritime patrol type derived from the VC10. Unlike HP's aim of achieving maximum BLC, as seen in the HP.117, BAC hoped to use the technique on the wings alone.

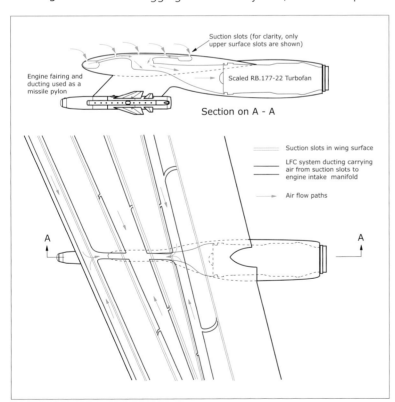

Suction slots (for clarity, only upper surface slots are shown)

Engine fairing and ducting used as a missile pylon

Scaled RB.177-22 Turbofan

Section on A - A

Suction slots in wing surface

LFC system ducting carrying air from suction slots to engine intake manifold

Air flow paths

A

A

have been evaluated…' or whether the Air Staff considered the VC10 operationally suitable or not. The DCAS continued, and pointed out that a shortlist of aircraft would be drawn up by the OR Branch and would not necessarily include the VC10 and that any premature moves on the VC10 for OR.357 might jeopardise any possible UK/Canada joint project.

The last months of 1963 saw Vickers address the main criticism of the VC10 – its size and inefficiency at low patrol speeds. Out of the Weybridge design office appeared an Optimum Aircraft for maritime patrol. The broad outline of this aircraft is laid out in a document that compares the VC10MR and the Optimum Aircraft and describes a four-jet aircraft with an AUW of 247,000 lb (112,037kg) or 243,000 lb (110,223kg) for a three-engined type. Engines were to be the RB.177 or RB.142 Medway, with both scaled to suit the airframe, mounted on the rear fuselage. Another engine considered was the RB.186 with an aft fan. The wing with a span of 130ft (39.6m) was to be 'Optimum for the transit speed required.', which was slower than the VC10 at 440kt (815km/h). So what was this Optimum Aircraft from BAC?

In the Brooklands Museum archive, a loose drawing of a very odd looking aircraft has been found. At first glance, it looks like a VC10 with its four engines faired into the rear fuselage. It has the graceful fin and T-tail of the VC10 and the forward fuselage of the maritime version of the Modular VC10. The tailplanes are reminiscent of the BAC One-Eleven, while the wings have the look of a VC10 but with reduced sweep and these last two characteristics may allude to this aircraft's identity. Could this be the Optimum Aircraft and could that have led to what would become the BAC 10-11? The drawing shows a flight refuelling probe ahead of the cockpit à la VC10, a MAD probe on the starboard wingtip and a searchlight on the port, with the wingtips sharing the curved sweep of the VC10. The planform of the aircraft is almost like the VC10 wing had been swung forward about its roots, reducing its sweep. As will be described below, BAC were not about to give up on the maritime patrol role in a hurry.

By the first week of January 1964 the press had become involved in the OR.357 debate, with an 'ill-informed' article in the Daily Telegraph '…extoling the virtues of a Trident conversion to meet AST.357.' Air Commodore R G Knott, DOR(b) was tasked with scotching such rumours in the press and drew up a spoiler to head off the article advocating the VC10 that was expected to follow.

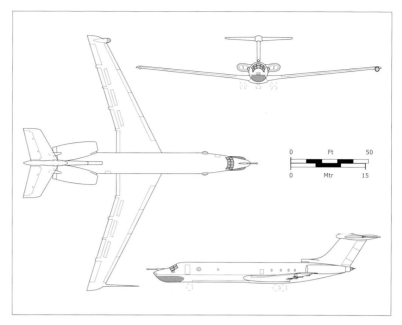

A New AW.681

One aircraft that the Air Staff and MoA spent quite a bit of time examining, and appeared quite keen on, was an update of another modified transport submitted and rejected for OR.350. The Armstrong Whitworth division of HSA re-submitted the AW.681 for OR.357, albeit in the updated form based on the definitive transport version to meet OR.351. Two variants were proposed, the MR.1, a minimum-change type based on the transport aimed at providing 'a reasonable maritime capability, but which fails to meet the requirement in several respects' and the MR.2. The AW.681 MR.1 retained the transport aircraft's wings and engines as well as much of the fuselage. The undercarriage fairings were extended to house weapons bays and the tail ramp replaced by a bomb bay while a pylon on each outer wing could carry the ASGW to meet OR.1168. The fuselage was to be stretched by 6ft (1.8m) to accommodate a 6ft (1.8m) diameter radar scanner in a chin radome, which also required the nosewheel to be relocated. The interior was to be fitted out with tactical control stations similar to those of the Avro Type 776, with a crew of twelve, including the pilots.

To achieve the required endurance, a large slipper tank was to be fitted under each outer wing, requiring strengthening of the wing structure, which with all the other modifications increased the AUW to 270,500 lb (122,697kg). The engines were to be Rolls-Royce RB.142 Medways with the thrust vectoring systems removed and a 32% reheat system fitted to achieve the OR.357 take-off criteria. Reheat was attractive as it allowed a smaller

This was BAC's 'Optimum' MR aircraft that used a VC10 wing swung forward to give a leading edge sweep of 22°, rather than 32½°, to improve transit cruise efficiency at 440kt while providing better gust response at patrol speed and low altitude.

The Armstrong Whitworth AW.681 MR.1 was pretty much the tactical transport with straight-through Medway turbofans and maritime kit installed. The MR.2 used a modified mid-fuselage tailored to the maritime role with a ventral weapons bay, rather than bays in the extended undercarriage sponsons. Note that the MR.1 and MR.2 designations were not official.

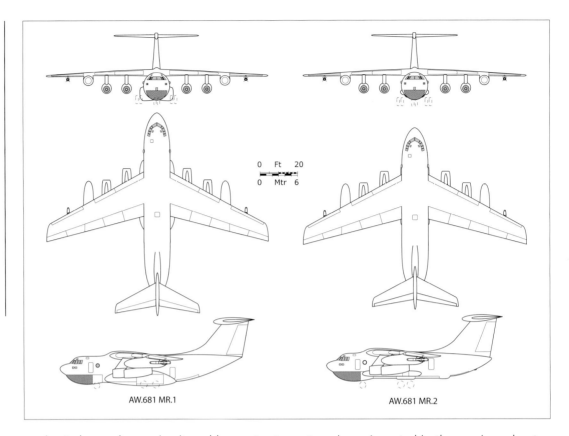

AW.681 MR.1

AW.681 MR.2

engine to be used, meaning it could operate at optimum power in the cruise and was therefore more economic on patrol. The AW.681 MR.1 could patrol for seven hours on two Medway engines, but this could be increased to eight hours if the Rolls-Royce RB.177-22 (a major redevelopment of the RB.142) was fitted. Additional fuel was carried in slipper tanks under the outer wings. As noted in the previous chapter, transports were not particularly suited to conversion to the MR role. Wing Commander R H Benwell of OR.13 expanded on this in a loose minute on 15th November 1961, stating '…it is considered that freighter aircraft physically shaped to provide maximum freight capacity and considering speed and range as secondary

The AW.681 MR.1 carried the usual equipment for ASW operations. Sections B:B and C:C show why converted transports fail to prosper in the maritime role – a great deal of internal volume remains unused. *Avro Heritage*

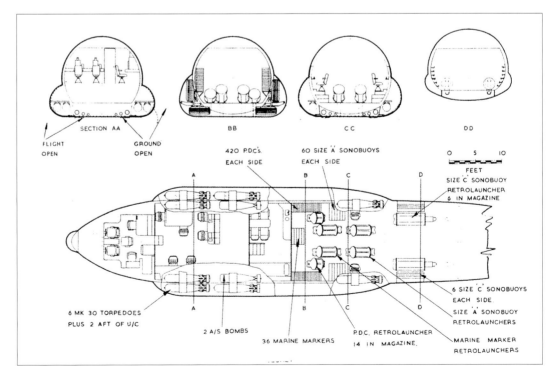

requirements, must inevitably possess inherent disadvantages in the maritime role…' This was quite obvious to all involved and Armstrong Whitworth took steps to address this dichotomy with the AW.681.

The AW.681 MR.2 was a different beast altogether, as AW had addressed the main drawback to converting a transport aircraft; the weight and drag penalty of the freight hold, not to mention the unused space within. Armstrong Whitworth opted to redesign the AW.681 fuselage, shortened it by 10ft 6in (3.2m), but retained the wings, engines and tail surfaces to produce the MR.2. Essentially the upper half of the existing fuselage was used, to preserve the same flight deck, wing and tail attachments, with a new 'pressure carrying floor' fitted. The tail ramp/cargo door was to be removed and replaced by a new faired-in rear fuselage. The redesigned lower half of the fuselage carried a pair of weapons bays and a new bay for a redesigned undercarriage, thus rendering the bulges on the fuselage flanks redundant and removed. The 6ft (1.8m) ASV radar antenna was installed under the nose and a visual observation/bomb aiming position located in the nose.

Power was to be provided by four Rolls-Royce RB.142 with 32% reheat or four Bristol Siddeley Pegasus 5-6A with 18% reheat and thus the AW.681 was to have had an AUW of 235,500 lb (106,821kg) and could meet OR.357 in full, but the search speed would need to be 230kt (426km/h) unless the flaps were deployed, for a speed of 180kt (333km/h). As with the MR.1, the MR.2 could patrol on two engines and carried extra fuel in two slipper tanks. Unfortunately the Air Staff's analysis showed that the Pegasus was '…unsuitable for the MR role.' And that if it was selected for the AW.681 transport, '…approximately £5m would have to be added to the engine R&D for the maritime version.'

The attraction in the AW.681 was that it had plenty of growth potential, would share development costs with the transport version and therefore be cheaper. Interestingly Armstrong Whitworth's assurance on meeting OR.357 in full was not borne out by the Ministry of Aviation's analysis. Air Commodore Fletcher summed up the AW.681MR thus: '…this aircraft would have – among a number of operational penalties – a search capability of about 4½ hours at 1,000nm radius of action.' Fletcher observed that this was about the same as the 'modernised Shackleton' and that the aim was to have 60 OR.357 aircraft to replace 93 Shackletons, this shortfall in patrol time would require

'more than 60 maritime AW.681s.' Unfortunately Fletcher doesn't mention which AW.681MR was being discussed, as according to Armstrong Whitworth, the MR.2 met the requirement with ease.

The AW.681, despite attracting the attention of Air Staff and Coastal Command, would not progress past the mock-up phase as it was cancelled as a tactical transport along with the BAC TSR.2 and Hawker Siddeley P.1154 in January 1965. Design authority and manufacture of the AW.681 had been transferred to Woodford where the designation was changed to HS.681 to reflect the new manufacturing group. The HS.681 V/STOL tactical transport would have been a difficult aircraft to develop given the perceived problems that an aircraft with such capabilities would have faced; the tales of a loud sigh of relief being heard around Woodford on the project's cancellation are probably not apocryphal.

The Vickers Vanguard was back in the running for OR.357, but not for long as the MoA considered it to be 'old-fashioned for an aircraft which would need to be in service from about 1970 until 1985'. The other problem, which had been identified by the OR.350 process, was that the Vanguard needed redesign and so many modifications that it would be prohibitively expensive and still not meet OR.357. The Vanguard MMC and GMRs would be particularly handicapped by having a weight penalty from carrying redundant civil variant components. Another criticism was that the Vanguard's weapons bay was tailored to the weapons in service in 1961, with no space to allow for larger weapons that might be in service in the 1970s.

BAC at Filton, formerly the Bristol Aeroplane Company, proposed a modified Britannia for OR.357, but since this design was even older than the Vanguard and being powered by Bristol Proteus turboprops, would no doubt require re-engining with Tynes. Canadair had based their Argus on the Britannia, but had replaced the Proteus with Wright R-3550 turbo-compound piston engines to attain the required endurance.

Avro also drew up a turboprop design for OR.357 in the shape of the Type 784, a conventional four-engined type not unlike the Vickers Vanguard in size and shape. Whether this was derived from the APG.1010A studies carried out for OR.350 is unclear, but the Type 784 was slightly larger, with a wingspan of 145ft vs the 135ft 6in of the APG.1010. The Type 784's fuselage was very similar to the earlier Type 776

While a propeller type was frowned upon by the Air Staff, Woodford drew up the Type 784. Slightly larger than the HS.1010, it was powered by four Tynes and, apart from the transit speed, capable of meeting ASR.357. The Type 784 has the Avro 'look' and was the last design study to carry an Avro Type Number. All studies after it were Hawker Siddeley types. Note that it carries ASMs on wing pylons, a new feature of OR.357. *Avro Heritage*

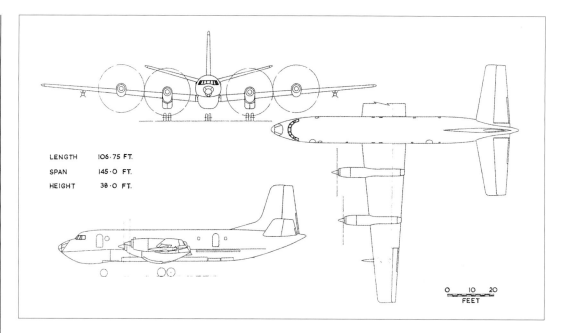

LENGTH   106·75 FT.
SPAN   145·0 FT.
HEIGHT   38·0 FT.

0   10   20
FEET

for OR.350, with the chin radome for the X-band ASV.21 antenna and glazed nose.

This brings the story back to the Avro Type 776, the Air Staff's preferred type and essentially the aircraft all the other studies had to beat. HSA's Hatfield plant drew up a design study based on the DH.121 Trident airliner. The MoA were not convinced and stated that Tri-

dent would require '…such extensive redesign in order to provide the necessary space and range that the result would be in effect a new aircraft.' The Trident, described as the MR.1 in the documents, utilised the fuselage, tail unit (including the intake and installation for the Number Two engine) and inner wing section of the Trident 1 airliner. A new, Woodford-

The poor-man's Type 776? The Trident MR.1 with new nose profile and on the MR.2, larger wings and new engines. The 'MR' designations were applied by Hawker Siddeley and were not RAF designations. Of note on the MR.2 is what appears to be an antenna for a sideways-looking airborne radar (SLAR) on the fuselage ahead of the wing.

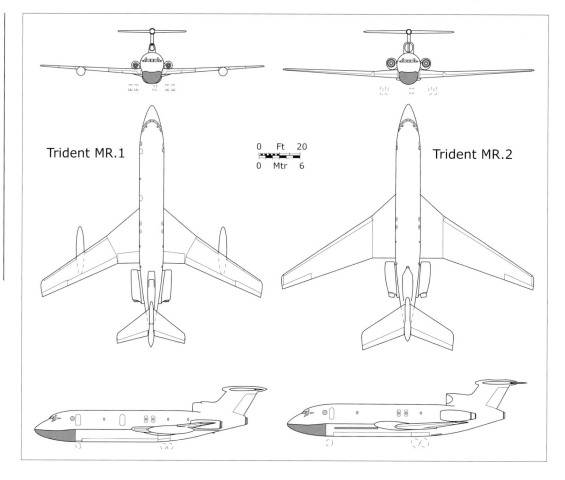

Trident MR.1

0   Ft   20
0   Mtr 6

Trident MR.2

designed outer wing, *à la* Avro 776, was fitted to the Trident MR.1 to give a wingspan of 112ft 2in (34.2m) and increase the wing area from 1,415ft² (131.5m²) to 1,800ft² (167.2m²). Since the Trident 1 airliner was considered to be 'short-legged' for the OR.357 requirement, the MR.1 was to carry extra fuel in slipper tanks at the 60% span point with hard points for AS.30 or AJ.168 missiles at mid-span. The Trident MR.2, using the 776's wings would carry more fuel within the wing, dispensing with the slipper tanks. The main change was the addition of a ventral pannier along almost the entire length of the fuselage, producing a double-bubble fuselage. The pannier carried a new weapons bay and an enlarged undercarriage bay to house a newly-designed eight-wheel bogie main undercarriage, which was becoming a bit of a Woodford signature, having been used on the Vulcan and the Avro 719 Mk.4 Shackleton. This undercarriage retracted to the rear and was required to handle the Trident MR.2's increased AUW of 210,000 lb (95,238kg) from the additional structure required for the pannier, the new outer wings and the increased fuel and weapons capacity. The oleos on the new undercarriage also had to be lengthened to provide ground clearance of 4ft 3in to allow a standard bomb trolley to be used for loading the pannier bomb bay. Essentially the Trident MR variants were more or less what the Avro 776, as proposed for OR.350, should have been.

The Politics of Patrol Aircraft

By January 1964 the timeframe for developing an aircraft to meet OR.357 was widening and its service entry further delayed. In a briefing paper dated January 1964, collaboration with the Canadians was looking less likely, due to much closer relations between the US and Canadian governments on the defence of the North American continent. A partnership with the French was unlikely 'There is officially no prospect of collaboration with the French whose requirement for a replacement for the Atlantic is aimed at later than 1973.' Then there is the first mention of a gap filling purchase: 'Unofficially it is thought that an interim buy of Atlantique by the UK might make collaboration possible.' This interim buy would be of 75 aircraft '...for about half the price of 50 aircraft to AST.357.' Lockheed's P-3 Orion was also considered as an interim, '...if the Americans deliberately lower the price.'

All-in-all the feasibility studies had shown that there was in reality no way of reducing the cost of the OR.357 aircraft without scaling back the requirement. What the Air Staff feared most was continual re-assessment of the proposals, effectively killing the programme though constant revision. The January 1964 paper stated '...we may find that the project is referred back for interminable re-evaluation after submission to WDC, as with the Hunter and Beverley/Hastings replacements (P.1154 and HS.681 respectively); or completely rejected.' What worried the Air Staff was that they would have to '...accept the Atlantic or P-3A.' Based on this the Air Staff thought that there was '...a good case for interim guidance on the course of action to be adopted.' It was time for an across the board reassessment of the Shackleton replacement and the future plans for the UK aircraft industry.

April 1964 saw Mr B C Fox, Permanent Secretary at the Air Ministry laying out the options for OR.357 as discussed at a meeting of the DCAS; the Controller, Aircraft and representatives of the MoA. Firstly it was felt that OR.357 was '...manifestly too expensive for this country to undertake alone...' and that the MoA and Air Staff were '...investigating possible interim replacements for the Shackleton Mk.2.' These included the Atlantic, and maritime conversion of the Trident and the VC10. Fox then pointed out that 'None of these however would have a capability approaching that specified in AST.357.'

Fox noted the costs involved and that these were not just financial but the case of the Atlantic '...political repercussions likely to flow from the suggestion that a further foreign aircraft should be bought...' As for the VC10 '...there is bound to be considerable political and industrial pressure to adopt this aircraft...the capital and running costs will present difficulties for the RAF.' The MoA were '...inclined to view more favourably the Trident solution...' that could be available by 1968/69 with a later upgrade bringing this type close to OR.357 standards. As far as the MoA was concerned 'This seemed... to offer a better solution than the purchase of Atlantics now and an entirely new aircraft in the late 1970s.' However, given the monies to be spent on the HS.681, TSR.2 and P.1154, there would be problems funding the purchase of Trident MRs from 1969.

In short, the MoA were against the Atlantic and for the Trident, but given the rapid deterioration of the Shackletons, a plan to refurbish, and ultimately, replace them before the end of the decade must be put in place. Thus was born OR.381.

'We intend to go for fixed cost, even if it means labels in French in the cockpit.'
Air Commodore R G Knott, DOR.2(RAF) 10th March 1964

'First, in considering the Atlantic we are dealing with something which physically exists, involves a number of our allies and is a certainty. A British alternative means embarking, by ourselves, on a venture which is far less certain both as to time and cost.' Hugh Fraser, Secretary of State for Air, 2nd July 1964

The true OR.357 type? The 'interim' Nimrod MR.1 would evolve via OR.381, upgrades and modifications to become by 1983 the MR.2P. The upgrade of the MR.1 was being planned while the type was entering service and would provide the capability sought by OR.357.
Graham Wheatley

As the powers that be continued with their attempts to replace the Avro Shackleton with an ever more complicated aircraft, there was a looming threat to the entire programme. That threat was time, which was not on the side of OR.357 and while cost was a factor in OR.357's demise, it was time that made the prospective Shackleton replacement an unsung victim of the Callaghan/Healey cancellations of 1965. The Air Staff was by late 1963 faced with a dilemma: stick with the Shackleton until the mid-1970s or bite the bullet and go for an interim type. They wisely opted for the latter course.

The Air Staff and Air Ministry were concerned that the cost of developing the OR.357 aircraft could not be borne along with the BAC TSR.2, Hawker P.1154 and Hawker Siddeley HS.681 projects (the 'Big Three'). This had prompted the deferment of OR.357 in 1964 and it is a distinct possibility that had the OR.357 programme continued, it too would have met the same fate as the 'Big Three' at the hands of Denis Healey. Added to that, it is quite probable that Healey would have come back from the States with the Lockheed P-3C Orion added to his shopping list of McDonnell-Douglas F-4K/M Phantom, Lockheed C-130K Hercules and General Dynamics F-111K. Oddly enough Coastal Command, the Air Staff and Air Ministry had foreseen this and opted to pursue an interim type.

For service in the Iraq and Afghanistan theatres the Nimrod was upgraded to MR.2P(GM) [Gulf Modification] standard and undertook missions in support of ground troops. Seen here on XV231 is the Wescam MX-15 electro-optical turret and a number of new antennae installed for the enhanced Sigint suite. Also visible are modular countermeasures pods (MCP) equipped with Terma Advanced Counter-measures Dispenser Systems (ACMDS) units. It was on one such mission that XV228 was destroyed.

Early March 1964 saw the tide turn against OR.357 and flow towards the Atlantic, despite the political and industrial pressure to adapt and adopt the Vickers VC10 or Hawker Siddeley Trident MR.2. BAC was quoting an in-service date of mid-1969 for a minimum-change maritime VC10, which effectively provided Shackleton MR.3 capability with the higher transit speed. The same date was quoted for the Trident, which, lacking the requisite range, needed more modification including additional tankage and a weapons bay. The interim Trident would cost up to double that of the VC10 at £20m a copy.

Air Commodore R G Knott, Director of Operation Requirements indicated that as an interim type, the only benefit a jet aircraft provided was to knock up to two hours off the transit on a long-range sortie and allow approximately 12% better coverage, but that extra capability would come at double, even treble, the cost of the Breguet Atlantic. The Atlantic could be available in 1966 and thus save the cost of the Shackleton refurbishments with further saving from buying the Atlantic off-the-shelf at a fixed price, in Knott's words, 'We intend to go for fixed cost, even if it means labels in French in the cockpit.' There was also the small matter of buying an aircraft that had been rejected by the RAF five years previously; something that would stick in the craw on both sides of the Channel and the French would no doubt exact a monetary and political price for that.

As for the VC10, it had the development potential the Atlantic lacked and in ten years' time (the intended service life of the VC10

Interim type), the Air Staff hoped that it could be re-engined with Medways, fitted with a bomb bay and brought up to OR.357 standard. However, Knott commented that two factors would govern the choice of MR aircraft; the first being an in-service date before 1970, the second the desire to have a fleet of 50 aircraft. The cost of the VC10 and Trident, plus the modifications to the Trident would make a fleet of 50 of the former and 1970 delivery of the latter unlikely.

A fourth aircraft, the Orion, was considered but apart from four engines (and presumably cockpit labels in English) differed little from the Atlantic in capability, although the Atlantic was fitted out to use the British Mk.1c sonobuoy system. Nor was price significantly different at £1.25m for the Atlantic, as the Americans had offered the RNZAF the Orion at £1.295m. Coastal Command estimated that 36 Atlantics (plus 11 for training, R+D, attrition) could replace the Shackleton MR.2 force. All-in-all if the interim type was required '…the Atlantic is preferred for financial and political reasons, there being little to choose between them operationally.'

So, by 10th March 1964 the preferred option was to buy 47 Atlantics, off-the-shelf, as an interim Shackleton MR.2 replacement pending the availability of an aircraft to OR.357. The Air Council was advised that to achieve a mid-1966 delivery date, the Atlantics would need to be ordered '…no later than July this year (1964)' and this would also allow the Shackleton MR.2 modernisation to be cancelled. If the interim type was procured to replace the Shackleton

MR.2 alone, Coastal Command would be carrying the logistics burden of two large MR types and the resultant drain on resources might even force the further deferment of OR.357 until 1980, which was tantamount to cancellation. Air Vice-Marshal R H E Emson, ACAS(OR) took the view that the whole Shackleton force should be replaced by 50 Atlantics in 1966/67, the Shackleton modernisation cancelled and OR.357 deferred until 1978 when the interim type would be replaced. This would mean that the expenditure would peak after the 'Big Three' aircraft projects, but too late to beat the rundown of the Shackleton force. Another factor in the deferment was that Canada would need to replace the Canadair CL-28 Argus and France its Atlantic aircraft in 1978 and the OR.357 type would be ready by then. In reality, by 1964 the heavier than predicted Cold War workload was eating up the Shackleton's fatigue life. The solution would be to fund the Shackleton replacement in 1966/67 before expenditure on the 'Big Three' peaked, with the OR.357 aircraft being funded after that peak.

The MoA and Air Staff were on a tight schedule to place the order for 50 interim MR aircraft before the perceived funding window closed after 1966. Another problem to be addressed was the procurement itself. There would be considerable pressure for domestic production under licence of any foreign aircraft to be placed with Avro, having had a relationship with Breguet since 1958, building the Atlantic and Shorts, for political reasons, building the Orion. Unfortunately such arrangements would lead to delay compared to buying the aircraft direct from the manufacturer as the relevant licenced builder set up production lines and got production under way. At this point, early 1964 the 'Big Three' were entering development, but one critical purchase had been made – the F-4 Phantom for the Fleet Air Arm instead of the naval Hawker P.1154– and the Ministries took the view that buying another American type would be politically untenable. Purchasing the French Atlantic would '…sweeten the atmosphere enormously and facilitate further collaboration…'

June 1964 saw the Atlantic in the ascendant, with Hugh Fraser, the Secretary of State for Air advising in a briefing paper that the Orion was more or less out of the running because ordering it so soon after the Phantom would be unattractive from a political viewpoint and it offered no operational advantage over the Atlantic. In its favour, the Atlantic used many British components such as the Tyne engine, whereas the Orion's British content was nil. Cost-wise the Atlantic was £57m cheaper to operate over a decade than the Trident MR, although over a twenty year period the Trident MR was £37m cheaper to operate. No matter how the cost of the Interim MR programme was examined, the final analysis was that the RAF could not possibly fund a fourth major aircraft programme in the late 1960s 'Indeed it is difficult to believe that the Royal Air Force could or should seek to absorb a fourth major aircraft project in this time scale.'

Quid Pro Quo

All-in-all the purchase of the Atlantic as an interim type would benefit all involved. The RAF would get a new MR aircraft, the financial burden would be spread and if a new type was to be developed to replace the Atlantic and the Shackleton MR.3s at a later date, it would benefit from all that new ASW equipment that was planned for the future. The new MR type would appear in service from around 1980 and be capable of taking on the next generation of Soviet nuclear-powered submarines that the intelligence analysts were describing.

That was the plan and all that was needed was a requirement to go with it and thus ASR.381 was drafted for an Interim Maritime Reconnaissance Aircraft to replace the Shackleton and this was submitted for discussion in March 1964. By July 1964 the draft requirement had been through the mill at the MoA and the Operational Requirements Branch with all concerned aware that ASR.381 was written specifically around the Atlantic as the interim type and it was endorsed by the Committee on 2nd July 1964. Then, just as the final OR was being completed, the 'buy British' aspect of the procurement reared its head. 'Suggestions have, however, been made within the last few days of other British aircraft conversions which might meet the requirement.' These were the BAC One-Eleven, the Comet and the Vanguard. Fraser was not particularly impressed by the appearance of these proposals so late in the process and stated that 'All experience with such conversions from civil aircraft…has shown that these are neither cheap nor quick projects.' Nor would these produce a Shackleton MR.2 replacement any quicker than the types previously proposed, such as the Trident.

The paper concluded by giving two choices: buy the Atlantic for an adequate capability at a known cost or develop a British type by converting a civil airliner at an unknown cost and timescale. In the latter case, the current estimates for cost and timescale would without doubt increase. Fraser was convinced that the Atlantic was the right choice, but there was trouble brewing at the Ministry of Aviation. Julian Amery, the Minister of Aviation had contacted Fraser, who described Amery as '...clearly very worried about the industrial impact of a decision to buy the Atlantic and, from his talks with Pierre Messmer (French *Ministre de la Défense*, Defence Minister), there seems to be no chance of an immediate quid pro quo.' This was a blow, as Amery had hoped that any deal involving the Atlantic would lead to more collaboration between the respective aerospace industries. While Amery considered it right to order the Atlantic on merit, he was at odds with Fraser with regard to buying it directly from the French. In fact the MoA had long held the view that British companies should be given the chance to offer alternatives to the Atlantic and had in June 1964 commented that it was '...unfair to have written AST.381 around the Breguet Atlantic.' The ACAS(OR) was briefed that the timescale meant the Atlantic was the obvious choice, the requirement should be written to fit its capabilities and there was a precedent for this in the Hawker P.1154, the HS.125 and HS.748 for the RAF.

Interestingly, from the industrial standpoint Breguet had previously approached HSA at Woodford about building the Atlantic centre section. This was being built by Fokker, but that firm had been having a problem with the 'honeycomb structural technique' used for the Atlantic structure. Breguet were dissatisfied with Fokker's work and wanted Woodford, who had pioneered the technique, to take over.

There was, in the early 1960s, a desire to cooperate with the Europeans, but not at any price. Amery's quid pro quo was most definitely not to be at any price and the various ministries associated with aviation were all in favour of it. In fact a June 1964 briefing for the Minister of Defence, Peter Thorneycroft, stated that '...joining the Atlantic consortium would do us a great deal of good, and indeed a wide programme of collaboration is probably the British industry's best hope of salvation...' This would be a long term, rather than just for the Atlantic for which there was '...no immediate quid pro quo though the French express an interest in a surface-to-air missile based on Bloodhound.'

The MoD briefing concluded with a paragraph that laid out the future of British co-operation with Europe, with the Atlantic being the

first step; 'Its purchase should be regarded not so much as a "foreign buy" but rather as rejoining a European consortium in which we had an interest and played an active part, before withdrawing in 1959, and for which we supply some of the parts. This can be turned to good advantage in many other projects, e.g. Bloodhound development, the trainer light strike aircraft, the tactical helicopter and possibly a future fighter, all of which the French and some other European countries are interested in.' So, the importance of European co-operation to the British aviation industry was appreciated from the early 1960s. Oddly, it appeared that on the military side of things the Atlantic, rather than Jaguar or helicopters, was the key.

Not that everyone was happy to see the French handed an order for fifty large aircraft in a role for which the same aircraft had been rejected five years before. Most notable in his opposition to the Atlantic procurement was Sir Arnold Hall. Director of the Royal Aircraft Establishment until 1955, Hall had since been the Technical Director of Hawker Siddeley Aviation and was most perturbed by the very idea that the RAF should be buying a French aircraft, especially after that same aircraft had been rejected in the past. This was not petty Francophobia, as Sir Arnold was serious and went further by requesting a visit to the MoD. On arrival at the Ministry, on the 6th July 1964 Hall made his objection to the forthcoming Atlantic deal clear to the Permanent Secretary at the MoD, Sir Henry Hardman. The next day Hardman in turn passed on Hall's views in a letter to his counterpart at the MoA, Sir Richard 'Sam' Way, and Frank Cooper, Deputy Under-Secretary of State at the MoD.

What Hall told Hardman was that HSA had just found out that ASR.381 had been written

Yet again the Breguet Br.1150 Atlantic reared its head in the Air Staff's search for a maritime patrol type. This time, for OR.381, the requirement was written around it with the intention that it be acquired come what may. Britain's aircraft industry would not take that lying down and Arnold Hall took on the ministries. This image shows two Dutch Navy Atlantics at the Breguet plant with a test rig for the Breguet Br.941 in the background. *Author's collection.*

around the Breguet Atlantic and that '…the parameters (as regards, for example, transit speed and load capacity), which had been set steadily higher in various Staff Targets from 1958, had now suddenly been lowered. For all practical purposes they were back again at the levels appropriate to the Atlantique.' Hall also pointed out that the Atlantic had been rejected as '…inadequate for the requirements of the RAF.'

Hall explained that HSA had been associated with the development of the Atlantic and six years before had to '…scrap a good deal of work that had been done on it as well as work on subsequently revised Air Staff Targets. All this was pure loss.' However, if the ASR.381 was a firm requirement, Hall suggested that Hawker Siddeley could supply '…a suitable aircraft without very much in the way of development in the shape of the Comet.' This would benefit from existing training and spares support and would probably cost around £100,000 per aircraft to develop and that the '…total cost of a maritime reconnaissance variant would certainly be no higher than the £1½m asked for the Atlantique.' At this point Hall must have adopted a somewhat suspicious tone with Hardman.

Hall asked if the Government had '…obligations to the French' or had decided that 'under no circumstances would they buy a British aircraft' and hoped that this Comet proposal would be considered. Hardman assured Hall that the Government had no commitment to the French and 'that I could not take seriously his implication that the Government would not bend over backwards to buy suitable British air-

craft.' Hardman also suggested that the change in the requirement might be 'A consequence of a realisation within the RAF that higher performance normally meant higher cost and that inadequate resources were available for anything approaching in performance to the earlier requirements.'

Hardman told Hall that if he thought HSA could '…offer a developed Comet at the same price as the Breguet, and in every other respect its performance was as good or better, I was sure that there were very good chances of the Comet being acceptable.' Hardman thought the right line should be that HSA begin talks with the MoA and the Air Force Department. He also laid out his view of Hall's visit, stating that 'Naturally Arnold Hall was striving to make my flesh creep. But his visit did indicate that the aircraft industry is very worried after the decision on the Phantom, about the suggestion of a Breguet purchase.' The Permanent Secretary thought that a decision on buying the Atlantic was unlikely before the next general election, to be held on the 15th October 1964 and won by Harold Wilson's Labour Party, elected to govern for the next five years. Hardman signed off by advising that the Comet proposal should be given 'careful consideration' and would discuss it with HSA as a matter of urgency.

When the Air Staff were told of the HSA proposal for the Comet, they were somewhat annoyed, because at that point in mid-July 1964, the only two British proposals that they were aware of were the Trident and the VC10. On making 'unofficial investigations with the firms', the officers 'uncovered a difficult situation'. Back in February the Air Staff had asked

A Comet 4 of Transport Command was in RAF service as a troop transport. Sir Arnold Hall was the driving force behind its transformation into the Nimrod and without his forceful representations, the British proposals might have foundered and the Breguet Atlantic be acquired by default. By late 1964 the RAE's Comet 3 was flying simulated maritime patrol sorties. *Author's collection*

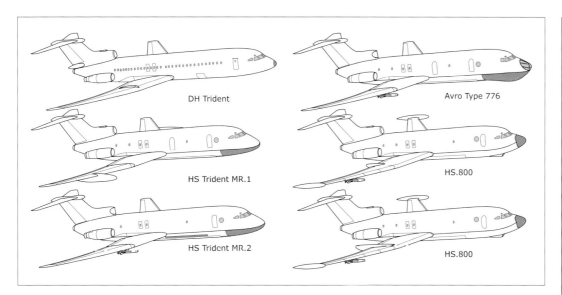

DH Trident

HS Trident MR.1

HS Trident MR.2

Avro Type 776

HS.800

HS.800

Woodford based a number of maritime types on Hatfield's Trident and these included the Trident MR.1 and MR.2 plus the HS.800. The latter would lead to the definitive HS.800 for ASR.381. The HS.800 was the successor to the Avro 776 and therefore the aircraft to beat in the design conference, but was deemed too expensive to develop.

the MoA for information on an interim MR aircraft based on an off-the-shelf type. The MoA insisted that only a British type was suitable and that the choice was between the Trident and the VC10, both with an in-service date of 1968, which completely ignored the Air Staff and Air Ministry's advice on avoiding the financial peak from the 'Big Three' projects by commencing deliveries in 1966. As noted earlier, the VC10 was too expensive and the Trident required so many modifications it would only be ready for service in 1968, if not 1969. This effectively ruled out both types as far as the Air Staff was concerned and therefore the only type that was cost effective and available in 1966 was the Atlantic.

So by 15th July 1964, Air Commodore R G Knott, Director of Operational Requirements (DOR.2) was advising the ACAS(OR), Air Vice-Marshal Denis Smallwood, on what his staff had turned up during their unofficial investigations. Essentially the Air Staff had the edge on the MoA, who were due to unveil their evaluation of the 'lesser proposals' around the 20th July.

What the Air Staff's 'unofficial investigations' revealed was a further two proposals from BAC at Weybridge and a third from HSA's Hatfield plant, the former home of the de Havilland Aircraft Company, now part of Hawker Siddeley Aviation. The MoA had ignored these three proposals and had persevered with the Trident and VC10 until the end of June, despite their obvious unsuitability, before calling in the firms to discuss the 'lesser proposals' at the beginning of July.

The two Weybridge proposals were based on the BAC One-Eleven and the Vanguard and had been included in the same brochure as the

VC10 for ASR.381 submitted to the MoA in July 1964. Strangely enough, the BAC One-Eleven MRA (Maritime Reconnaissance Aircraft) is on page one and describes a BAC One-Eleven airliner fitted with a pair of Rolls-Royce RB.177-22 (developed Medway) as intended for the HS.681 transport aircraft. Waiting for the Medway could delay the BAC One-Eleven MRA's service entry, so other engines were being examined. BAC advised that this new type had plenty of 'stretch' and potential for growth in the maritime role. It would be larger than the airliner, which in turn opened up the opportunity to develop the maritime airframe as an airliner in its own right. BAC were correct, since the BAC One-Eleven MRA did not particularly look like a standard BAC One-Eleven as it utilised a completely new wing centre section carrying fuel tanks and outer wings with large nacelles on the wings that carried fuel tanks

One of Weybridge's last design studies for OR.381 was this highly modified BAC One-Eleven. The lack of four engines would have met with disapproval from the Air Staff. It is unclear from the documentation whether this pre- or post-dates the BAC 10-11 hybrid.

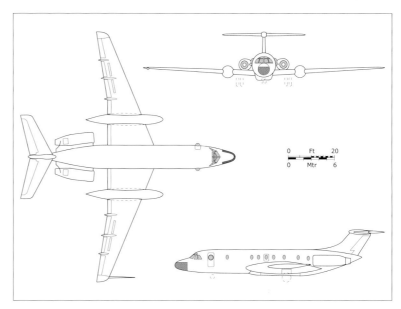

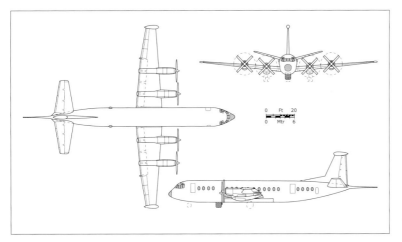

Vickers also tendered the Maritime Vanguard for OR.381, but as before the Air Staff was looking for an aircraft with high transit speed. The pod on the fin carried ESM equipment while the port wingtip contained a Yellow Duckling or Clinker IR linescan and the starboard tip pod, a searchlight.

De Havilland's Comet 4 would provide the basis of what ultimately became the Nimrod. Main changes included being re-engined with Speys, the fuselage shortened and adding a ventral pannier for weapons. All these modifications would provide as good, if not better capability than the Trident MR, but at reduced cost. Despite its use of Shackleton systems, it was a quantum leap in the RAF's anti-submarine capability.
Author's collection

and the undercarriage bays. The undercarriage had to cope with an increased AUW of 155,000 lb (70,295kg) and was taken from the Vickers Vanguard. These large nacelles also carried an infrared linescan system (derived from Yellow Duckling or Clinker) and a searchlight. The rear fuselage was modified to carry the RB.177s and up front a large radome for the ASV.21 scanner was faired into the existing forward fuselage. A retractable radome could be fitted under the forward fuselage, but this would entail penetration of the pressure cabin. Weapons and ASW stores totalling 10,000 lb (4,535kg) were to be carried in a 19ft 4in (5.9m) long weapons bay in a ventral fairing. The BAC One-Eleven MRA for ASR.381 had a patrol time of five hours at 1,000nm (1,852km) with a transit speed of 415kt (769km/hr) and BAC estimated that a fleet of 40 aircraft could be acquired for £1.9m each, with a development cost of £17m.

Rather than quote the Duke of Wellington for a second time, suffice to say the Vanguard was

back in the running yet again. The aircraft was similar to the civil Vanguard apart from the fuselage and wing centre section that were to be modified to house a fuel tank and a 32ft (9.75m) long bomb bay, while the Tyne engines were to be fitted with 15ft (4.6m) diameter propellers. The ASV.21 radar was fitted under the forward fuselage in a retractable radome, with the original weather radar in the nose removed and its radome replaced by the transparency for a bomb aimer's position. As for previous submissions to OR.350 and OR.357, access to this required a redesign of the cockpit and instrument panel. The ASR.381 Vanguard MRA was to have an AUW of 166,000 lb (75,283kg) with a 10,000 lb (4,535kg) ASW stores capacity, with a transit speed of 355kt (657km/hr) and an endurance of five hours at 1,000nm (1,852km). Equipment pods were fitted to the fin tip (ECM antennae), port wingtip (IR linescan) and a searchlight in the starboard tip pod. As with the BAC One-Eleven, the cost per aircraft was to be £1.9m.

The Mighty Hunter from Hatfield

'The root of this original underestimation lies with the technical staffs' initial misconception that the Nimrod was only a Comet fitted with Shackleton black boxes. The extent of this lack of appreciation of the complexity of a modern ASW avionics system is most surprising, particularly in view of the numerous AST.357 studies which preceded Nimrod.' Brief for DCAS, R+D Board, Monday 8th April 1968.

De Havilland had conducted a design study based on the Comet for OR.350, but this was

not submitted on an official basis, presumably because the Comet design was older than the Shackleton it was to replace. De Havilland stated that the Comet 4, fitted with the Rolls-Royce RB.168-1 Spey, rated at 9,850 lbf (43.8kN), as fitted to the civil Trident, would '…give a performance not far short of the Trident with far less structural alteration.' By fitting the Spey turbofan, the much improved specific fuel consumption (SFC), 0.7 lb/(lbf/hr) (19.8g/(kN/s)) of the engine over the Rolls-Royce Avon turbojet, 0.9 lb/(lbf/h) (28.9g/(kN/s)) allowed the Comet MR to meet the ASR.381 endurance requirement on existing fuel capacity. This in turn allowed the existing undercarriage to be used as the AUW would not be increased by as much, estimated at 60%, as the Trident MR had over the Trident airliner. The major structural modifications to the Comet were a reduction in fuselage length of 6ft (1.8m) if Comet 4Cs were used plus new air intakes and engine bays that required redesign of the inner wings. The addition of a ventral pannier with a positively massive 46ft (14m) long weapons bay completely changed the Comet's profile and a hard-point for a pylon was added to each outer wing. The windscreen was extended to improve visibility and eyebrow windows added to help when making hard manoeuvres while prosecuting a target. Unlike the Trident, the existing undercarriage was tall enough to allow bomb trollies under the aircraft.

An ASV.21D scanner was to be installed in the nose or in a retractable 'dustbin' under the fuselage, but this would require the pressure cabin to be penetrated. ASV.21D could detect a *schnörkel* from an altitude of 500ft (152m) up to a range of 30nm (55.6km) in sea state 2. The ASV.21 antenna was 32in x 24in (81cm x 61cm), but a 72 x 24in (183cm x 61cm) antenna would improve the medium altitude surveillance performance. This would need to be installed in the nose, restricting coverage to 240° rather than the 360° of the 'dustbin'. A SLAR could be incorporated, with the antennae fitted along the fuselage at the forward freight hold.

A glazed bomb aimer's position was to be installed in the nose under the radar, to allow accurate bomb delivery during 'Internal Security' missions or depth change dropping. As for performance, at an AUW of 162,500 lb (73,696kg) the Comet MR could transit to its patrol areas at 410kt (759km/h) and patrol for eight hours at a distance of 600nm (1,111km) or almost six hours at 1,000nm (1,852km).

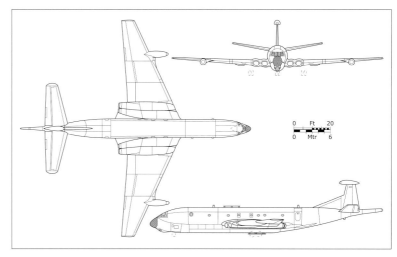

Once the AUW had reduced to below 150,000 lb (68,027kg), two of the engines could be shut down for the remainder of the mission, ideal for a dash to the seas north of Norway. The AUW was governed by the undercarriage, which by being manufactured in steel could be beefed up for a higher weight, but above 167,000 lb, the Spey RB.168-1 engines would need to be replaced with engines of higher power. One artist's impression of the early Nimrod shows a version with six engines – one in each wing root, presumably Speys, and another two on pylons under each wing. These look like small turbofans, and given the era, these could possibly be RB.172 turbofans, the forerunner of the Adour turbofan for the SEPECAT Jaguar and HS Hawk. The caption says that the drawing shows a version proposed for

One of the earliest studies for the Maritime Comet (which might even date back to OR.350) shows a different nose profile that included a bomb-aimer's position in the chin. The deletion of the 'internal security' role and inclusion of ASGWs such as Martel meant that the bomb-aimer's position could be dispensed with.

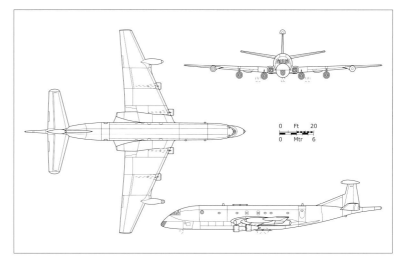

Possibly the oddest Nimrod proposal was the six-engined variant that appears to have been drawn up for the Royal Canadian Air Force. Details remain scanty, but the engine type appears to have been chosen on the maximum diameter to maintain the same undercarriage. Therefore to keep the same engine thrust, six engines were required. A later version with underwing refuelling pods and strengthened airframe was called Export Nimrod.

the Royal Canadian Air Force, but no further information has been found.

Interested, the Air Staff decided to investigate the Comet as a potential Shackleton replacement. Since they had already conducted test flights of the VC10 and Trident during which they were flown on maritime mission profiles, the sorties were repeated using a Comet. The RAE's Comet 3, XP915, was used for this and it performed well, particularly in the low-altitude and low-speed search phases. The verdict was that the Comet could form the basis of an interim maritime patrol type and the Air Staff thought that a Comet MR had growth potential, through re-engining at a later date or strengthening the undercarriage for increased AUW. They did take the view that larger engines in a buried installation might prove problematic, of which more later, with subsequent delays and cost increases. As for overall costs, well, Sir Arnold Hall had, at their meeting, pretty much told Sir Henry Hardman that HSA would, building on work for OR.357, develop and produce the Comet MR for a fixed price to match, if not better the Atlantic. The next designation in the Hawker Siddeley series was 801 and that was what was applied to the Comet MR, which became the HS.801.

By November 1965, the Chief Scientist, Solly Zuckerman and the Weapons Development Committee were examining the Maritime Comet proposal for the Minister of Aviation, Roy Jenkins. The Chief Scientist advised that the financial authority for the project definition phase ran out at the end of the month and so a quick decision was required. Zuckerman pointed out that in some areas the Comet '...does not match the formal ASR (and falls short of the performance claimed when the decision was taken to adopt the Comet) but the Operational Requirements Committee have confirmed that the changes are acceptable.' Zuckerman describes some of the changes that were needed before full development could be sanctioned, but in general he was happy with the Maritime Comet. In his conclusions for Jenkins, Zuckerman said that the Committee '...recommend that full development of the maritime reconnaissance Comet should now proceed...' and that the Air Force Department and the Defence Secretariat '...concur in this proposal...and accept the production commitment. I would be glad of your authority to proceed on this basis.' And so the HS.801 went into the development phase as the interim Shackleton replacement. Two Comet 4Cs would be used for development and returned to HSA for full HS.801 conversion at a later date. Former Transport Command Comets with less than 5,000 hours would be suitable for conversion, but surplus BOAC airframes with 20,000 hours were deemed unsuitable as the Comet had a fatigue life of 30,000 hours.

By July 1964 the Atlantic was looking less likely, not due to new British proposals but because the Atlantic itself was becoming unattractive from development, operational and political viewpoints. The C-in-C Coastal Command Air Marshal A D Selway wrote on 9th July 1964 to the DCAS Air Marshal C H Hartley saying he considered the Orion 'operationally superior', that the Atlantic '...has even less "stretch" than we supposed'. Selway poured further scorn on the MoA's VC10 proposal by

The Big Three. These were the three aircraft projects whose funding peaks the Air Staff wanted to beat by issuing the Instruction to Proceed for OR.381 before the end of 1964. All three; P.1154, TSR.2 and HS.681, were cancelled by Denis Healey who turned to the USA for the F-4K, F-111K and C-130K.

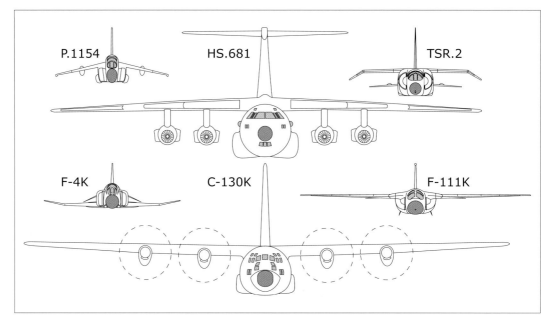

stating that it would '…never be anything but a scandalous inefficient waste of money' but was positive on the Trident '…potentially a very attractive proposition if it comes anywhere near meeting the time-scale.' Selway continued by hoping that a decision would be made soon, as the General Election was scheduled for October 1964, which could see any decision '…shelved for a year…'.

Rounding off his letter to Hartley with what is more or less a *précis* of the entire Shackleton replacement saga, Selway wished Hartley success in making a decision and it is worthwhile reproducing this in full. 'I sincerely trust that you will be successful, but if not, I can hardly suppose we shall have anyone but ourselves to blame. It is now almost exactly four years since OR.350, itself preceded by protracted discussions, was issued on 18th July 1960. This has been supplanted by two further requirements, consideration of the problem has been continually postponed in the supposed interest of other projects, and various paper studies have been made. As far as I can see, however, there has been no significant progress towards a solution.'

Of course a straight buy of the Atlantic could never happen, there had to be some quid pro quo to keep the politicians and the trades unions happy. The fact was that the MoD, with Peter Thorneycroft as Secretary of State had just ordered 140 McDonnell Douglas F-4K Phantom FG.1 fighters and associated kit such as Sparrow missiles and that had not gone down well with the UK taxpayers, never mind the aircraft industry. At various times in 1964 Air Vice-Marshal Reginald Emson exchanged letters with the DCAS Air Marshal Christopher Hartley on what the French could buy from the British. The main hope was that they would buy the Hawker P.1154 rather than their own Mirage IIIV, but Emson hoped to interest them in the Bloodhound II SAM and the forthcoming AEW aircraft (OR.387 for a land-based AEW aircraft was finally issued in 1966). In August 1964 Air Commodore Reginald Knott, Director of Operational Requirements advised that he had arranged, via the Air Attaché in Paris, for Canberra PR.9s from 39 Sqn to visit Toulon as a goodwill gesture. Knott added that the visiting crews had been advised that the French '…may be shown and be fully briefed on…' the Canberra PR.9 and that the French Air Force had '…apparently spoke in terms of a possible buy of 3 to 6 aircraft.'

Time was running out for the Atlantic and the pursuit of a quid pro quo from the French was going nowhere. The funding for ASR.381 had to be allocated before the General Election in October 1964 to ensure that it avoided a monetary clash with the 'Big Three'. Once the TSR.2, HS.681 and P.1154 projects got under way, the maritime patrol aircraft would be left high and dry once again.

The OR.381 Alternatives

Having examined the politics behind the selection of the Comet as the interim maritime patrol type, the other types under consideration require examination. The Atlantic had been a constant presence in Coastal Command's search for a Shackleton replacement. However, when OR.381 was under consideration, the Atlantic had become the benchmark, much to Arnold Hall's annoyance. Despite its place in the OR.381 story, there is little to be gained by describing the Atlantic again, so the alternative types submitted by BAC and HSA should be described. Like the Atlantic they have been discussed above, but in the case of OR.381, the shock of discovering that the Atlantic was the front-runner prodded the companies into action.

Trident

The Trident, as proposed by Woodford for OR.381 was more or less that proposed for OR.357. However, by 1964 Hawker Siddeley Aviation was the name of the entire group and designations of new designs gained the HS prefix. Since the Type 784 was the last of the Avro line, Woodford decided to start the HS series at 800 and the Maritime Trident became the HS.800. Its service name would still be Trident (what better name for a maritime weapon) and Woodford were describing the version for OR.381 as the Trident MR.2 and was outlined by the MoA's DFS Mr T V Somerville at a meeting at the Ministry on 26th June 1964

The Trident MR.2 was more or less the same as the MR.1 but slightly longer at 125ft 9in (38.3m) and had an increased wingspan of 127ft 7in (38.9m) with more area and hence increased lift, with the centre section beefed-up to carry the larger wings. The slipper tanks were dispensed with, as was the twin-wheel undercarriage, with four-wheel bogies adopted. Two other variants of the HS.800/Trident MR were examined and these were more or less standard Trident airliners with the ventral pannier and tip tanks to carry the necessary extra fuel. The four-wheel bogie undercarriage with extended oleos was installed in large pods on the trailing edge of the wings, lending these HS.800 Trident MRs

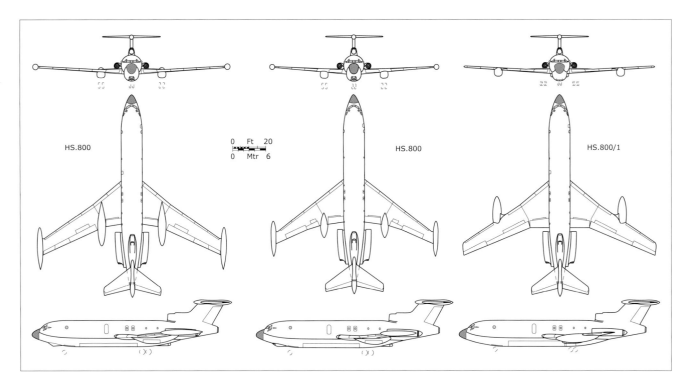

HS.800   HS.800   HS.800/1

0   Ft   20
0   Mtr   6

This drawing shows the various guises of the HS.800. The earliest versions have wingtip tanks and fairings for the bogie undercarriage. Also of note is the bomb aimer's position in the nose. The later variant, identified as the HS.800/1, has pinion/slipper tanks and an enlarged wing box to house the undercarriage. Note the lack of a bomb-aimer's position.

a somewhat Soviet look. These pods also carried fuel in their rear portions, with one variant having much larger pods that were faired into the wing across most of its upper surface.

In addition to the tip tanks these two variants retained the glazed bomb aimer's station on the chin between the ASV radar and the nose undercarriage bay. This was deleted from the definitive OR.381 Trident MR.2. The other main change from the standard Trident was the installation of the more powerful Rolls-Royce RB.168-25R turbofans with upgraded generators. In this configuration the Trident could carry 11,000 lb (4,988kg) for six hours at 1,000nm (1,852km), an improvement on the Atlantic's

6,000 lb (2,721kg) for four hours.

As with the Avro proposals for OR.357, a radar using FASS was initially proposed for the Trident MR.2 but this became less likely as the word from Lockheed was that such a system was being problematic on the P-3 Orion. HSA opted to fit the ASV-21 scanner with a 6ft (1.8m) antenna in the Trident's nose, but this was limited to 240° while the requirement called for 360° cover for the surveillance role. Fitting an antenna larger than 6ft (1.8m) across would require structural changes to the Trident nose, producing the deep, aquiline nose that would soon become familiar. Familiar indeed, but on a different aircraft.

This Hawker Siddeley drawing shows the internals of the HS.800/3 and illustrates the unused space in the sonobuoy dispenser bay. The HS.800 was the first Woodford design to use the 'HS' prefix. Note the weapons carriage in the ventral pannier, a new feature transferred to the HS.801.
Avro Heritage

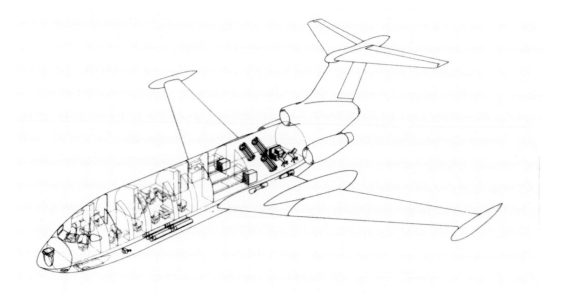

VC10 Again

The VC10 was 'identical to the civil Standard VC10 in service with BOAC' suggesting that the ten VC10s that BOAC wanted rid of would be convertible. The weather radar was to be replaced by an ASV.21 set with a 6ft (1.8m) antenna in an extended nose radome with 270° coverage (or a shorter radome over a 200° arc). The extended nose also carried an intake for the Autolycus Mk.3 equipment. Additional glazing to improve the pilots' field of view and the conversion of the forward freight hold as a sonobuoy launcher completed the changes to the forward fuselage. The Super VC10's fin with integral fuel tank was to be fitted and the uprated Conway R. Co.43 engines were to be installed in the existing nacelles. Apart from the large freight door, this was more or less the configuration of the Type 1106 that would enter service with RAF Transport Command as the VC10 C.1 in December 1966. Both wings were to be modified with strong points under the inner wing for carrying the spare engine pod, rather than just on the starboard wing of the airliner. The spare engine pod was to be converted into a weapon cocoon, two of which could be carried, but this affected endurance. With an AUW of 312,000 lb (141,497kg), including 10,000 lb (4,535kg) of stores, the ASR.381 VC10 was to have a transit speed of 470kt (870km/h), with an endurance of six and a half hours at a distance of 1,000nm (1,852km). As ever, the VC10 was most expensive at £2.725m per copy, with development costs of £6m spread over 40 aircraft.

Goldilocks – The Weybridge Hybrid

Having criticised the Vanguard as too slow, the VC10 for being too big and expensive, and the BAC One-Eleven as too small and twin-engined when the Air Staff preferred at least three, if not four engines, BAC at Weybridge sought to address all these aspects in one aircraft. This proposal led to the BAC Ten-Eleven, which appears to have been a development of the Vickers Optimum Aircraft described in the previous chapter and the BAC One-Eleven MRA outlined above.

The configuration of BAC's Ten-Eleven proposal, as submitted for ASR.381 in October 1964, closely resembled that of the BAC One-Eleven MRA put forward by BAC the previous July. It was, in effect, an enlarged One-Eleven MRA with four engines. BAC were quite clear

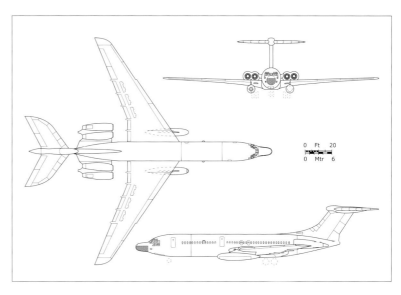

Not the final maritime patrol VC10, but the last for almost two decades. This variant to meet OR.381 carried its weapons in pods under the inner wing. These pods were a modification of the 'fifth engine' pods that allowed the airliner to transport spare engines under the starboard wing. This example carries a pod with a searchlight in the front, while the other pod may have housed a weather radar.

BAC artwork showing the BAC Ten-Eleven on patrol. The study was derived from Vickers/BAC's lack of an aircraft in the same AUW class as the Avro 776 or HS Trident and took components from the BAC One-Eleven and VC10. Note the MAD probe on the nose and the four Speys, mounted VC10-style. The Ten-Eleven had an unusual engine system: rather than being a four-engined type, BAC advised that it was twin-engined with two auxiliary engines. *via RAF Museum*

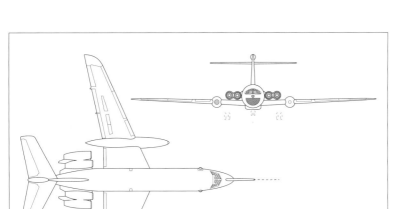

the Ten-Eleven was a hybrid aircraft that used components from the BAC One-Eleven and VC10 but saw it as a stepping stone toward '...an "ultimate" Maritime Reconnaissance Aircraft...' as planned in OR.357.

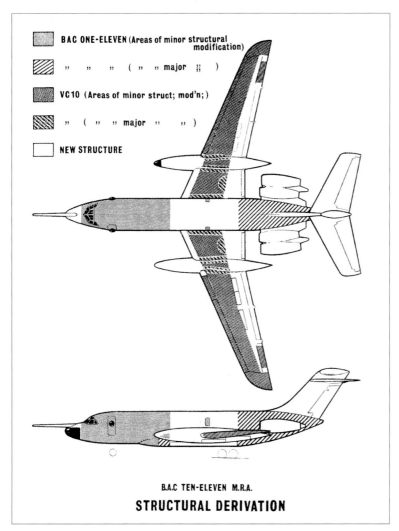

BAC ONE-ELEVEN (Areas of minor structural modification)

" " " (" " major ")

VC10 (Areas of minor struct; mod'n;)

" (" " major " ")

NEW STRUCTURE

B.A.C TEN-ELEVEN M.R.A.

STRUCTURAL DERIVATION

The Ten-Eleven was to be powered by four Rolls-Royce RB.163-25 Spey turbofans rated at 11,400 lbf (50.7kN) as used in the Trident and One-Eleven airliners, addressing the Air Staff's preference for four engines for aircraft involved in long endurance patrols. As with the other four-engined MR aircraft proposed for the various requirements, patrols on two engines would be possible once the aircraft's AUW permitted. The Speys were mounted VC10-style in pairs on the rear fuselage, that according to BAC '...makes readily possible a change of engine diameter or type, should this be required in the further development of the aircraft.' With an AUW of 210,000 lb (952,454kg) the Ten-Eleven was at the ideal weight for OR.381 and had enough growth potential to reach the requirements of OR.357, if that was still the ultimate aim. It did indeed meet the basic endurance and transit speed requirements with eight hours at 1,000nm (1,852km) and a transit speed of 420kt (778km/h).

If a ventral dustbin for the ASV.21 was used, fitted just ahead of the wing box, it would shave 'minutes per hour' of patrol of the endurance if deployed at 50% extension for patrols. As per the Shackleton MR.2 and MR.3, the dustbin was only fully extended for attacks. The existing ASV.21 used a 32in (81cm) wide antenna, but a larger 72in (183cm) antenna would improve the radar's performance in the medium altitude surveillance role. This could only be fitted in an extended nose radome, with 240° coverage.

BAC went to great lengths to assure the Air Staff and MoA that the Ten-Eleven wasn't some rehashed old airliner or a completely new untried design. Their brochure stated that it was '...derived from two second-generation civil airliners which utilised the most up to date structural and systems design knowledge.' These guaranteed a long fatigue life and this would be applicable to the maritime environment as the VC10 and One-Eleven used the latest corrosion protection techniques. BAC were hopeful that if the Ten-Eleven were selected for ASR.381, it would '...continue the production lives of the VC10 and One-Eleven civil airliners, and increase the sales potential of both aircraft by reductions in price.' They also hoped that the availability of a new MR aircraft, effectively the next generation, would make the Ten-Eleven '...attractive to a number of other countries.' All this would be music to the MoA's ears, but the major spanner in the works as far as the Air Staff was concerned would be timing: the Ten-Eleven would not be available until 1970 at the earliest, smack in the middle of the expenditure peak for the 'Big Three'.

Structurally, the Ten-Eleven used the wing torque box, slats and ailerons from the VC10 with the sweep reduced from 32.5° to 22°. The nose and forward fuselage are derived from the One-Eleven 300 Series, while the rear fuselage, geometrically similar to the One-Eleven airliner, was to be strengthened to cope with higher weights. A completely new fuselage centre section incorporating a 25ft 4in x 7ft 4in (7.7m x 2.2m) bomb bay was to be fitted below the wing torque box. The tail was based on the One-Eleven tailplane with a new fin with integral fuel tank as used on the Super VC10. If the addition of an ARAX/ARAR.10 ECM pod on the fin tip was required, the tailplane could be moved down to a position on the fin, similar to that on the Vickers Valiant bomber. A new four-wheel bogie main undercarriage was to be fitted, retracting outboard into the wing nacelles that also carried fuel tanks, with the starboard pod housing a searchlight and an IR linescan unit. The nose undercarriage was to be based on the Vanguard's.

Internally, the cockpit was based on the One-Eleven but modified for four–engined operation. The tactical, crew rest and sonobuoy areas were pretty much laid out as per the other MR aircraft dating back to NBMR.2 in 1958. For that all-important trooping role, up to 58 troops could be carried if the sonobuoy and crew rest equipment was removed and replaced with seating, with the light kit carried in the freight holds aft of the nose undercarriage bay. There was provision for a visual bomb aimer's position

in a chin fairing if required, but by late 1964 visually-aimed weapons delivery was limited to the co-pilot guiding ASGWs from the cockpit. Unusually, the Ten-Eleven carried its MAD kit in a telescopic boom extending forward from the nose, possibly due to limits on the aircraft angle of rotation on take-off if a tail boom was fitted.

The Ten-Eleven's wing was the outboard section of the VC10 swung forward through 10.5° and fitted with VC10 slats and new flaps based on those of the VC10. Inboard, the wing was to be a standard VC10 torque box, with a large nacelle, *à la* Lockheed JetStar forming the junction between the two wing sections with the nacelles containing the undercarriage bays and fuel tanks. The wing torque box structure also acted as a series of integral fuel tanks.

When it came to the engine installation BAC explained that to incorporate as many One-Eleven systems as possible, the Ten-Eleven should not be thought of as a four-engined aircraft, but as a twin with two booster engines. So it had two sets of engine controls each commanding an engine per side ie. No.1 and No.4 or No.2 and No.3. Unlike the VC10, the Ten-Eleven was not fitted with a thrust reverser system to shorten its field length requirements, but a drag chute was to be fitted in the tail cone.

What's In a Name?

On 19th March 1965 Director of Operations (Maritime, Navigation and ATC) Air Commodore R H C Burwell wrote to the three ACAS on the subject of naming the HS.801. The Air Commodore had pointed out that the name

Opposite page: BAC's primary submission to ASR.381 was the BAC 10-11 that was a hybrid of BAC One-Eleven and VC10. One novel feature of the BAC 10-11 was the nose-mounted telescopic MAD boom. A tail-boom, when combined with the 10-11's short undercarriage legs, would have affected the aircraft's rotation on take-off.

BAC diagram showing the structural derivation of the Ten-Eleven. The type was much more complex than a mere mating of VC10 wings to a BAC One-Eleven fuselage. The new wing pods and centre section housed fuel, undercarriage and sensors. *via RAF Museum*

Left: Maritime Comet? Slessor? Cormorant? What should the HS.801 be called? Nimrod was without doubt the finest choice for the type, not only the Biblical mighty hunter but the name of Ernest Shackleton's ship on his 1908 South Pole expedition. This fine portrait of the first production MR.1 shows the main changes from the Comet 4. *Avro Heritage*

Maritime Comet was being used increasingly for the Shackleton MR.2 replacement. He considered the expression 'cumbersome and confusing' and suggested the name *Plymouth* be adopted. Since at that time the formal contracts on the *Maritime Comet* had yet to be signed, it was highly unusual to apply a service name to an aircraft so early in its procurement. This came to the attention of a Mr B A Rawet at the Ministry who suggested that to prevent the name *Maritime Comet* becoming the service name by default, '...it is perhaps opportune to select a suitable name.' Rawet continued, outlining the 'rules' for naming RAF aircraft with reconnaissance aircraft named after 'British historical names, and flying boats (which have largely performed MR duties) after coastal towns and seaports of the British Commonwealth.' Mr Rawet promptly ignores all that and suggests *Osprey* and *Cormorant*! This loose minute from B A Rawet initiated one of the most fascinating and heated exchanges of paperwork in the postwar maritime patrol story. For a start one recipient of the minute, DD(ORs), has scribbled a note on it with an asterisk next to *Cormorant* and the note reads 'I would have thought not. Any brain waves?'

First out of the traps was Air Vice-Marshal Emson, ACAS(OR) who agreed that the use of the name *Comet* might be confusing and so advised that the aircraft should be referred to by its manufacturer's designation, HS.801. Emson, obviously unable to resist temptation, put in his suggestions and has a dig at Rawet at the same time and, possibly mindful of the recent events around TSR.2, P.1154 and the Royal Navy, stated 'I am not sure whether you had tongue in cheek when proposing *Plymouth*, which suggests to me either a gin and/or Naval connection.' Emson continued and suggested that an 'outstanding personality' from the RAF might be apt and suggests '*Slessor*, a former C-in-C of Coastal Command or *Trenchard*, though the latter perhaps ought to be reserved for a supersonic aircraft.' Here, Emson alludes to 'Boom' Trenchard's nickname from his time at the Central Flying School.

By the end of April Air Marshal Sir John Davis, the Air Member for Supply and Organisation and thus the officer responsible for procurement, had become involved and his Permanent Secretary, Mr A G Rucker, delivered a dismissal of *Osprey* by pointing out that it was a name given to a Hawker Hart variant in service with the Fleet Air Arm' whose service was 'not understood to have been so meritorious as to call for revival of the name, quite apart from the fact that the real osprey habitually folds its wings and dives headlong into the water...' and also states that the cormorant is also a diving bird. Rucker then states that the AMSO '...favours the name *Albatross*, which is a graceful bird distinguished by its ability to undertake safely long flights over the sea: it is of course traditionally unlucky to shoot an *Albatross*.' AMSO also suggested *Drake* because it combined the name of a bird with the 'Elizabethan sea-dog.'

ACAS (Ops) Air Vice-Marshal Denis Smallwood questioned the need for a new name as *Comet* was fine, but favoured *Albatross* rather than *Drake* as this was the name of '...the RN barracks in Devonport.' The DCAS Air Marshal Sir Christopher Hartley favoured the names of gods such as were already in use for *Vulcan* and *Thor* and so suggested *Venus*, which has maritime associations or *Apollo*, did admit to liking *Albatross*, but pointed out that this was the name for the Grumman amphibian in service with the Royal Norwegian Air Force.

By the 4 of May 1965 Wing Commander G G Beaugeard, OR23(RAF) had taken on the collection of names (perhaps he was also running a book on this?) and had sixteen names, of which he favoured *Albatross*, with the last on the list being *Dolphin*. Three days later Beaugeard had a further seven names, courtesy of a colleague, including *Cabot*, for the explorer in a continuation of the Shackleton theme, and *Churchill*. By the 24th March 1966 the intermittent debate was still under way and Wing Commander Loveland pointed out that the British Army had used *Churchill* for a tank, but *Winston* was a possibility as was *Woodford* as that was Churchill's parliamentary constituency and where the aircraft were being built. The next day and after a year of suggestions and no real progress on the name, Air Marshal Paul Holder Air Officer Commanding-in-Chief, Coastal Command, became involved with naming the HS.801, which by this time was being referred to as the *Coastal Comet*. Holder felt that the time was right to give the HS.801 a name and that name should '...have a direct connection with the role of the aircraft or be traditional.' but that all the more 'magnetic names' had been used for RN ships or shore establishments. Holder observed that there was one '...namely NIMROD, meaning any great hunter, which has not been used.' and that this was the name that Coastal Command preferred with *Trenchard* as an alternative, but '...NIMROD would be a more suitable choice because it reflects the role of the aircraft, is short, easy to spell, unlikely to be confused and not at present in use for any other ship, aircraft or weapons system.'

A couple of weeks later the ACAS(OR) Air Vice-Marshal W D Hodgkinson put his thoughts on paper. Hodgkinson was set against any mention of Comet as it would '…give the impression that this aircraft would seem to be old before it came into service than if it were given a new name now.' Hodgkinson also pointed out that while he liked the name *Nimrod*, it had been '…used for the carrier version of the Hawker Fury before the war (albeit flown by the RAF).' In fact Air Vice-Marshal A A Case, Chief of Staff HQ Coastal Command was set against any mention of *Comet* because 'Unfortunately, the press and others insist on referring to the "*Maritime Comet*" rather than the "HS.801", and the misguided are under the impression that we are converting aircraft handed down from BEA and Transport Command.' D J Pearson, Permanent Secretary to DCAS Air Marshal Hartley, described how the HS.801 was Coastal Command's first new aircraft in years and the service was '…very proud of it…' and so wanted no association to be made with the Comet. Pearson also advised that some NATO nations had described the HS.801 as '…merely a Comet which was very good in its time but which is hardly suitable for the 1970s.'

By 2nd June, Wing Commander Loveland suggested that the name *Scott* should be adopted for the HS.801 because *Scott* was a famous explorer like Shackleton. The reply from Wing Commander G A Chessworth was non-committal and advised that the CAS Air Chief Marshal Sir Charles Elworthy had dictated that the HS.801 was not to be named until the first example had flown. The rationale for this was to avoid the impression that the aircraft's service entry has been delayed. However, by 5th September 1966 Pearson advised that the DCAS had written to AOC-in-C Coastal Command 'asking for his recommendations' and that the '…selection of a name for the HS.801 should go ahead without delay.' Pearson advised ACAS(OR) that he should inform the MoA and Hawker Siddeley '…on the choice of name prior to the preparation of a draft paper for the Air Force Board.'

By 22nd September Mr Rawet, DS9, was writing to all concerned that the HS.801 would be named and that a shortlist be prepared before the MoA and HSA were consulted. The Shortlist included *Trenchard* (with a comment about it being reserved for a supersonic aircraft), *Albatross* and five that had been 'con-sidered but rejected earlier' and of course, Nimrod. Even at this late stage more names were appearing including '*Avenger, Bowhill, Tempest* and *Tornado*', suggested by AOC-in-C Coastal Command in case *Nimrod* was rejected. Interestingly the copy in the National Archives at Kew has 'My choice is *SCOTT*' scribbled on it by DDOR8. Five days later another two suggestions, *Scorpion* and *Explorer*, arrived from the office of Air Commodore R G Knott DOR2(RAF). The AMSO had also changed his mind, suggesting *Calshot* as it had been a flying boat base with a long association with maritime patrol.

On the 28th May 1967 HS.801 prototype XV148 took to the air, flown by Group Captain John Cunningham, who had been the pilot of the DH Comet's first flight. Twelve days later Mr F Anderson, of the MoA Air B1 Mods, issued a short statement 'It has been decided that the name *Nimrod* shall be given to the HS 801 Maritime Reconnaissance Aircraft which is hereby designated *NIMROD MR Mk.1*.' There had been forty names in the hat, and oddly enough AOC-in-C Coastal Command Air Marshal Paul Holder's favourite had been selected. But did Wing Commander G G Beaugeard have his money on Nimrod?

Colour – The Customer Isn't Always Right

Concurrent with the name debate, a similar cast of senior officers was discussing the colour scheme for the HS.801. On 17th September 1965 DOR2(RAF) Air Commodore R G Knott sent a letter to ACAS(OR), VCAS, ACAS (Ops) and Director of Operations (Maritime, Navigation and ATC). Air Commodore Knott was inviting suggestions on the colour scheme for the HS.801 and was in a spot of bother, caught between Coastal Command and the OR and Ops staffs at the MoD.

Back in 1941 Professor Sir Thomas Merton of the Coastal Command Research Branch '…had established that, at least in the North Atlantic, maritime aircraft under and side surfaces should be painted white…and the tops of the wings and fuselage should be painted dark slate or dark sea grey.' This scheme persisted until 1956 when the scheme was changed to dark grey with a white top '…because it proved impossible to keep the aircraft clean'. Knott's problem was that Coastal Command wanted to return to Merton's scheme while the MoD wanted to maintain the current scheme. Knott was concerned that operational effectiveness

This Vickers Warwick GR.II is sporting Prof. Merton's recommended dark grey over white colour scheme. When this proved difficult to keep clean, it was changed to the reverse. This Warwick is spick and span because it is being photographed for 'Research & Development Technical Publications'. *Author's collection*

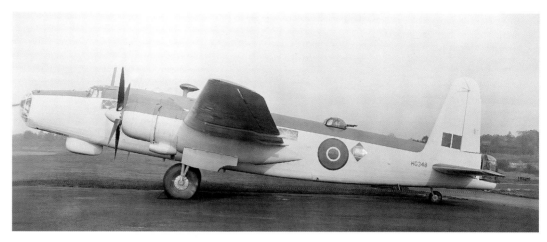

Not looking so spick and span is this Boeing Fortress II showing signs of being worked hard over the Atlantic and Bay of Biscay. Perhaps Air Marshal Holder had reached such heights in the Air Staff that he rarely saw aircraft except on inspections?

would be compromised by 'extraneous factors' such as cleaning. Knott observed that the pannier with limited ground clearance on the HS.801 would pick up dirt and that the salt spray would provide a sticky surface for 'day-to-day grime' to adhere to. The white undersides would show any contrasting staining, which might prove more durable after exposure to flight conditions. A further complication was the need to fit antennae and dielectric panels

on the underside, which in the 1960s, were invariably black '…so providing a more noticeably patchy effect against a stark white background.'

Air Commodore Knott mentioned that the US Navy had also considered these points and '…decided on Gull Grey and White as the new colour scheme for their Orion aircraft.' Apparently the best 'aesthetic impression' was desired and that the colour scheme for the Dominie had been revised for this reason. Knott concludes with 'In view of the conflicting views between MoD staffs and Coastal Command, perhaps you would consider discussing this matter with the VCAS.'

Coastal Command preferred an all-over white finish similar to the V-Force and their rationale was that, like the Merton scheme, the white undersides would be less conspicuous against a bright sky, while white topsides would reflect heat. Unfortunately this scheme was undermined by the black aerials and any accumulated grime and so by the 1st October 1965 the VCAS, Air Marshal Sir Brian Burnett and took the view that the DCAS was right on the grey undersides and gave his approval for the DCAS to'…progress this decision accordingly, after speaking with AOC-in-C Coastal Command.' However, this met with dissent at HQ Coastal Command at Northwood and Air Marshal Holder voiced his displeasure in a letter to the DCAS, stating that 'We in the Command all feel strongly that that our aircraft should be white all over, as are Bomber Command's Victors and Vulcans.' Holder continued, advising that since the underside was indeed lower the dirt would be more difficult to see and that approved washing plant would be available at the stations. This tends to suggest that Holder was more concerned with the look of the aircraft when lined up on the apron than when airborne. AOC-in-C Coastal Command concluded with 'I hope you will agree to an all-white finish. Such a decision will, I know, make the Command happy. After all we are the customers!'

Air Marshal Hartley, the DCAS, on 8th November 1965 drafted a letter in reply to Holder's plea for all-white HS.801s. Firstly Hartley made the point about the dirt being visible in the air 'Your point on the dirt being unlikely to be noticed is surely only valid in the ground case and it would no doubt provide an unwelcome contrast where you least want it, in the air.' Next Hartley describes the effect of the dark dielectric panels on the underside, and possibly pandering to Holder's perceived wish for parade ground turn-out, describes how these would be '…providing an untidy and operationally untidy patchwork effect against a stark bright background.' Concluding, Air Marshal Hartley advises that the 'The unanimous advice from the MoD Staffs is that the HS.801 should be painted light grey with a white fuselage top and both VCAS and I agree that this will be the most aesthetic solution both operationally and aesthetically.'

Bashing Johnnie Foreigner

One role that the Shackleton replacement was meant to fill as part of ASR.381 was 'internal security' (IS), which in the early 1960s meant operating against insurgents/freedom fighters in various parts of the shrinking British Empire. The Shackleton was admirably equipped for this, with a pair of 20mm Hispano V cannon in the nose and a bomb bay that could carry a variety of bombs, from 25 lb (11.3kg) to 1,000 lb (454kg). Shackletons had operated in support of British forces on a number of occasions,

Opposite, bottom: Resplendent in the DCAS's preferred white over grey, a Nimrod MR.1 makes a turn over a *Moskva*-class helicopter cruiser. *Moskva* caused a stir when it entered service in 1967. Assigned to the Black Sea Fleet, they were prime targets for NATO maritime patrol and Elint aircraft whenever they ventured into the Mediterranean and Atlantic. *Author's collection*

Left: For the 'Internal Security' role a bombing capability was written into the initial OR.381. This drawing shows an early HS.800 carrying 21 1,000 lb (454kg) bombs. Interestingly the lower drawing might allude to another role as it shows the bays occupied by four '2,000 lb bombs' whose shape matches the Red Beard tactical nuclear weapon. *Avro Heritage*

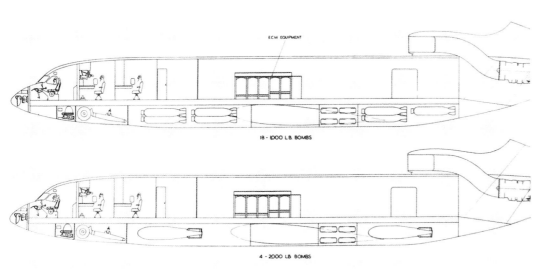

ECM EQUIPMENT

18 - 1000 LB. BOMBS

4 - 2000 LB BOMBS

HS.801 prototype XV148 fitted with TV Martel ASM under the port wing. The store was specified for the OR.381 type, but HSA soon discovered that the Martel affected the aircraft's stability, particularly after one had been launched. The weapon was soon dropped from the HS.801 weapons suite and the bomb option was re-examined. *via Avro Heritage*

Nimrod XV231 cleans up its landing gear as it climbs out on another long oceanic sortie. The Nimrod, despite carrying much of the same equipment as the Shackleton MR.2 and MR.3, revolutionised Coastal Command's operations. Within two years of entering service, Nimrod crews were winning the Fincastle Trophy. *Author's collection*

notably in Aden where the Shackletons of No.37 Sqn dropped over three thousand bombs in the Radfan.

This of course required the guns and a bomb-aimer's station and although the early drawings of the Maritime Comet show the glazing for a bomb-aimer's position, these soon disappeared as the withdrawal from Empire continued. In early May 1967, the ACAS(Ops) Air Vice-Marshal Lewis M Hodges wrote to the VCAS Air Marshal Sir Brian Burnett on the matter of the medium-level bomb sight for the HS.801. Hodges took the view that '...it was most unlikely that this highly sophisticated maritime aircraft would ever be used in the IS role.' To this end Hodges had discussed the matter with the ACAS(OR) and ACAS(Pol) who agreed that

'...the requirement for the bomb sight and any other equipment which would only be used for the IS task, be cancelled.' By 5th May 1967, the requirement for IS equipment on the HS.801 had been cancelled.

And so the HS.801 became Nimrod, the Mighty Hunter of the Book of Genesis, and on 2nd October 1969, sporting a smart white over grey colour scheme, XV230 moved to RAF St Mawgan and 236 Operational Conversion Unit (OCU). The next summer 201 Squadron at RAF Kinloss received its first Nimrod and in a milestone mission on 30th August 1970 a Nimrod MR.1 detected and tracked its first Soviet submarine. For the next forty years Nimrods patrolled the seas around the United Kingdom. Then they were gone.

10 Replacing the Maritime Comet

'The Working Party has weighted its efforts more on the ASW role because it believes that the submarine poses a greater threat than the surface ship, and because the current Nimrod sensor fit appears weaker for ASW than for anti-shipping purposes.'
Annex B, Nimrod Refit Working Party Paper No.68, MinTech, August 1969

'Would you like to see our Comets? They're on the way to the canteen.' I traipsed after George into the main hangar. Within was a hive of activity and, arranged down the length of the hangar, were four Comets, with engineers and technicians going about their business. This was not way back in the 1950s or 60s, this was 2010 and the hangar was BAE Systems' Woodford factory where the Nimrod MRA.4 was being produced. Stripped of the panniers and fairings that maketh Nimrod, the distinctive shape of the Comet was revealed, but within a year these airframes would be reduced to scrap. Author's visit to BAe Systems, Woodford, March 2010

With the Maritime Comet on track for a 1970 service debut as Nimrod, the Air Staff looked back at ASR.357 and longed for its sensors and weapons capabilities. Could the Nimrod be refitted at a later stage with equipment to approach ASR.357 standard? The answer was yes and the result would be the MR.2, but it was still a Forties airframe with Sixties kit. The late Seventies and into the Eighties saw a number of new studies for a New Generation Maritime Reconnaissance aircraft, but these fell by the wayside. The Nineties were typified by the repetition of the early 1960s contests, albeit with fewer runners. In the end, the search for a Nimrod replacement was another Nimrod – Nineties kit in a Forties airframe – and that would be its undoing.

Hawker Siddeley's Nimrod to meet ASR.381, certainly until the cancellation of ASR.357 in August 1969, was always viewed as an interim type, built to fill a gap that was operational and financial. Finance was a great concern of the Wilson Government, with the UK economy in crisis and a need to make savings. The Chancellor of the Exchequer, Jim Callaghan sought

Nimrods dominated the Fincastle Competition for many years after their introduction. Seen here on the ramp at RNZAF Whenuapai for the 1978 event is a pair of 404 Sqn RCAF Argus Mk.2, two Nimrod MR.1s from 201 Sqn RAF plus P-3C Orions of 5 Sqn RNZAF and 11 Sqn RAAF. The trophy was won by the RAAF and its P-3C Orions.
Terry Panopalis Collection

to balance the books and Defence Secretary Denis Healey was charged with reducing the defence budget. High cost military projects such as the 'Big Three' aircraft projects were cancelled in the budget speech in April 1965, with the CV(A).01 aircraft carrier in 1966. Nimrod was the sole survivor of the Healey/Callaghan cull that had cancelled the 'Big Three', the most advanced aerospace projects outside the United States. Had Arnold Hall not walked in on the Air Staff with his Maritime

Comet proposal and the Ministry of Aviation had held sway, then a Maritime VC 10 would have joined the 'Big Three' – TSR.2, P.1154 and HS.681 – on the British aviation scrapheap. What would become Nimrod escaped the attention of Healey, not merely because its 'interimness' had kept it simple and cheap; it had been perfectly timed.

By late 1969 the Nimrod MR.1 was entering service with the Operational Training Unit at RAF St Mawgan and was soon bringing a new

Nimrod MR.1 XV243 monitoring a *Hotel*-class SSBN. The MR.1's sensor suite was criticised for its lack of capability against the Soviet nuclear-powered boats. This prompted the Air Staff to issue Specification MR.286 and the development of the MR.2. Nimrod XV243 was upgraded to MR.2 in 1984 and was later converted to an MRA.4 in 2004. *Author's collection*

sense of *ésprit de corps* to Coastal Command, the Growler was gone and the Nimrod took the maritime reconnaissance world by storm. An example of this was the interservice competitions that build expertise and teamwork between air forces. The Fincastle Competition saw the ASW squadrons of the RAF, RAAF, RNZAF and RCAF compete in a series of challenges in the anti-submarine, anti-shipping and surveillance roles. As of 1970 Coastal Command had won the Fincastle Trophy only once since its inception in 1960, but swept the board from 1973 until 1977 and by 2012, the RAF had won the trophy seventeen times.

As ever, as soon as a type entered service, its replacement was being planned by the Operational Requirements Branch of the Air Staff. Having essentially fitted Shackleton kit in a jet, the aircraft's performance allowed Nimrod crews to conduct operations that the Shackleton crews could only dream of. Some Shackleton crews considered their MR.3s superior in submarine hunting, but the Nimrod with its high-speed transit at high altitude made for much better crew comfort and consequently higher efficiency. This could only be improved by providing state of the art systems and possibly even bring the Nimrod MR.1's systems close to the level of ASR.357. This was well under way by 1969, but the initial movement towards refitting the Nimrod dated back to 1965 with discussions within the Air Staff and Royal Aircraft Establishment (RAE). To progress a potential refit, a Working Party was set up at the RAE with the aim of examining ASR.357 and potential refitting of the Nimrod to a capability approaching ASR.357. This AST.357 Cost/Effectiveness Working Party reported its findings in a First Interim Report in early 1966 and advised '…that studies of AST.357 were unrealistic with the effort available…' and that they would '…advise on the prospective advantages and cost of equipment changes proposed for a possible refit of the HS.801 aircraft.' The report's writers explain that 'Cost/Effectiveness' should be read as 'cost *and* effectiveness' because the differences in sensors for the Nimrod (essentially of the Shackleton era) and AST.357 were such that direct 'like with like' comparisons could not be made.

In this field the Working Party would '…evaluate the improvement which such changes might make to the operational effectiveness of the aircraft and its systems…' and indicated areas for particular examination. These were the detection and tracking of submarines, classification and localisation of targets and finally

'The attack of ships and of submerged and surfaced submarines.' The Working Party was to compare the current systems to be fitted in what would become the Nimrod MR.1 with systems then in development that could be used in any future refit of the aircraft.

As ever the first port of call was to assess the threat and so the Defence Intelligence Staff was consulted for their best estimates of the Soviet submarine fleet and capabilities 'of the different types of submarine likely to be encountered during the life of the Nimrod (assumed to be 1970-1990).' They noted that the current sensor fit of Nimrod '…appears weaker for ASW than for anti-shipping purposes.' and that the sensor fit for detecting diesel/electric submarines differed from that for nuclear types. Having assessed the needs for a future upgrade, the Working Party drew up Specification MR.286 for an upgraded Nimrod with new radar to ASR.846, navigation system to ASR.872, acoustic processor to ASR.875 with sonobuoys to ASR.879, new ESM to ASR.833 and MAD to ASR.876. The Specification for the upgraded Nimrod was issued on 6th May 1975 as MR.286 and the result was the Nimrod MR.2.

Sensors

For the refit a suite of three main sensors was drawn up: radar, ECM/ESM and sonar, all of which were available on the MR.1, but with the emphasis on anti-submarine operations rather than anti-surface ship. Sonar, the 'wet kit', was seen as the primary sensor whereas radar, the 'dry kit', would be secondary in hunting nuclear submarines. The Royal Aircraft Establishment produced a number of reports on the Nimrod refit and the Radio Department became particularly involved in the selection of sensors and processors. The sonar system was to use the usual suite of sonobuoys including Jezebel, Miniature Jezebel, DIFAR and *Barra*, with Ranger and CAMBS also available, as were the usual bathythermal buoys.

The radar was to meet requirement ASR.846 that called for a new radar to detect snorting submarines and fast patrol boats, as well as the usual surface ships. Experience had shown that the full capability of the X-band ASV.21D could not be utilised due to the '…complexity of the task facing the operator.' The project to replace the ASV.21D radar had been under way at the Royal Signals and Radar Establishment at Malvern who were working with EMI on project P1149 to develop a new maritime surveillance radar. The result was called Searchwater,

177

an X-band set that used a narrow beam and short pulse length to minimise the clutter from sea returns and allow classification of targets, in what was described as 'searchlight mode'. Searchwater could detect a snorkel at 28nm (52km) and a fast patrol boat at 60nm (11km) and the basic Searchwater formed the basis of a number of radar sets in a variety of roles on a range of platforms including the Sea King AEW.2.

Other sensors included Autolycus (although this was deleted from the final upgrade), the MAD system and the passive ECM suite to meet ASR.833. The Air Staff had a choice in the electronic support measures (ESM) systems, with MEL proposing an arrangement based on the Royal Navy's Abbeyhill intercept receiver, as used on warships. The other new system was a suite under development by GEC and AEI and differed from MEL's by relying heavily on computer power, but fully integrated with the new radar to meet ASR.846, whereas ASV.21 would need antennae fitted to its scanner. The third option was a modified ARAR/ARAX, a French system, as fitted to the existing Nimrod MR.1 and Breguet Atlantic.

The Air Staff and RAE preferred the GEC proposal as '…MEL is an adaptation of a shipborne system, whilst the GEC is designed entirely for the Nimrod, The GEC system is therefore likely to provide better performance, and in the opinion of the RAE, does so.' Further to this, the RAE were happy that the GEC system would allow the existing ARAX/ARAR system to be retained and being computerised and integrated with the radar, provided more flexibility than MEL's.

Having set out the basic sensor suite for an anti-submarine aircraft, The Working party turned its attention to more advanced sensors such as ORADS, the laser radar system described in Chapter Two. The finding, tracking and ultimate destruction of an enemy submarine is, as noted in the Introduction, a problem that involves the integration of a number of inputs from different sensors, all providing data to produce a solution. This may appear to be a mathematical problem and essentially that is what it is, although the final decisions are taken by the crew, the information needs to be collated and displayed quickly and clearly. That is where the electronics revolution of the 1960s met the age-old problem of locating something in the ocean, with 'in' now being the operative word. Sonar processing by computers would allow the signals from sonobuoys to be analysed quickly, picking out the sound of the machines from the sounds of the sea. Similarly, radar signal processing allowed new radars to detect and classify targets amongst the clutter of the surface returns or the electronic warfare suite to identify or locate a transmitter. Any aircraft that was intended to see service in the 1980s would benefit from the next, digital, revolution but as with any revolution, it was easy to be in the wrong camp – choice of computer was fundamental to the Nimrod upgrade programme.

Processing Power

The Elliott 920B computer in the Nimrod MR.1 was operating at its limit and so for the forthcoming Nimrod refit more processing power was required. American airborne computing doctrine was to use a '…single large central computer on the grounds of greater cost effectiveness and efficiency…' in an aircraft. In the UK, '…current thinking in this country leans towards the multicomputer configuration…' with each computer tailored to specific tasks. The British view was that the system load would be spread over several processors and have an element of redundancy, avoiding aborting a sortie if one computer failed. This concept was intended for ASR.357, with the sensor processing spread across the various sensor operators, each an expert in their particular field, with their analyses sent to the Tactical Navigator, also likely to be the captain of the aircraft. The Working Party identified two types of computer: the 'management' and the 'sensor processor', with the management computer '…concerned with routine clerical and housekeeping matters associated with displays and interaction with the crew.'

The Nimrod MR.1's working environment was a vast improvement on the Shackleton. Shown here are the workstations of the Routine Navigator on the left and Tactical Navigator on the right. Note the 24in (61cm) diameter tactical display.
Terry Panopalis Collection

The Sensor Processor's main task was '...to transform its raw output into a form which is meaningful to the crew or management computer.' with each '...custom built to meet its own individual requirement.' As ever, the real reason the British opted for the separate management computer and sensor processor was cost, as the sensor data processor could tolerate restrictions that would be unacceptable in a general purpose computer, such as processing power and particularly 'working storage', now called memory. In the mid-Sixties this accounted for '...significant proportion in the price of a present day computer...'. Candidate British-made computers were the Elliott 920M, 920C or 102C, Ferranti Argus 400 or 500, the Ferranti FM1600 or FM1200 and the Decca Omnitrac IIB. Ultimately the Elliott 920ATC computer was selected for the upgrade and two were used for the acoustic processing, one for the ESM and all interacted with a fourth used for the Central Tactical System (CTS). A single Ferranti FM1600 was used for the Searchwater radar, again feeding data to the CTS.

What finally emerged was the Nimrod MR.2, with new systems and sensors, including the AQS-901 acoustic processor bringing the Nimrod close to the capability that had been laid out in ASR.357 back in 1963. The first MR.2, XV147, took to the air in April 1977, with XV236 becoming the first MR.2 to enter service on 23rd August 1979. In the MR.2, No.18 Group, Strike Command, the inheritors of Coastal Command's mantle, finally had what they had outlined in 1963. The MR.2 upgrade

programme lasted until late 1984, but in the meantime, there had been a flurry of activity that produced a number of enhancements and also saw the return, in a sense, of the Internal Security role.

The Falklands Conflict of May and June 1982 prompted a number of modifications to the Nimrod MR.2s, most visible being the refuelling probe that made the aircraft the Nimrod MR.2P. The probes, intended as a temporary measure, came from recently retired Avro Vulcans and Woodford fitted out an initial batch of 16 Nimrods with probes and plumbing. The probes affected stability, so in addition to a ventral fin under the rear fuselage, small finlets were added above and below the tailplanes. Ultimately all MR.2s and the AEW.3 Nimrods were fitted with probes. The other highly visible addition was the restoration of the wing pylons, but this time carrying a pair of AIM-9L Sidewinder IR-guided AAMs. Some wag suggested that the Nimrod MR.2P was the biggest fighter in the world, but the AAMs were more for self-defence, something that had also been included in ASR.400 for the AEW Nimrod.

The Falklands Conflict also saw the return of the free-fall bomb capability that had been deleted from the original ASR.381 along with the other 'Internal Security' equipment. Interestingly, in *Dam Busters* fashion, the bombsight was a rudimentary affair developed by the A&AEE at Boscombe Down. Apparently it produced good results in the right hands. Another 'crowd-pleaser' for the Nimrod MR.2 crews was the adoption of the McDonnell Douglas

The Mighty Hunter's Achilles heel was the air-to-air refuelling capability that the Nimrods acquired from 1982. When combined with original heat exchangers and cooling pack for systems, the system failed over Afghanistan in 2006 with disastrous results. Nimrod XV238 is seen here in contact with Victor XH672.
via Ron Henry

Shown here in red, the external clues to the Nimrod MR.2P included new one-piece radome for the Searchwater radar and Yellow Gate ESM in the wingtip pods. Most of the upgrade affected internal kit; new computers and acoustic suite brought the Nimrod into the 1980s and much-improved the Mighty Hunter. The Falklands Conflict added the refuelling probe and associated ventral and tailplane fins required to maintain stability, underwing AIM-9L Sidewinder AAMs, 1,000 lb bombs and AGM-84 Harpoon ASM.

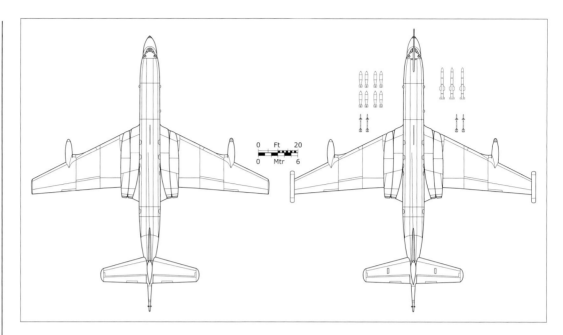

AGM-84 Harpoon anti-ship missile that gave the Nimrod over-the-horizon anti-ship capability. With up to three Harpoons carried in the weapons bay, this was a vast improvement on the earlier 'disappointing' Martel whose tenure with the Nimrod had been very short. In the ASW role, the Stingray torpedo was cleared for use by Nimrods on patrols in the South Atlantic. The Falklands Conflict precipitated many of the improvements on the Air Staff's wishlist, improving capability against airborne, surface and submarine targets. Unfortunately one of these enhancements, the in-flight refuelling probe, quite possibly sowed the seeds for the ultimate demise of the entire Nimrod fleet.

The MR.2 fleet was soon sporting cylindrical pods in their wingtips that carried Loral Yellow Gate ESM systems, replacing the ARAX/ARAR equipment in the pod on the fin tip. They also appeared on the Nimrod R.1s used for elint (or in the official parlance of the 1980s 'radio and radar calibration') but what the pods on the R.1 carried is still open to discussion. The Nimrod MR.2, what could be described realistically as the ASR.357 aircraft the Air Staff longed for, entered service in the early 1980s and with constant upgrades to systems, communications and weapons remained in service until 2011. Its role changed and Nimrods took on tasks such as sigint and comint missions over Afghanistan and Iraq. Specialist equipment fitted under Project *Broadsword* for the wars of the Noughties included the Wescam MX-15 electro-optical turret that allowed Nimrods to send real-time imagery to commanders on the ground. It was on such a mission that the last chapter of the Nimrod story opened.

The Mighty Fallen

In my view, XV230 was lost because of a systemic breach of the Military Covenant brought about by significant failures on the part of all those involved. This must not be allowed to happen again.' Charles Haddon-Cave QC, 28th October 2009

On 6th September 2006 twelve RAF personnel, a British Army soldier and a Royal Marines Commando died when Nimrod MR.2P XV230 on a surveillance mission in support of British ground forces suffered a catastrophic fire and explosion following in-flight refuelling over Afghanistan. The Coroner's report, released in May 2008, was damning and essentially stated that the Nimrod was 'not airworthy' and never had been since the refuelling probes had been fitted. The Coroner called for the entire Nimrod fleet to be grounded forthwith.

The writing was on the wall for the Nimrod when leading barrister Charles Haddon–Cave QC produced an independent review that was highly critical of all involved in the Nimrod programme and subsequent legal proceedings revealed a catalogue of failures in the oversight of the various Nimrod modification programmes. The *Haddon-Cave Review* identified three items in particular: Firstly, a cross-feed duct installed in the original MR.1 build and used to feed compressed air from the Marton auxiliary power unit (APU) to the Spey engines for starting or to restart engines that had been shut down on patrol. This had been identified as a 'major potential source of ignition'. Secondly, a Supplementary (air) Conditioning Pack 'which increased potential for ignition' had been added during the MR.2 upgrade. The last item on Haddon-Cave's list was the post-Falklands installation of air-to-air refuelling equipment 'which increased the risk of an uncontained escape of fuel'.

Haddon-Cave was particularly scathing of the Nimrod's safety case. A safety case is 'a structured argument, supported by a body of evidence that provides a compelling, comprehensible and valid case that a system is safe for a given application in a given environment'. This sets out how and why an installation or piece of equipment is safe to use or that the risk is as low as reasonably possible. Safety cases came into widespread use in the UK after the Piper Alpha disaster in July 1988 but they originate in the aftermath of the fire at the Windscale nuclear plant in 1957. Mr Haddon-Cave QC also advised that the '…generic problems associated with aged and "legacy" aircraft are addressed' by the RAF and MoD who had, in his view, failed to fulfil their part of the Military Covenant. This is an informal agreement whereby it is understood that the needs of British service personnel will be properly addressed in return for serving in the Armed Forces. Since 2001 the Military Covenant has acquired a pseudo-legal status in the eyes of public, military and politicians alike, despite its having no basis in law. The Independent Review concluded with Mr Haddon-Cave QC stating 'In my view, XV230 was lost because of a systemic breach of the Military Covenant brought about by significant failures on the part of all those involved. This must not be allowed to happen again.' In March 2009 the MoD admitted that XV230, the first Nimrod to enter RAF service, was 'not airworthy' on the day it foundered. On 9th March 2009 all Nimrods that had not received an urgent safety upgrade were grounded.

Clean Sheet or Conversion?

Thirty years before the tragedy over Afghanistan the process of replacing the Nimrod had begun. The engineers involved at that time would be long retired before the Nimrod replacement flew, some would be surprised by its identity but the old hands merely shook their heads in disbelief. There were many attempts to replace the Nimrod, for a variety of reasons, but airframe age never seemed to be the driving force, particularly when the ultimate replacement for a Nimrod was another Nimrod.

Hawker Siddeley at Woodford were mindful of two aspects of the maritime patrol aircraft market – new rules and old aircraft. The new rules were the Exclusive Economic Zone (EEZ) legislation that was under discussion in the mid-1970s and due to be implemented in the early 1980s. The second was the observation that the majority of maritime patrol aircraft were based on old aircraft whose service life would be coming to an end in the 1980s. While the short-range, minimum capability market was well catered for by the likes of Embraer with maritime variants of the EMB-111A Bandeirante, HSA identified at least seven countries with a requirement for a more capable 'coastline protection' type that could replace or supplement the existing ASW aircraft in service around the world.

Woodford's sales office in 1976 began their studies by compiling a table showing the existing ASW aircraft and their time in service, noting that there were many Neptunes in service around the World, Shackletons in South Africa and Lockheed Constellations in India, all of which had more than twenty years' service under their belt. Even the purpose-built Atlantic had eleven years on operations and the Orion averaged twelve years. And then there was the Nimrod, with eight years which wasn't bad for an interim design (unless it was for the RAF, whose interim kit served for decades).

So HSA began by looking at the various aircraft, current and projected, in comparison with what they described as the 'Optimum MR' aircraft. The designation for these maritime studies was HS.829 and this covered a number of types in various configurations. The first necessity for HS.829 was an airframe to use as the basis for a maritime patrol aircraft, or if no suit-

By the mid-eighties, fuselage volume was considered the best basis for examining a type's suitability for the MR role. When cross-referenced with range, some surprising results were obtained. This graph has additional data on the Airbus A320 and Boeing 737-800 that were proposed in the early 21st Century. *Based on Avro Heritage data.*

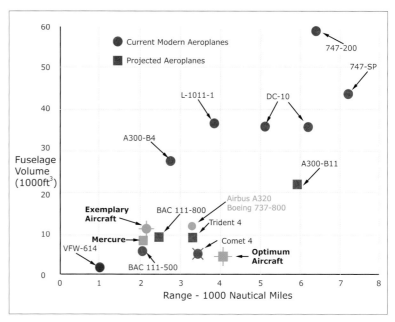

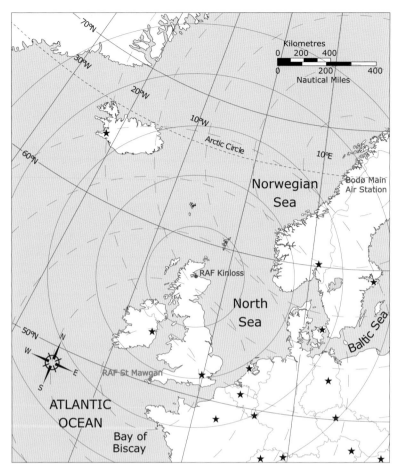

Woodford's maritime patrol studies examined the needs of anti-submarine patrols over the north east Atlantic. Using St Mawgan, Kinloss and Bodø as bases, an aircraft with a radius of action in the order of 1,000nm (1,852km) could cover the area of operations from Kinloss. This had been the baseline endurance requirement for the Air Staff's various requirements since 1960.

Sitting midway between the Exemplary and Optimum aircraft on Woodford's graph, the Trident 4 would have been a good compromise. Unfortunately two factors were against it: 1) it remained an unbuilt study and 2) the Air Staff would not compromise on the number of engines. The Dassault Mercure existed and had growth potential, something the Trident 4 probably lacked.

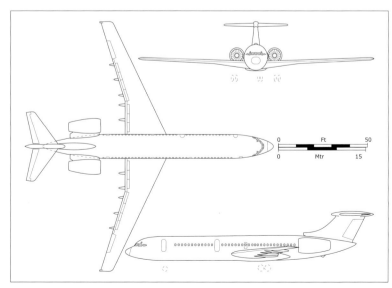

able aircraft was available, whether a clean-sheet design was necessary. What qualities were needed in an existing aircraft and how could its suitability be assessed? Back in the early 1960s when the OR.357 selection process was under way, the ideal fuselage diameter was deemed to be 12ft (3.7m) and HSA had also noted that despite having that diameter, types such as the VC10 were far too big. The Woodford maritime patrol study group opted to base their work around fuselage volume, which they related to range. In addition to the 'Optimum Aircraft' they also identified an 'Exemplary Aircraft' and this revealed a somewhat surprising result – the closest type to the optimum aircraft was the Comet 4! Of course these were all parametric studies and ignored some of the practicalities of producing a maritime patrol aircraft.

Woodford's Mercure

As for the exemplary aircraft, well, there was a further surprise. Hawker Siddeley in 1975 had no airliner in the Comet class, although the putative Trident 4 came close. This comprised a slightly stretched Trident 3B fuselage, fitted with a new wing based on A300B technology, and a pair of CFM56 (or Pratt and Whitney JT10D, both known as 'Ten Ton' engines due to their thrust in the 20,000lbf/8kN class) on the rear fuselage and a buried RB.162. Similarly BAC had no extant aircraft in that class, but again a design on the drawing board based on the BAC One-Eleven airliner, the -800, was considered. The surprise came in the form of a French airliner, the Dassault Mercure that was under development as a second generation airliner to follow the Caravelle.

Mercure holds the dubious honour of being amongst the least successful airliners ever, with ten built for the French carrier Air Inter, plus one of the prototypes brought up to production standard as the eleventh in service. The Mercure was intended to compete with Boeing's 737-200 and was powered by the same Pratt and Whitney JT8D low-bypass turbofan. Unfortunately the Mercure lacked the range and, more significantly was launched at a time when the US Dollar was devalued, making American airliners cheaper than European types and when this was combined with the economies of large-scale manufacturing, customers bought American.

Hawker Siddeley's Woodford and Chadderton design teams had a long-standing relationship with Breguet through their work with on the Atlantic back in the late Fifties and early Sixties. The Weapons Research Department at Woodford had liaised with Dassault through their work on stand-off weapons for l'Armée de l'Air, and by 1971 Breguet had been absorbed into Avions Marcel Dassault-Breguet Aviation. It wasn't only Britain that merged its aircraft companies; France also saw fewer, larger aircraft builders as the way forward. This was the high point of the Anglo-French project era, with the fruits of the collaborative programmes such as Concorde, Jaguar and Lynx approaching service entry. The Mercure would have made a suitable addition to the range of projects, and could have replaced the Nimrod and the Atlantic, as well as many of the Neptunes around the world.

While the Mercure-100 fitted the bill in fuselage size, with a diameter of 13ft (4m), its Achilles heel was lack of endurance if used in the ASW role. Firstly the engines would need to be changed, for a more efficient modern turbofan and of course, if it was aimed at the RAF, the Mercure would need four engines. The RAF's preference for four engines had two reasons. Firstly there was the standard procedure of shutting down engines as the aircraft weight came down as the mission progressed, with most four-engined MR types being able to patrol on two engines. The other reason was that a four-engined aircraft could continue its mission after an engine failure. Endurance could also be extended by fitting a wing that was tailored to the mission profile, possibly by extending the wing and increasing its aspect ratio.

Dassault by 1975 had also looked into improving the Mercure, mainly by re-engining it with the CFM-56 high bypass ratio turbofan and stretching the fuselage to produce the Mercure-200. Along with the new engines, a new wing with higher aspect ratio and greater span was to be fitted and Woodford based their maritime patrol studies on the Mercure-100 and -200. These Hawker Siddeley studies were given the designation HS.829 and at least five design studies were carried out; four based on the Mercure and one using the HS.146.

These HS.829 studies involved re-engining the Mercure, by doubling the number of engines or fitting newer technology powerplant. The engines were the CFM-56 high bypass ratio turbofans rated at 25,000lb (112.2kN) or scaled RB.415 turbofans, rated at 12,500lb (55.6kN). The CFM-56 has been a mainstay of the civil aviation propulsion market since 1978 when it was selected to re-engine

the first generation of jet liners including the Douglas DC-8 and Boeing 707. It gained a mass market in new-build airliners when it was selected for the Boeing 737-300. On the military side it was installed on the Boeing KC-135R Stratotankers and the Boeing E-3D and E-3F Sentry, operated by the RAF and l'Armée de l'Air. The RB.415, despite its RB prefix, had its roots with SNECMA, the French engine manufacturer who had developed the M45 low bypass ratio turbofan in collaboration with Bristol Siddeley Engines Ltd. The M45 was based on

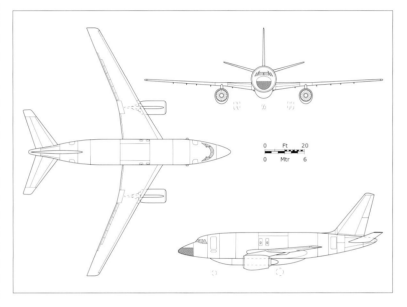

Dassault were examining an improved Mercure, the Mercure 200, with stretched fuselage, new wings and CFM.56 turbofans. Woodford took that design, modified the wing and added a ventral pannier for stores to produce the HS.829B. The undercarriage was lengthened to improve ground clearance for the larger engines.

The HS.829C2 retained the Mercure 100 wing, but was fitted with four scaled RB.415 turbofans rated at 12,500lbf (55.6kN) apiece. The ventral fairing housed an unpressurised weapons bay and the extended nose was fitted with the antenna for the Searchwater radar.

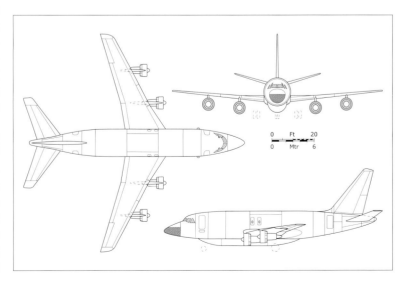

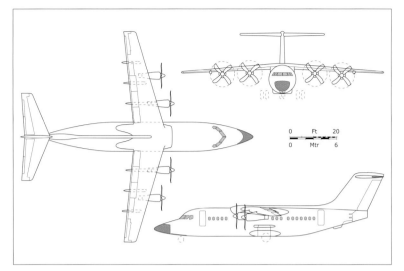

The BAe 829D was based on the BAe.146 feeder-liner airframe that was still on the drawing board in 1977. With extended nose for Searchwater radar, weapons bay and most obviously, four RB.401 turboprops, the HS.829D met little interest at the time.

the high-pressure section of the BS.116, a scaled version of the Olympus 320, as used on the BAC TSR.2. Bristol's expertise in big fan technology made the logical next step to be a joint effort to produce a high bypass ratio engine. When Rolls-Royce bought out Bristol Siddeley in 1968, the project transferred to their books as the RB.415.

The four variants based on the Mercure differed in planform and engine configuration, but all were fitted with a ventral pannier to carry 8,000 lb (3,629kg) of weapons and other ASW stores. An extended nose cone housed an ASV radar, no doubt a variant of the Searchwater series and the underfloor freight bays were converted to carry fuel tanks, further extending the type's range. The HS.829A had an AUW of

152,000 lb (68,946kg) and sported an extended wing to enhance its endurance, particularly in the high-speed/high-altitude transit phase, and was powered by four scaled RB.415s. The HS.829B had completely new wings, each fitted with a CFM-56 turbofan to produce an aircraft with an AUW of 158,000 lb (71,668kg). The HS.829C came in two flavours, depending on the engine fitted, but sharing the same wing as the Mercure-100. The C1 used a pair of CFM-56 while the C2 was fitted with four scaled RB.415s.

The HS.829D was derived from a British type that was, at the time, still on the drawing board – the HS.146. This was a small high-wing, four turbofan airliner, with a 12ft (3.7m) fuselage diameter, designed at the former de Havilland plant at Hatfield. The project became the BAe.146 upon the nationalisation of the British aviation industry in 1977 and was a fairly successful airliner. Powered by a quartet of Lycoming ALF-502 turbofans carried on pylons under the wings, the BAe.146 was marketed as the 'Whisperjet' on account of its low noise footprint. For the maritime patrol role as the HS.829D, the -200 variant with longer fuselage was used and re-engined with four RB.401 turboprops rated at 3,680shp (2,744kW) and driving 13ft (4m) diameter four-bladed propellers. At an AUW of 108,000 lb (48,988kg) the HS.829 could carry a stores load of 8,000 lb (3,629kg). As with the other HS.829 studies, the Searchwater radar was installed in an extended nose, with the stores carried in the

From the Optimised ASW type, Woodford drew up the HS.830 Optimum Maritime Reconnaissance Aircraft. Two versions were proposed: A turbofan version powered by four RB.415s with an AUW of 137,000 lb (62,412kg) and a turboprop variant with four RB.401 rated at 4,200shp (3,132kW) and an AUW of 117,500 lb (53,070kg).

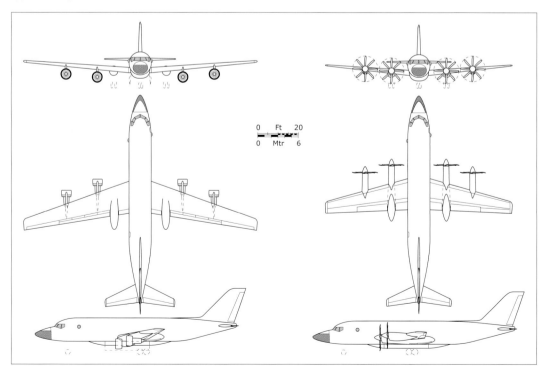

lower fuselage, but there was no MAD sting, no doubt because by the mid-1970s the detectors were less susceptible to magnetic interference from the aircraft itself or a SQUID system was installed under the wing.

Having based their HS.829 studies on what was deemed the exemplary aircraft; Hawker Siddeley now turned their attention to the optimum aircraft and for this opted to produce a clean-sheet design, the HS.830. Three variants of the HS.830 were examined, the differences being powerplant and planform, with all three sharing a fuselage with a diameter of 11ft 3in (3.4m) and a total length of 104ft 7in (31.9m), with the Searchwater radar antenna housed in an extended nose radome. The planforms were dependent on the powerplant, one with turbo-props and two with high bypass-ratio turbofans, while the fuselage has the look of the BAe.146 about it. The stores and weapons were carried in a 22ft 6in (6.9m) long bay within a pannier under the wing box.

The turboprop version had an AUW of 117,000 lb (53,061kg) and was powered by four RB.401s rated at 4,200shp (3,132kW) driving 12ft 1in (3.7m) diameter eight-bladed propellers. The wing, which spanned 93ft 11in (28.6m), had a moderate sweep of 15.5° at the leading edge and 2.5° on the trailing edge, with the engine nacelles projecting from the leading edge. The four-wheel bogey undercarriage retracted in nacelles that projected far behind each of the inboard engine nacelless. The first of the turbofan variants weighed in at 137,170 lb (62,209kg) and had slightly more sweep to the wing, 18°, with the four scaled RB.415 turbofans, rated at 10,138 lbf (45.1kN) mounted on pylons ahead of the wing leading edge. This version had a wingspan of 116ft 11in (35.6m) and a higher aspect ratio. In this case the undercarriage was housed in a pod extending aft of the trailing edge midway between the fuselage and the inboard engines. The last, and largest, variant had an AUW of 162,000 lb (73,469kg) and a moderately swept (18°) high aspect ratio (11.7) wing with a wingspan of 136ft (49.3m). Again the undercarriage retracted into trailing edge pods between the fuselage and inboard engines. The engines were RB.410-03 scaled to 0.85 and mounted in long cowlings, unlike the short cowls of the RB.415.

These paper studies showed that BAe were more than prepared to replace the Nimrod and all that was required was a requirement from the Air Staff and the MoD Procurement Executive that had been set up in 1971 to co-ordinate weapons development and finance. However,

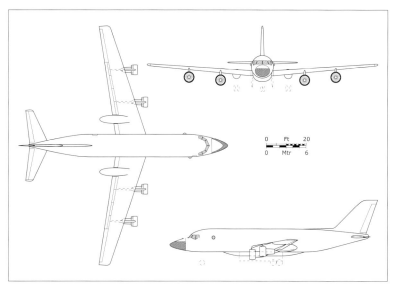

nothing was forthcoming and so BAe continued with numerous paper studies, but in the real world the Nimrod upgrade was ongoing and one other former Avro aircraft was taking up the maritime role.

At Last a Maritime Vulcan – MRR

'First, we have provided reconnaissance aircraft so that N.A.T.O. can be continuously aware of the precise position of Soviet ships at all times…' Denis Healey, Secretary of State for Defence, 4th March 1969

While Woodford examined the next generation of maritime patrol aircraft other aircraft were keeping watch over the seas around the United Kingdom and beyond, scanning swathes of the ocean from medium and high altitude. These involved two Handley Page types, Victor and Hastings, but the role also saw Avro's Vulcan employed in the maritime role, but not quite as Woodford had hoped back in 1960 with OR.350. The application of these types in the maritime radar reconnaissance (MRR) role was more a by-product of the H2S Mk.9 bombing and navigation system these types carried rather than the types' suitability in the conventional maritime role. However, this was a different aspect of maritime operations and the Victor and Vulcan's high-altitude capability made them ideal for it.

The medium-altitude patrols were conducted by the HP Hastings T.5 that had been used by Bomber Command Bombing School and after April 1968, Strike Command Bombing School and finally Radar Flight, 230 Operational Conversion Unit. The Hastings also conducted

Based on their examination of the various existing maritime patrol types plus existing and projected aircraft, Woodford had identified the characteristics of an Optimum and Exemplary aircraft. From this information they drew up their Optimised ASW type. Powered by scaled RB.410 turbofans, it had an AUW of 162,000 lb (73,482kg) and spanned 136ft (41.5m).

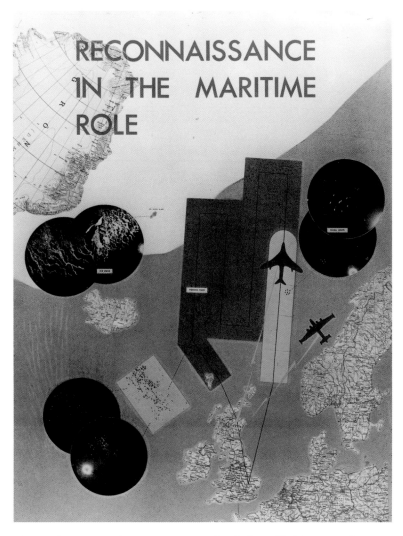

RECONNAISSANCE IN THE MARITIME ROLE

Above: This image shows the extent of coverage provided by the maritime radar reconnaissance Victors of 543 Sqn. The information gleaned from the radar sorties would be passed to base for analysis, then transmitted to the maritime patrol aircraft, in this case a Shackleton MR.3. *Author's collection*

ture' of shipping movements. Victors of 543 Sqn had this role until 1974, with the Vulcans of 27 Sqn taking over when the Victors were converted to K.2 tankers. This role was assigned to Supreme Allied Commander Atlantic (SACLANT), as part of the UK's NATO commitment to the security of maritime operation in the North Atlantic. The radar reconnaissance surveillance operations involved the Victors, then Vulcans, conducting patrols to monitor shipping movements and if necessary direct Nimrods onto targets for a close-up inspection or, in wartime, Buccaneers for an anti-shipping strike. It was this capability that Healey had in mind on commenting that he knew where the Soviet ships were, adding that since they lacked air cover, they could be dealt with swiftly.

The Victor briefly returned to the MRR during the 1982 Falklands Conflict when three MRR missions were undertaken by Victor K.2s (which retained the H2S Mk.9 radar) in the run-up to Operation *Paraquet,* the recapture of South Georgia. Known to the Royal Marines as Operation *Paraquat* after the weedkiller, this operation saw the island back under British control on 25th April 1982.

The Nimrod AEW.3 was intended to replace the Vulcan in the maritime radar reconnaissance and anti-shipping strike co-ordination roles from the mid-Eighties. One of the reasons that the MoD persisted with Nimrod AEW.3 was that the Boeing E-3 Sentry was deemed unsuitable for ASR.400 because, unlike the Nimrod AEW.3, the Sentry's radar could not produce the required 'surface picture'. This was the undoing of the Nimrod AEW.3 and once that dilemma had been resolved and the Sentry selected as the AEW platform, BAe Woodford proposed using the redundant Nimrod AEW.3

Tapestry patrols of the oilfields using the H2S Mk.9 radar and would also be deployed in the Third Cod War (see Appendix 2). On a more martial note, HP Victor SR.2s and Avro Vulcan B.2MRRs were employed on radar reconnaissance missions to provide a radar 'surface pic-

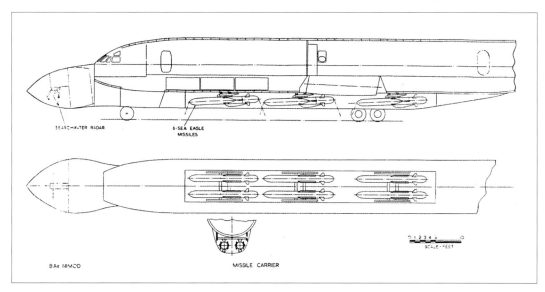

airframes in the anti-shipping role. The Marconi AEW radar was to be removed and a Searchwater antenna fitted in the nose radome, with the armament comprising six BAe Dynamics Sea Eagles in the weapons bay. However the proposal was not taken up as by the late Eighties all Nimrod MR.2s were equipped with the AGM.84 Harpoon.

BAC's Seventies Maritime Types

BAC's lack of an 'optimum-sized' aircraft, between the BAC One-Eleven and the VC10 prompted them to look elsewhere for sales of an ASW aircraft based on existing airliners. The VC10 in the mid-Seventies was long gone from production lines but the BAC One-Eleven was in production and was being presented to the RAF not only as a potential AEW type, but as a MR aircraft. While the RAF were interested in the AEW aircraft, fitted with the FASS system or the AN/APS.120 rotodome system from the Grumman E-2C Hawkeye, they were less than impressed by the maritime patrol version upon which it was based. As ever with the smaller airliner conversions, the Air Staff's main gripe was that the BAC One-Eleven only had two engines.

With the BAC Ten-Eleven proposal for OR.381, Weybridge had added engines outboard of the BAC One-Eleven's existing engines, making the BAC Ten-Eleven appear to carry a quartet of engines in a VC10-style installation, although the 'original' and 'additional' engines were on separate fuel and control systems. The BAC Ten-Eleven, and the BAC One-Eleven for OR.381, were radical redesigns and amalgamations of two types. Such hybridisation wasn't feasible in the 1970s as the VC10 was out of production, so in September 1974 BAC opted to modify a new version of the One-Eleven and add engines under the wings. However, this was not for the RAF but for the Japanese Maritime Self-Defence Force (JMSDF). BAC were promoting the One-Eleven-700, a longer, re-winged version as a replacement for not only the JMSDF's NAMC YS-11 transports but their Kawasaki P-2J Neptunes, and it is this variant that is of interest here.

BAC's One-Eleven 700 series was to be fitted with a pair of Rolls-Royce Spey Mk.606 turbofans, the so-called Quiet Spey, rated at 16,900 lbf (75.2kN) and fitted on the standard mountings on the rear fuselage. The Spey 606 was a refanned Spey with higher bypass ratio '...which reduces both fuel consumption and noise.' Compared to the -500, the -700's fuselage was lengthened by 8ft 4in (2.5m) with a new fuselage plug inserted ahead of the wing

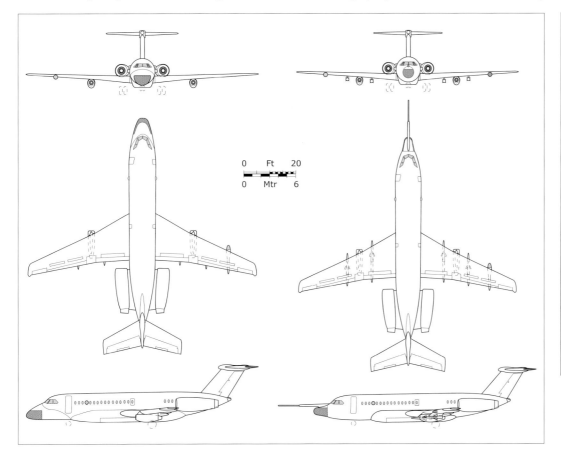

Above: In the mid-1970s BAC at Weybridge tendered a couple of BAC One-Eleven maritime patrol variants to the Japanese Self-Defence Force. These were based on the BAC One-Eleven 800 and carried a Rolls-Royce/Turbomeca Adour turbofan under each wing. Much of the equipment was from the ongoing Nimrod Upgrade that became the MR.2. On the right is a study dating from 1974 while the study on the left dates from 1976.

0 Ft 20

0 Mtr 6

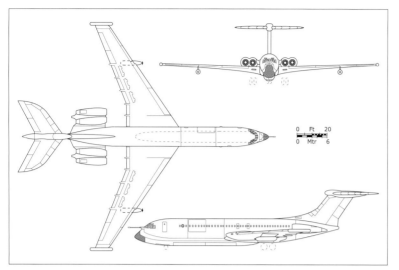

Another Post-Falklands proposal was to graft the pannier from a Nimrod onto a VC10. This would provide weapons bay and radar installation while minimising modifications to the structure of the VC10. This would provide two benefits for the RAF, the first being provision of a long-range MR capability in support of 'Fortress Falklands' operations while the second was to free up Nimrod MR airframes for conversion to AEW.3. This was the last of a long line of Maritime VC10 studies.

Two VC10MRs on a Tapestry patrol over the Beryl Field in the East Shetlands Basin. Such patrols became a regular part of RAF maritime operations in the 1980s. The VC10s are climbing to a suitable altitude to refuel Tornado F.3s operating in the area. *Adrian Mann*

box. The new 1280 sq ft (119 sq m) wings were intended to allow the BAC One-Eleven-700 to operate from the same class of airfield as the YS-11 and would utilise '…new wing sections, still more advanced leading and trailing edge high lift systems and possibly major application of carbon fibres.' In the ASW role these high-lift devices would assist low-speed handling at low altitudes on patrols when making attacks using the MAD. BAC hoped that in the -700 they would have a '…completely modernised aircraft with outstanding airfield performance and low noise and growth potential.' BAC saw a huge worldwide market of up to 1000 for the BAC One-Eleven-700.

As a Neptune replacement, BAC proposed the One-Eleven ASW with its fuselage 13ft 4in (4.06m) shorter forward of the wing than the proposed -700 airliner and modified around the wing centre section to accommodate a

weapons bay that extended forward of the wing box as a short pannier. The most obvious change to the BAC One-Eleven was the addition of two Rolls-Royce Turbomeca RT.172-06 Adour turbofans under the wings. The Adours were already being built in Japan as the TF40-IHI-801A for the Mitsubishi F-1 fighter and T-2 fast jet trainer and so would be attractive to the JMSDF. The Adours would be mounted on short pylons under the wings which were identical to the -700 airliner's but with structural strengthening to support the Adours, two underwing stores pylons each side and a higher AUW of 152,000lb (68,946kg).

BAC took the view that the Spey/Adour combination was '…particularly well suited to the multiplicity of mission performance requirements for the maritime reconnaissance/AEW roles. The mixed powerplant arrangement provides some of the advantages of both twin-engined and four-engined aircraft.' The One-Eleven ASW would take off and climb on all four engines, transit to the mission area on the Speys (or if higher speed was required, run all four engines at optimum thrust) then conduct patrols on the two Speys. A failure of one engine, even a Spey, would allow the aircraft to continue its sortie, which was the rationale behind the RAF's 'minimum of three engines' requirement for over-water patrols. The One-Eleven ASW had more than adequate endurance considering its size, with four and a half hours at the RAF's 1,000nm (1,850km) yardstick, courtesy of the two 900 imperial gallon (4,091 litre) fuel tanks installed in the cabin above the wing box. To cater for the increased AUW, the undercarriage was to be beefed-up

and larger tyres fitted. The nose undercarriage leg was to be extended to improve access to the weapons bay in the pannier. A Searchwater ASV radar was to be installed in the nose, which also sported a long probe a carrying the MAD detector, rather than carrying the sensor in a tail sting that was too close to the engines and thus susceptible to magnetic interference and would restrict the aircraft's rotation on take-off.

BAC had by February 1976 honed the One-Eleven ASW by enlarging the nose to house a larger antenna for the Searchwater and extending the weapons pannier forward to the nose-wheel bay and aft beyond the wing trailing edge. The nose-mounted MAD boom was deleted from this variant and replaced by a towed AN/ASQ.81 'MAD bird' trailed behind the aircraft and stowed in a bay between the Speys. The One-Eleven ASW mission avionics were to be based on those being developed in the UK for the Nimrod refit that became the MR.2 and sensors such as low-light television and even Autolycus could be fitted. Weapons-wise the aircraft could carry four ASMs such as Martel or Sea Eagle on the wing pylons and up to six Stingray torpedoes in the weapons bay, along with space for depth charges and mines. The rear fuselage housed storage for 100 Type C sonobuoys, deployed via a pair of rotary launchers, with an available payload of up to 7,700 lb (3,493kg) available. With this stores load and fully fuelled, this second variant was slightly lighter, with an AUW of 146,392 lb (66,402kg).

By the early Eighties, BAC at Weybridge were also looking at replacing, or supplementing the Nimrod. In the wake of the Falklands Conflict, the RAF and the MoD realised that there was a requirement for more support aircraft: Tankers, Elint, AEW and Maritime Reconnaissance. BAe at Weybridge and Filton proposed yet more versions of the VC10 for all four roles, based on ex-airline airframes. On the maritime patrol front, Filton and Airbus Industrie offered maritime versions of the A320, even proposing three and four engine versions of the A320, for which they had high hopes as a Nimrod replacement for the RAF and to replace Neptunes and Orions around the world. Unsurprisingly, Weybridge trotted out yet another, if not rather radical, transformation of the VC10. BAe were keen on a short term gap-filler that would need replacement in the Nineties, thus allowing the time for BAe/Airbus to develop a new maritime platform.

BAe proposed what was more or less the grafting of a Nimrod MR.2 nose, radar and weapons bay pannier onto a Super VC10. The Air Staff had long criticised Vickers and BAC for

pushing the VC10 as an MR type, but in the Post-Falklands climate, a large MR type with long endurance was quite attractive, even more so if the original airframe was cheap. Add to that the VC10MR's ability to act as a refuelling tanker and BAe thought they were onto a winner. The main problem with the VC10 in the MR role was that there was '…insufficient airframes available to provide a single–type fleet…' and so would not be a practicable programme.

New Engines in Old Airframes

And what about the truly optimum aircraft? Woodford looked into re-engining the Nimrod as part of their maritime patrol studies in the Seventies. One reason for the success of Boeing's 707 and Douglas's DC-8 was their pod-mounted engines. By mounting the engines externally, rather than buried in the wing roots, new and bigger engines could be fitted allowing for development and increased power as a buried installation limits the diameter of the engine that can be fitted. Another consideration is the limited size of the intake, and therefore mass flow, that the installation can allow without the need for structural changes. An earlier study towards a Canadian Argus replacement requirement produced a six-engined Nimrod that retained one of the Speys in the wing roots, but added a pair of pylon-mounted turbofans under each wing. These were installed in individual pods, allowing for growth or a change of engine type during the aircraft's service life.

One project study examined re-engining the Nimrod and saw the buried Speys removed and the inner wing structure modified to remove the intakes, trunking and engine bays. Four RB.410 turbofans were fitted on underwing pylons, which as might be imagined, proved problematic. The main and most obvious prob-

One Woodford proposal to increase the capability and efficiency of the Nimrod was to rework the wing and fit four RB.410 turbofans on underwing pylons. Parametric studies showed that the Nimrod's patrol time at 1,000nm (1,852km) would be increased from five and a half hours to just under eight hours. The proposal also required a new, longer undercarriage.

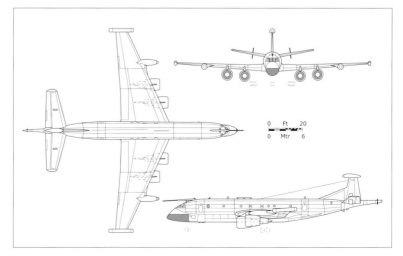

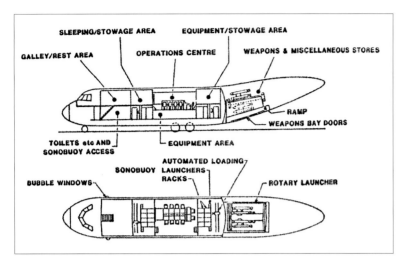

This maritime patrol FIMA variant carries a rotary dispenser on the modified tail ramp. No radar is shown on this drawing, but a corresponding German study shows a nose-mounted ASV radar. *via Avro Heritage*

BAe Woodford's Future International Military Airlifter formed the basis of a number of Multi-Role Support Aircraft proposals including this version combining the AEW and maritime patrol roles providing surveillance of air and surface traffic. As with earlier converted transports, the loading ramp housed the weapons bays. *via Avro Heritage*

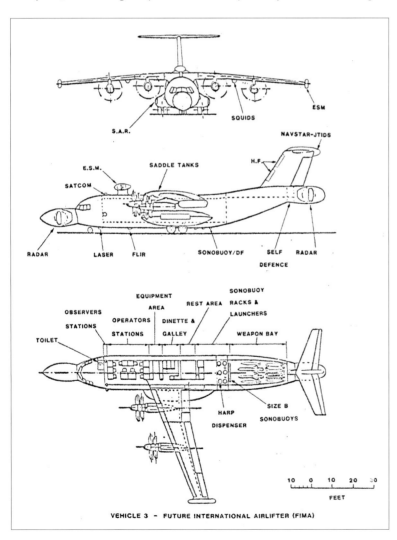

VEHICLE 3 - FUTURE INTERNATIONAL AIRLIFTER (FIMA)

lem was ground clearance and the existing Nimrod had limited ground clearance even without the new engines. Woodford's engineers addressed the problem and extended the length of the undercarriage legs. This lent the Nimrod a somewhat stalky look, while the engines detracted from the Nimrod's sleek lines. It just did not look right.

Multi-role MR

The Labour Government of Jim Callaghan in 1977 nationalised the UK aviation industry, merging BAC and HSA to form British Aerospace (BAe). The major project on BAe's books in the late Seventies was the Panavia Tornado Multi-Role Combat Aircraft, the best example of the new trend towards multi-role aircraft. The Woodford Division's main work was the Nimrod MR.2 upgrade and the Nimrod AEW.3, but as ever, future projects were being studied and by 1982 Woodford was involved in developing a Multi-Role Support Aircraft (MRSA) as a tanker, transport, AEW and of course maritime patrol platform. A further collaborative project involved the European countries in a series of project studies for a replacement for the C-130 Hercules under the designation Future International Military Airlifter (FIMA). FIMA had its origins in a 1982 agreement that led to Aérospatiale, BAe, Lockheed and Messerschmitt-Bölkow-Blohm joining forces to develop a replacement for the C-130 Hercules and C-160 Transall. Lockheed departed in 1989 to develop the C-130J, but CASA and Alenia joined the group, which was renamed Euroflag (European Future Large Aircraft Group). The basic configuration remained the same throughout the study, although the powerplant changed from turbofans to turboprops with contraprops and finally to turboprops with eight-bladed propellers. It was this last configuration that ultimately entered service as the Airbus Military A400 Atlas.

Woodford produced a number of studies under the BAe.841 designation, drawing up a high-wing, four-engined transport reminiscent of a Boeing C-17 Globemaster III. BAe also examined using the FIMA as a multirole platform capable of filling a number of roles including AEW and maritime reconnaissance. The MR type was powered by four General Electric GE34 turboprops driving 12ft (3.7m) diameter contraprops and retained the fore-and-aft scanner system (FASS) that was the trademark of British AEW types. This configuration, using a radar optimised for the maritime role would

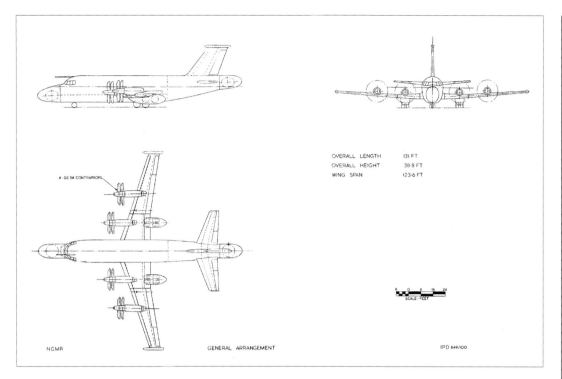

OVERALL LENGTH   131 FT
OVERALL HEIGHT   38·8 FT
WING SPAN   123·6 FT

SCALE - FEET

NGMR   GENERAL ARRANGEMENT   IPD 849/100

By 1984 Woodford's maritime patrol studies had evolved into the BAe.849 New Generation Maritime Reconnaissance aircraft (NGMR). This was tailored to the export market as well as the RAF and its origins in the HS.830, particularly the wings, can be seen in this drawing. Its GE34 engines sport the multi-bladed contraprops that were all the rage in the early 1980s. The BAe.849/100 shown here was a large aircraft that might also have had a AEW role given that it was fitted with a FASS radar system. *Avro Heritage*

allow 360° coverage, something that had prompted the ASR.357 design studies to be fitted with tail radomes. The capacious cargo hold provided plenty of room for workstations, rest areas and sonobuoy storage and dispensers. Additional ASW equipment included the horizontal array random position system (HARP) for sonobuoy monitoring, forward-looking infrared (FLIR), a laser system such as ORADS under the forward fuselage and SQUID under the wing between the No.1 and No.2 engines.

Alongside the FIMA studies, Woodford also examined a number of design studies for a New Generation Maritime Reconnaissance (NGMR) aircraft. This, the BAe.849 from 1984, would be a clean-sheet mid-wing design with a double-bubble fuselage cross-section, the upper lobe being 13ft (4m) in diameter with the lower lobe, containing fuel and weapons bays, 10ft (3m) across. The BAe.849 was to be powered by four General Electric GE34 turboprops driving contraprops mounted on a moderately swept wing with an aspect ratio of 9.48 and a planform reminiscent of the Airbus A320. Three variants have come to light, the /100 and /104, plus a smaller version with no designation save

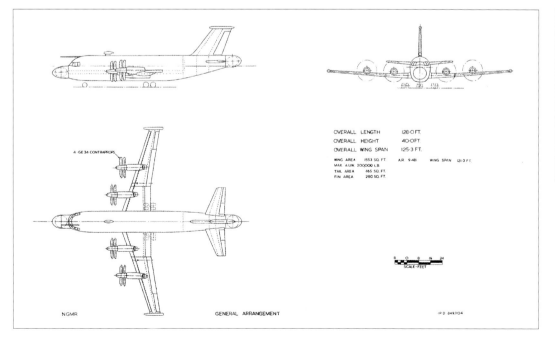

OVERALL LENGTH   128·0 FT.
OVERALL HEIGHT   40·0 FT.
OVERALL WING SPAN   125·3 FT.

WING AREA   1553 SQ. FT.   A.R. 9·48I   WING SPAN   121·3 FT.
MAX. A.U.W. 200000 LB.
TAIL AREA   465 SQ. FT.
FIN AREA   280 SQ. FT.

SCALE - FEET

NGMR   GENERAL ARRANGEMENT   IPD 849/104

The BAe.849/104 was slightly larger than the /100, but the main characteristics were the same. The dual role is probably a result of Woodford's Multi-Role Support Aircraft (MRSA) studies. *Avro Heritage*

BAe.849. The first two were fairly large aircraft with an AUW in the 200,000 lb (90,703kg) range and fuselage lengths of 131ft (40m) and 128ft (39m) respectively with both sporting FASS radar antennae for 360° coverage. The BAe.849/100 had a wing span of 123ft 8in (37.7m) with ESM pods at the tips and large fairings aft of the inboard engines into which the four-wheel bogie undercarriage retracted. The /104 had a span of 125ft 4in (38.2m), with similar ESM pods and a similar undercarriage that retracted into the lower fuselage/wing box, which would restrict the size of the weapons bay.

The third study was slightly different, mainly due to having its radar in a retractable dustbin under the rear fuselage. Like the /100, this variant carried its undercarriage in wing pods, leaving the entire underside available for a long weapons bay. The Nimrod was famed for its long unrestricted weapons bay that allowed the largest of stores to be carried and pods for the undercarriage would continue this feature in the NGMR. While the various design studies were under way at Woodford, there was another possibility from across the Channel in the form of the C160ASF Transall. This was more or less a replay of the conversions of tactical transports that had fared so badly in the OR.350 process twenty years before. It suffered from the same failings as the earlier proposals and of course, being twin-engined and only fitted with four stores pylons rather than an internal bay, it would not have been considered by the Air Staff. Unfortunately for the aircraft companies, there was no requirement for a Nimrod replacement in the mid-1980s, so the NGMR projects remained as design studies, but the lessons from NGMR played a part in the last and final act of the British maritime patrol drama. The story of the MRA.4 is, if nothing else, dramatic.

This variant of the BAe.849 appears to be a dedicated maritime patrol version that dispensed with the FASS radar and used a dustbin-type retractable radome à la Shackleton. The deletion of a possible AEW role allowed a more compact airframe.

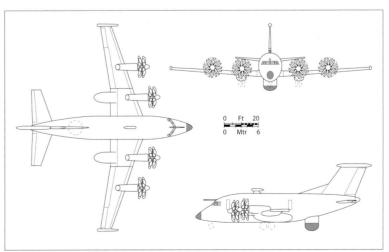

Our Comets – Nimrod MRA.4

The description of the author's visit to Woodford at the start of this chapter gives a number of clues to the cause of one of the biggest wastes of defence expenditure seen in Britain to date and Britain is a country with an unenviable track record in defence procurement disasters.

On the 9th November 1989, the Government of the German Democratic Republic opened the gates of the Anti-Fascist Protection Rampart, known in the West as the Berlin Wall and the Inner German Border. This event heralded the last phase of the Cold War, which finally came to an end when the Soviet Union was dissolved in 1991. With the end of the Cold War came a draw-down of the armed forces of NATO and the Warsaw Pact, which rapidly collapsed as the Soviet Empire disintegrated. The Russian Navy became a shadow of its former Soviet self and that, and the end of the Cold War commitments, removed much of the more obvious threats to British maritime activity in the North Atlantic and North Sea. Many saw this as an opportunity to save a spot of cash, in fact this became known as the Peace Dividend, and the NATO Allies in general began to reduce the strength of their armed forces.

From the standpoint of the maritime patrol forces, there would be fewer enemy submarines to find and track and some commentators considered the role to have been rendered obsolete by the apparent disappearance from the seas of the Soviet submarine fleet. However, there was still a requirement for the Royal Navy to operate a nuclear deterrent and that deterrent took the form of the four *Vanguard*-class SSBNs and their Trident D5 missiles. The RAF's Nimrod force provided ASW protection for the *Vanguard* boats, a NATO commitment tasked to SACLANT that dated back to OR.350 and thus the RAF's maritime patrol role was saved not by the need to destroy Russian boats but the need to defend British boats. In addition, there was the prospect of 'second rate' navies acquiring the latest Soviet submarines in the grand sell-off that was occurring in Russia in the early 1990s. Seventy years of Marxist-Leninism hadn't stripped the Russian people of the basic tenets of capitalism.

This argument for the maritime force was important to the next stage in the story of UK maritime patrol aircraft and had it not been for the deterrent protection role, the RAF's maritime patrol fleet might have been reduced to a basic EEZ patrol and SAR asset. As noted above, the Nimrod had transformed from an interim

The US Navy was also examining a new maritime patrol aircraft in the late 1980s. The Lockheed P-7, a modernised Orion was the front runner, while Boeing proposed their 757ASW. One of the more interesting proposals was the P-9D from McDonnell Douglas. Derived from the MD-91 airliner, this utilised the new technology of the propfan. This manufacturer's model shows how the P-9D would have looked in US Navy service. *John Aldaz Collection*

type in the shape of the MR.1 to a fully-capable upgraded MR.2 in the early Eighties. By the end of the Eighties it had become obvious that the MR.2s would need to be replaced in the mid to late Nineties and to this end the Air Staff drew up a Staff Requirement (Air) for a Replacement Maritime Patrol Aircraft (RMPA) which became SR(A).420, issued in November 1992 and tenders invited. Up to 25 aircraft would be required and there was a preference for a minimum development aircraft (the Air Staff had apparently learned something from the earlier ASR.350 and ASR.357 – like suits, bespoke aircraft were pricey) and the usual RAF MR mission of eight hours at 1,000nm (1,850km) applied. It was to be equipped with state of the art communications and radar systems plus the latest sonar and signal processing. Nor were the RAF interested in converted twin-jet airliners as they were adamant that four engines were required. By 1992 the world's airlines had shown that twin-engine airliners could fly long sectors over oceanic routes under Extended-range Twin Operations (ETOPS) rules whereby an aircraft could fly a route that allowed it to be 60 minutes flying time, on one engine, from a diversion field. So why should the RAF not opt for a converted twin-jet airliner? The Air Staff pointed out that a steady cruise at 35,000ft (10,668m) was somewhat different to a high-speed transit followed by an eight hour patrol at low altitude in turbulent, salt-laden air in the middle of the Atlantic.

Meanwhile across the Atlantic the US Navy had been looking to replace its P-3C Orion force and drew up a requirement for a Long-Range Air ASW-Capable Aircraft (LRAACA), and had selected another Lockheed product, a new-build Orion called the P-7, in late 1988. This was essentially an Orion with advanced systems and new engines, General Electric/Lycoming T407s, rated at 7,600shp (5,667kW) and driving five-bladed propellers. The competitors to the P-7 had been the Boeing 757ASW based on the 757 and McDonnell Douglas' P-9 that was derived from the MD-91, a propfan-powered airliner. Unfortunately for Lockheed, problems with the mission suite systems plagued the project and with the end of the Cold War there was a severe cutback in US military spending and so in 1990 the P-7 was axed. Not that the P-7 was the only type with systems problems as a new acoustic processor for the Nimrod to meet SR(A).909 and SR(A).910 for a Central Tactical System for the Nimrod MR.2 had run into development problems and both were cancelled in 1990. Unfortunately, the cancellation of the P-7 put the RAF in a bad position as they had been eyeing the P-7 as a Nimrod replacement, since it was the only western maritime patrol type that would fit the RAF's needs, as acquiring the Atlantic was unthinkable.

The Air Staff hadn't exactly been lax in looking into a Nimrod follow-on type and had since 1986 been examining a variety of conversions and new designs, one of which was for the Airbus A310. That type most definitely exceeded the 12ft (3.7m) optimum fuselage diameter, not to mention the fuselage volume guidelines. With a requirement for less than 25 aircraft a new design wasn't economic and conversion of an airliner in the relevant weight class wasn't

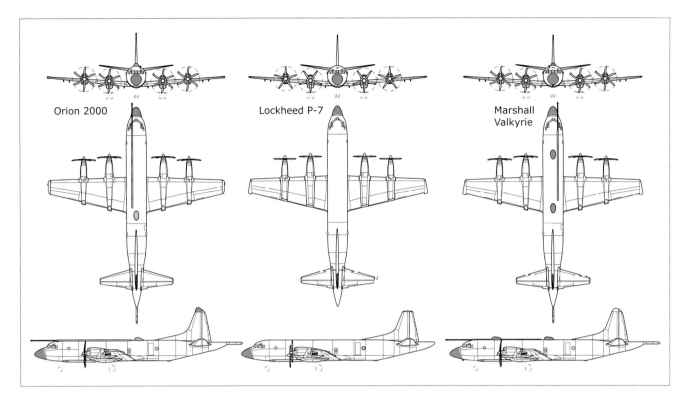

Orion 2000   Lockheed P-7   Marshall Valkyrie

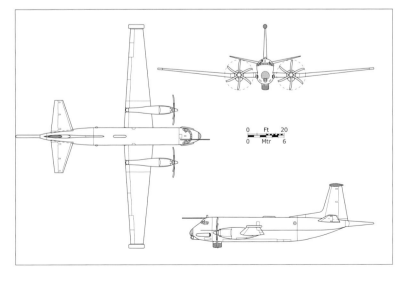

viable as most were twin-engined. The Lockheed P-7 was showing promise and looked like an ideal solution until it was cancelled by the US Government in 1990, prompting a new flurry of activity in Whitehall. This resulted in the Replacement Maritime Patrol Aircraft programme (RMPA) to meet a new requirement SR(A).420, issued in 1993.

So, in the true spirit of the late Eighties, a competition had to be held to select a RMPA. The graph from the 1970s that showed the Comet 4 as the optimum aircraft must have come in handy in BAe's presentations. BAe project managers expounded the view that the Nimrod was the best understood airframe in the RAF, possibly because it and the Comet had

been around so long. By 2010 that statement was looking a bit rash.

Despite the re-engined Nimrod looking like the cheapest option for the MoD, other proposals had to be examined. Lockheed invariably tendered a variant of their P-7 with AE2100H engines and renamed the Orion 2000. These were to be new aircraft and Lockheed entered into a joint venture with GEC-Marconi, who became the British lead contractor in the Orion 2000 bid. These aircraft had benefitted from lessons learned from Lockheed's work, particularly its internal layout, on the CP-140 Aurora that replaced the CL-28 Argus in the Canadian Armed Forces. A second Orion-derived tender came from Loral, fronted in Britain by Marshalls of Cambridge, who also proposed refurbished US Navy P-3A and B Orions under the name Valkyrie.

Dassault entered the fray with the Air Staff's favourite reject, the Atlantic, this time the new-build Atlantic 3 with glass cockpit, new systems and AE2100H engines with six-bladed propellers. The final bid was for a radically upgraded Nimrod MR.2 from BAe called Nimrod 2000. So, the choices were between two newly-built types and two refurbished types, from four companies until in 1996 Lockheed Martin bought Loral and opted for an each-way bet on RMPA by keeping the new Orion 2000 and the refurbished Valkyrie in the competition. As usual the Atlantic suffered from the fact that it had two engines, although as ever, a four-engined variant was proposed with pylon-mounted underwing jets. Dassault were advised by the Air Staff that they were wasting their time and money and the French company withdrew from the tender process in January 1996. One other aircraft was discussed for SR(A).420, but it was never a real contender, mainly because it came from Russia. The Beriev Be-40P Albatross (NATO reporting name *Mermaid*) was a jet-powered amphibian that had been proposed to the Russian Navy as an ASW type to replace its Beriev Be-12 *Mail*. Ostensibly a two-engined aircraft, the Be-42 actually had four engines, two Soloviev D30KPV turbofans rated at 26,460 lbf (117.7kN) plus two Kolesov RD36-35 turbojets rated at 5,170 lbf (23kN) used as auxiliary boosters on take-off. Whether this would have satisfied the Air Staff is not known and whether there was an official tender from Beriev is unclear, but the Air Staff stated that (as they had feared would happen back in the early Fifties) they had indeed lost the art of operating flying boats.

On the face of it, this looked like the plucky Brits up against the big Yanks and less than a decade after the Westland affair, a row that had torn the Thatcher government apart, there was a strong political case for buying British. In addition to the political aspect of the decision, there was the small matter of the Nimrod 2000 being the most cost-effective of the bids. In this case cost-effective meant cheapest, but that would

soon be shown to be a flawed assumption. All that was in the future and in December 1996 the contract to produce the Nimrod 2000 was granted to British Aerospace, with the first delivery to the RAF to be made in the spring of 2003. The service designation was to be Nimrod MRA.4, the A standing for 'Attack', reflecting its much improved capability to carry offensive weapons including the Boeing AGM-84 Harpoon, Raytheon AGM-65 Maverick and the MDBA Storm Shadow stand-off missile in addition to the usual Stingray torpedoes and mines.

The work to produce the twenty one Nimrod MRA.4s from existing MR.2s involved what was effectively a complete rebuild. The wings were cut off and a new inner wing and centre-section with larger engine bays and intakes were built. The Nimrod MRA.4 demonstrates why engines buried in the wings failed to prosper: fitting a new, larger engine requires a major reworking of the wing structure. The engine selected for the MRA.4 was the Rolls-Royce Deutschland BR.710 turbo fan, rated at 15,550 lbf (69.2kN) against the 12,000 lbf (53.4kN) of the Spey 250. This is also used by the RAF's Sentinel surveillance aircraft, but in that case is mounted externally. The BR.710's diameter was approximately 30% greater than the Spey and being installed in the wing roots, this had forced the complete redesign of the inner wing and engine bays. The fuselage was to be completely

Opposite page:

Orions galore! In the early Nineties when SR(A).420 was circulated, three Orion variants were examined. From left to right: Orion 2000, Lockheed P-7 (the Air Staff preference) and the Marshall Valkyrie. None of these prospered, the Air Staff opting for a reworked Nimrod, no doubt seeing a propeller type that predated the Nimrod as a retrograde step after the high-speed Nimrod.

The Dassault Breguet team were back with the Atlantic 3 with glass cockpit, new systems and Rolls-Royce/Allison AE2100H turboprops with six-bladed propellers. The Air Staff advised Dassault Breguet that they had no prospect of gaining the contract and should withdraw. The advice was heeded.

Left: BAe artwork from the early Nineties showing Nimrod 2000 as bid for SR(A).420. At this point the type was effectively a re-engined MR.2P with new systems. That would soon change.
BAE Systems

stripped to the bare metal and refurbished. New cockpit systems derived from Airbus Industrie's latest 'glass' cockpits from the A340 airliner were to be installed and the cabin was refitted for the new mission systems, with the primary 'dry' sensor being the Searchwater 2000 radar. The 'wet' sensors were to be co-ordinated by the Ultra Electronics AQS.970 acoustic processor.

The original programme called for Flight Refuelling Ltd (FRL) to carry out the refurbishment, with the wingless fuselages transported in Antonov An-124 *Condors* from Kinloss to FRL's base at Hurn Aiport. FRL and BAe soon discovered that the original Comet airframes upon which the Nimrods were based had not been built '...to a common standard' and so BAe took over the work and moved the project to Woodford. Woodford discovered that there were a number of flaws in the wing, which entailed a completely new design. All this delayed the programme and as costs mounted, the number of airframes on order was reduced to eighteen in 2002, sixteen in 2004, twelve in July 2006 and ultimately, nine in 2008. The first production Nimrod MRA.4 flew in September 2009 but just over a year later the entire programme was cancelled.

In the fallout from the affair, the *Financial Times* stated that the MRA.4 was '...still riddled with flaws...' which included a number of safety concerns including the fuel pipe that proved to be the downfall of the MR.2. The UK aviation magazine *Air Forces Monthly* also published information on the background to the cancellation, specifically dealing with the flight characteristics of the Nimrod MRA.4. The magazine stated that there were '...significant aerodynamic issues and associated flying control concerns in certain regimes of flight meant that it was grounded at the time of cancellation and may not have been signed over as safe by the Military Aviation Authority.' The gist of this was that the RAF did not consider the aircraft a viable type to accept into service, the MoD accepted this view and thus after spending £3.6bn on the programme the airframes were stripped of kit and 'reduced to produce' and Her Majesty's Ministry of Defence recouped £1m from the scrappie. Perhaps, Mr Haddon-Cave QC's comments on the operation of elderly or legacy aircraft had a hand in this cancellation, but it must be suspected that the 'flaws' in the Nimrod MRA.4 were more than enough to justify cancellation. Not one complete aircraft or large remnant of the airframe remains. Despite the furore that followed the cancellation of the MRA.4, there may come a time when it will be seen as the correct decision. The RMPA programme was an example of bad procurement, with good defence budget money being spent on a project that was flawed from the start.

Presented in July 1996 with a range of maritime patrol types to replace the Nimrod under SR(A).420, the Ministry of Defence opted to repeat the Shackleton experience of four decades before and update the current service type. Within fifteen years the Royal Air Force's maritime patrol capability would not even be a shadow of its 1945 self; it was gone. The RAF could not confidently operate an aircraft that had been deemed unsafe in a coroner's court, and after forty years of operations, the Nimrod owed no debt to the British people. Seven of the surviving Nimrod MR.2s were sent to museums while components of the MRA.4s were open to bids in online auctions. This was an ignominious end to a fine aircraft that had served its country well. The Nimrod saw off its potential, and its actual, replacements but fell afoul of a temporary modification that had further increased its capability.

New wing, engines, cockpit instruments, cabin floor, mission systems and radar made the Nimrod MRA.4 a brand new aircraft. Unfortunately the prospect of merely re-engining and fitting new systems to the MR.2s foundered for a number of reasons, not just the bespoke nature of the original Nimrods identified in the popular press.

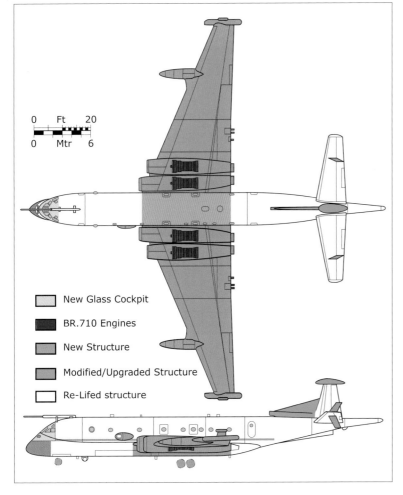

0 Ft 20

0 Mtr 6

New Glass Cockpit

BR.710 Engines

New Structure

Modified/Upgraded Structure

Re-Lifed structure

'If, at relatively short notice, we thought that we needed to get back into having a dedicated capability we could put something together or buy something off the shelf pretty quickly. Would it be of comparable complexity to Nimrod? No. Would it be capable of performing a maritime surveillance function because we perceived the need to get back into that urgently? I think it would.' Nick Harvey MP, the Minister for the Armed Forces, House of Commons Defence Committee report on Future Maritime Surveillance, September 2012

On 26th May 2010, Nimrod MR.2 XV229 took off for the final official Nimrod flight and landed at Kent International Airport (formerly RAF Manston) to end its days as a training aid for the Defence Fire Training and Development Centre. A few more 'last flights' of Nimrod MR.2s took place as the remaining Nimrods were delivered to museums around the UK. Banging on a desk at the MoD, would Sir Arnold Hall have thought that he was to be outlived by the aircraft he was proposing as an interim type? Doubtful, but that's exactly

what happened. Sir Arnold Hall did a great deal for the United Kingdom's aerospace industry; he also gave us a superb aircraft in the Nimrod.

The loss of a dedicated maritime patrol aircraft in 2010 left a gap in the UK's armed forces. The role was taken up by the Royal Navy's surface ships and anti-submarine helicopters while SAR support was supplied by C-130J Hercules fitted with a 'strap-on' surface search radar and Lindholme gear while the on-scene co-ordination role would be carried out by the E-3D Sentry AEW.1, whose AN/APY-2 radar can also provide a 'surface picture' of marine traffic for SAR co-ordination.

What of the future? As far as personnel were concerned, relinquishing the role did not necessarily mean losing the experienced aircrew from the Nimrod units. On the retirement of the Nimrods, the RAF quickly established Project *Seedcorn*, to retain a core cadre of experienced personnel that could form the basis of a maritime patrol unit if (or when) the RAF regained the role. These personnel had from late 2011 been flying on maritime patrol missions with the Australian, Canadian and New Zealand air forces plus the US Navy. When Malaysian Air-

A Boeing P-8 Poseidon of VX-1 'Pioneers' shows off its considerable sensor suite, weapons bay and CATM-84D Harpoon ASMs. If the RAF were to bite the 'silver bullet' and pay the cost of procuring the Poseidon, it would regain the formidable maritime capability it lost when the Nimrods were retired. ©*Graham Wheatley 2014*

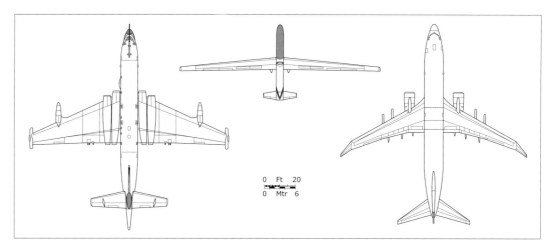

lines' flight MH370 (Boeing 777, 9M-MRO) disappeared in March 2014, coverage of the resulting search in the British media included interviews with a few RAF personnel, seconded to the RAAF and RNZAF under Project *Seedcorn*.

Platforms were a different matter and a number of options were available for the RAF to return to the maritime reconnaissance role. These included resurrection of the Orion projects intended to meet SR(A).420; the full-blown anti-submarine requirement from the early Nineties. The RAF could modernise the new force and adopt the US Navy's Boeing P-8 Poseidon/Northrop Grumman MQ-4C Triton double-act or merely opt for a limited type for EEZ surveillance and SAR support. In the former case, the twin-engined Boeing P-8 does not approach the RAF's long-held requirements for 8 hours patrol at 1,000nm (1,852km), more like 4-5 hours, so that lack of endurance could be a consideration in its procurement. For the latter case, the Airbus Military C295ASW, a maritime patrol version of a twin-engine tactical transport could be examined, especially if the Triton provided the surveillance capability. Airbus Military have proposed a number of maritime reconnaissance variants of the A320, although the RAF would probably be the launch customer for a new variant of this aircraft and its development would take time. Then there are the options to strap kit onto exiting types such as the RAF's C-130J Hercules or modify the Raytheon Sentinel R.1, although this type was slated for withdrawal in 2018. There is however, an aircraft that could meet the Air Staff's long-held wish for a maritime patrol aircraft with more than two jets: The Dassault Falcon 900MPA. Another option could be to follow the lead from 51 Sqn, the RAF's Elint unit: acquire refurbished KC-135s and install the kit intended for Nimrod MRA.4. Boeing had in 1973 proposed a maritime version of the 707-320 for the Canadian Long Range Maritime Patrol Aircraft, which got as far as a Boeing 707-720 being used for mission profile trials.

One of the more realistic proposals involved

keeping the C-130J Hercules for SAR duties and the monitoring of the EEZ around the Falkland Islands while the UK-based maritime patrol role could be filled by the Airbus C295ASW, Bombardier Q400MP or the SAAB 2000MPA. Airbus Military have been attracting interest by developing an ASW variant of the CN-295 military transport, developed via the CN-235MPA Persuader, an EEZ patrol type of which two are currently in service with *an tAerchór* (Irish Air Corps). The Airbus C295ASW is equipped with the latest anti-submarine sensors and can be armed with ASMs and torpedoes, the preferred weapon for use against submarines. Bombardier's Q400 already operates in the maritime patrol role around the world and was by 2014 being touted as a possible maritime platform for the RAF. There was also a consortium of UK firms including Ultra Electronics, Selex and Marshalls Aerospace involved in the proposal to modify the Canadian aircraft. In addition to the usual Searchwater radar, the Q400 also carried the Eagle electronically-scanned wide-area radar which Selex claimed offers similar '…air-surveillance capabilities offered by the RAF's E-3D Sentry…' if that was required in addition to its maritime application. The SAAB 2000 airliner is in airline service in the UK, so support would be available and BAE Systems have close ties with SAAB through their building of major components for the JAS-39 Gripen. The 'kit' is available off the shelf in the form of the Sea-

spray 7500 radar, Ultra Electronics acoustic systems and one of the many high-definition FLIR systems. The name for this type? SAAB call it the Swordfish, a name that would be more than suitable for a maritime aircraft and no doubt one that Wing Commander G G Beaugeard would have approved of.

As of September 2014, the Breguet Atlantics, repeatedly written off by the Air Staff since 1958, continue to operate and approach their half century in service. In September 2014 there had been no movement towards the RAF returning to maritime patrol, no doubt related to the considerable uncertainty about the outcome of the referendum on Scottish independence on 18 September 2014. In the event, the Scottish electorate voted to remain in the United Kingdom, thus preserving the Royal Air Force in its current form and maintaining the Royal Navy's Trident facilities at Faslane on the Clyde. The United Kingdom has the fifth largest Exclusive Economic Zone in the World and the largest area of responsibility for search and rescue in Western Europe and so the RAF must have a maritime patrol aircraft of some description. With the uncertainty of the last four years gone, a decision on a Nimrod successor cannot be far away. Whether that will be the P-8 Poseidon, SAAB Swordfish, Bombardier Q400MP or one of the many types available, is a decision yet to be made, but the men and women of *Seedcorn* should come home.

The SAAB 2000MPA Swordfish is another possibility if the RAF was to regain the maritime patrol capability it lacks. SAAB and BAE Systems have a close relationship, the SAAB 2000 airliner is in service in the UK, so the support and experience in their operation is available.

Postscript

The Russian Navy battlegroup? It moved off as the storm subsided and the only evidence of its presence was some garbage that had been allegedly dumped over the side, which caused a furore in the Scottish media. Changed days indeed.

Glossary of terms

A&AEE	Aircraft and Armament Experimental Establishment, renamed Aircraft and Armament Evaluation Establishment in 1992
ACAS	Assistant Chief of the Air Staff – Since 1969 second most senior military officer in the RAF.
ADE	Armaments Design Establishment
AEW	Airborne Early Warning
Afterburner	see reheat
Air Staff	The cadre of senior officers that runs the RAF
Amphibian	A flying boat or float plane with retractable undercarriage and thus can operate from land or water.
APU	Auxiliary Power Unit. Used to power an aircraft's electrical systems when on the ground and for starting the main engines.
ARI	Airborne Radar Installation
ASE	Admiralty Signals Establishment
ASM	Armstrong Siddeley Motors
ASR	Air Staff Requirement – outline of what the subject has to do. The basis of the development programme.
ASR	Air/Sea Rescue – replaced by SAR
AST	Air Staff Target – formed the basis of a feasibility study for the Requirement.
ASV	Air-to-Surface Vessel – generally refers to radar.
ASW	Anti-Submarine Warfare
Avro	A.V. Roe
AWA	Armstrong Whitworth Aircraft
BABS	Beam Approach Beacon System.
BAe	British Aerospace plc, became BAE Systems
BAC	British Aircraft Corporation
BSEL	Bristol Siddeley Engines Ltd
CA	Controller, Aircraft – Senior RAF officer charged with clearing an aircraft for service.
CAS	Chief of the Air Staff – the most senior military officer in the RAF.
DCAS	Deputy Chief of the Air Staff – until 1969 second most senior military officer in the RAF.
DH	De Havilland – aircraft and guided weapon company
DHP	De Havilland Propellers – guided weapons branch of De Havilland that became Hawker Siddeley Dynamics.
Flying boat	Fixed-wing aircraft whose fuselage comprises a boat-like hull that can operate from a body of water.
ECM	Electronic CounterMeasures
EEA	English Electric Aviation
EECo	English Electric Company, parent company of The Marconi Company Ltd, English Electric Aviation Ltd and D. Napier and Sons Ltd.
Elint	Electronic Intelligence

BAE Systems Nimrod MRA.4 ZJ518 shows its weapons bay and the type's trademark engine housings at the wing roots.

Nimrod ZJ518, formerly MR.2 XV234, was the second to fly and carried the full mission suite. ©*Graham Wheatley*

ERAPS	Expendable Reliable Acoustic Path Sonobuoy
ESM	Electronic Support Measures
EW	Electronic Warfare
FAA	Fleet Air Arm – Royal Navy's air branch
FLIR	Forward Looking Infrared
HSA	Hawker Siddeley Aviation – aircraft division of Hawker Siddeley Group
HSD	Hawker Siddeley Dynamics – guided weapons division of Hawker Siddeley. Formerly de Havilland Propellers.
Hunter/Killer	A platform that can both find and destroy a target.
JATO	Jet Assisted Take Off – original acronym, now RATO.
Mach Number	Aircraft speed divided by local speed of sound.
MAD	Magnetic Anomaly Detector – sensitive magnetometer to measure changes in the Earth's magnetic field that might indicate the presence of a submarine.
MoA	Ministry of Aviation – took over the Ministry of Supply's interests in military aviation
MoS	Ministry of Supply – the ministry in charge of everything from 'boot laces to atoms'. Responsible for aerospace contracts and management thereof. Dissolved with the creation of the MoD in 1958.
MoD	Ministry of Defence
NDB	Nuclear Depth Bomb
NGTE	National Gas Turbine Establishment
NPL	National Physical Laboratory
OR	Operational Requirement
RAE	Royal Aircraft Establishment
RAF	Royal Air Force
RATO	Rocket Assisted Take-Off
Reheat	Thrust augmentation by injecting neat fuel into the exhaust flow between the turbine and the propelling nozzle. This fuel burns thereby increasing the temperature and expands the gas as it leaves the nozzle, increasing thrust. British term, American equivalent is afterburner.
RN	Royal Navy
RPV	Remotely Piloted Vehicle. Now called UAV or UAS
RRE	Radar Research Establishment/Royal Radar Establishment
SAR	Search and Rescue. This term replaced Air/Sea Rescue
Schnörkel	Device to allow diesel/electric submarines to operate submerged by drawing in air for the engines and crew. Also known as a 'snort' in the UK.
SLAR	Sideways-Looking Airborne Radar
SLBM	Submarine-launched ballistic missile
Sonobuoy	Air-dropped, anti-submarine sensor used to locate and track submarines. Can apply to active or passive acoustic, thermal or bathymetric sensors.
SSK	Diesel/Electric-powered submarine
SSN	Nuclear-powered submarine
SSBN	Nuclear-powered submarine that carries SLBMs
SST	Supersonic Transport
Step	A break in the line of the keel on a flying boat hull that reduces water drag on the hull and aids take-off. This could be a sharp step, but more advanced boats had a curved step.
STOL	Short Take-Off and Landing
STOVL	Short Take-Off Vertical Landing
TRE	Telecommunications Research Establishment, became Radar Research Establishment in 1953, but renamed Royal Radar Establishment. Became Royal Signals and Radar Establishment in 1980 when Signals Research and Development Establishment and RRE were merged.
UAS/UAV	Unmanned (or Uninhabited) Air System or Vehicle
VCAS	Vice-Chief of the Air Staff – Senior assistant to the CAS. Post abolished in 1985.
V/STOL	Vertical/Short Take-Off and Landing
VTOL	Vertical Take Off and Landing
Wing loading	Total aircraft weight divided by wing area

Selected Bibliography

Books

Avro Aircraft Since 1908, A J Jackson and R Jackson, Putnam 1989
Avro Shackleton, B Jones, Crowood Aviation, 2002
British Secret Projects: Jet Bombers since 1949, Tony Buttler, Midland Publishing, 2003
British Secret Projects: Hypersonics, Ramjets and Missiles, Buttler and Gibson, Midland Publishing, 2007
Cambridge Aerospace Dictionary, Gunston, Cambridge University Press, 2004
From Sea to Air, Tagg and Wheeler, Crossprint, 1989
Nimrod: The Centenarian Aircraft, W Gunston, The History Press Ltd, 2009
Project Cancelled, Wood, Janes Publishing, 1986
Vickers Aircraft since 1908, C F Andrews and Eric B Morgan, Putnam Publishing, 1988
Vickers VC10 – AEW, Pofflers and other Unbuilt Variants, Chris Gibson, Blue Envoy Press, 2009
Vulcan's Hammer, Chris Gibson, Hikoki Publications, 2011
Avro's Maritime Heavyweight: The Shackleton, C Ashworth, Aston Publications Ltd, 1990

Shorts Aircraft Since 1900, C H Barnes, Putnam, 1989
The Short Sunderland, C Bowyer, Aston Publications Ltd, 1989
Supermarine Aircraft since 1914, C F Andrews and E B Morgan, Putnam, 1989
Vickers Aircraft Since 1908, C F Andrews and E B Morgan, Putnam, 1988

Journals and magazines

Air Forces Monthly, various issues, Key Publishing
Flight International, various editions, Reed Publishing
International Air Power Review, various issues, Air Time Publishing
World Air Power Journal, various editions, Aerospace Publishing Ltd
Wings of Fame, various editions, Aerospace Publishing Ltd
Aeroplane, various issues

Websites

www.FlightGlobal.com
www.secretprojects.co.uk
www.thinkdefence.co.uk

Appendix 1
Requirements & Specifications

The following is a list of the postwar Air Staff Requirements and Targets and Operational Requirements relevant to the maritime patrol role in the RAF.

ASR.381   Interim Maritime patrol aircraft to Spec MR.254. Written around the Breguet Atlantic but produced Nimrod MR.1.

ASR.420   Replacement Maritime Patrol Aircraft (RPMA) Nimrod MRA4 became SR(A).420.

ASR.549   Maritime crew trainer for Nimrod MR.1

ASR.600   Magnifying sight for Nimrod MR.1 incorporating an airborne simulator for the AS.12 operator.

ASR.602   Passive night-viewing equipment For Nimrod.

ASR.618   Mission simulator for Nimrod.

ASR.631   Electro-optical surveillance aid For Nimrod.

ASR.827   Radar for the ASR.357 maritime reconnaissance aircraft but replaced by ASR.846.

ASR.828   Low frequency receiver equipment for Nimrod MR.2.

ASR.831   IFF interrogator for maritime aircraft.

ASR.833   Surveillance receiver for Nimrod AEW and MR.2.

ASR.842   Short-range active sonobuoy to detect and localise submerged submarines.

ASR.846   ASV/ASW Radar to detect fast patrol boats. Searchwater radar intended for Nimrod MR.1 but ultimately fitted to Nimrod MR.2.

ASR.872   Aircraft modifications including upgraded datalink, navigation and tactical displays for Nimrod MR.2

ASR.875   Acoustic processor for Nimrod MR.2

ASR.876   Magnetic anomaly detection for Nimrod MR.2

ASR.879   Long-range passive directional sonobuoy for Nimrod MR.2

ASR.883   Maritime crew training for Nimrod MR.2

ASR.1211   Short-range ASM for maritime aircraft leading to the AS.12 on Nimrod

ASR.3557   Part of OR.357/OR.381 covering equipment for the detection of submerged submarines that became NAST.807.

ASR.3613   Mk.1c homing system fitted to Nimrod MR.1

AST.350   Maritime reconnaissance Staff Target leading to Avro 776 maritime patrol aircraft.

AST.357   Maritime reconnaissance Staff Target that ultimately led to Hawker Siddeley Nimrod.

AST.1186   Improved air-launched, anti-submarine homing weapon.

AST.3583   Airborne anti-submarine detection system. Joint Staff Target with Admiralty's AW.144 for which the Red Setter Q-band SLAR proposed as part of the solution.

CAOR 2/46   Specification for naval anti-submarine helicopter, leading to Bristol 173.

GD/A.101   Admiralty requirement for an air-to-surface missile that was a joint Requirement with OR.1168 which led to Martel

GDA.5   Requirement for Bullpup ASM to arm the Scimitar

and Buccaneer

GDA.10   Requirement for a Red Beard replacement, initially called Red Flag, ultimately became WE.177.

MR.218   Specification for an advanced maritime reconnaissance aircraft related to OR.350/357. Developments of Avro Shackleton and Vickers VC10 proposed

MR.254   Specification for an advanced maritime reconnaissance aircraft to meet OR.381 that led to HS Nimrod MR.1.

MR.286   Specification for Nimrod MR.2 issued 6th May 1975

NASR.563   Assisted sea survival system.

NASR.7709   Mk.31 air-launched anti-submarine torpedo.

NAST.6632   Anti-surface ship missile.

NAST.7511   Active/passive anti-submarine weapon that replaced OR.1163. Air-launched torpedo leading to Stingray.

NAST.7709   Improved Mk 31 air launched torpedo.

NR/A.32   Lightweight carrier-borne anti-submarine aircraft to Spec M.123 that became the Shorts Seamew

NSR.6134   Acoustic location device for NGASR.3335 helicopter

NST.6169   Improved air-launched anti-submarine weapon leading to Mk.31 air-launched anti-submarine torpedo

NSR.6418   Improved weapon system for Buccaneer S.2 that included Martel on Buccaneer.

NSR.7511   Stingray Lightweight torpedo

NST.6623   Helicopter-launched ASM based on AS.12.

OR.179   Air Sea Rescue aircraft leading to the Avro Lancaster ASR.III.

OR.183   Amphibious Air Sea Rescue aircraft. Requirement met by Consolidated Catalina

OR.200   Modification of Lincoln for maritime reconnaissance to Spec R.5/46 that ultimately led to Shackleton.

OR.231   Maritime reconnaissance flying boat to Specification R.112D, formerly R.2/48

OR.320   Long-range maritime reconnaissance aircraft leading to Avro Shackleton

OR.326   Maritime reconnaissance helicopter for the RAF. Bristol Type 191 proposed

OR.347   Patrol Missile Carrier known in Ministerial circles as a Poffler.

OR.350   Maritime reconnaissance aircraft to Spec MR.218. See also OR.357.

OR.357   Maritime reconnaissance requirement. Superseded by OR.381

OR.381   Maritime reconnaissance aircraft to MR.254 that led to Hawker Siddeley Nimrod.

OR.1009   Uncle Tom 1,000 lb anti-ship rocket

OR.1015   Sight for use with Bootleg tossed torpedo.

OR.1023   A 250 lb anti-submarine bomb with hydrostatic & impact fuzing.

OR.1035   Practice round for Dealer acoustic torpedo

OR.1048   1,000 lb/2,500 lb bomb/torpedo carrier.

OR.1054	Anti-ship blind bombsight for strike aircraft.
OR.1058	Anti-submarine homing weapon Zeta (later called Pentane). Joint naval requirement with AW.59 but was too big and heavy for small-ship helicopters. Admiralty favoured embarked helicopters for anti-submarine operations and so fixed-wing naval ASW aircraft were retired. Its cost was prohibitive for RAF as sole user.
OR.1060	Bootleg BA.290 high-speed, airborne, tossed torpedo.
OR.1061	Tossed torpedo sight, possibly for Bootleg. See OR.1015
OR.1156	Nuclear depth bomb. Met by Mk.90 Betty NDB and Mk.101 Lulu NDB.
OR.1163	New air-launched active/passive torpedo that ultimately led to NASR.7511 and Stingray.
OR.3458	Aircraft systems for Jezebel and Julie sonobuoy systems
OR.3501	British directional sonobuoy and receiver.
OR.3507	Anti-snort radar for maritime aircraft.
OR.3512	British non-directional sonobuoy.
OR.3513	Sonobuoy trainer.
OR.3546	Mk.10 Passive directional sonobuoy system
OR.3548	Joint requirement with AW.348 for a long-range, low-frequency directional sonobuoy system to be in service by 1965. Similar to the American Julie/Jezebel.
OR.3555	Anti-submarine radar detection device for OR.326 helicopter.
OR.3558	Air Staff Requirement for an underwater area search device for MR aircraft.
OR.3562	Autolycus exhaust gas detector fitted to Shackleton and early Nimrod MR.1s to 'sniff out' diesel-electric submarines.
OR.3569	Active directional sonobuoy system
OR.3583	Airborne anti-submarine detection system ASV with better *schnörkel* detection than ASV.21
OR.3586	Anti-submarine and surface vessel radar leading to ASV.21. Joint Naval requirement with AW.298 then AW.339 for use on Gannet, Buccaneers and Shackleton with Green Cheese.
OR.3589	ASV.21 synthetic radar trainer.
OR.3613	Mk.1c sonobuoy improvements for ASR.357
OR.3629	Sensor such as SLAR and IR and optical linescan for ASR.357
R.2/48	Specification for a maritime reconnaissance aircraft to OR.231. Saunders-Roe P.104, Shorts PD.2 and PD.3, Supermarine 524 proposed,
R.5/46	Specification for a maritime reconnaissance aircraft that became Avro Shackleton.
R.42/46	Specification covering the MR Liberator's replacement by Avro Shackleton
R.112D	Maritime reconnaissance flying boat to OR.231
SR (A) 420	Requirement for Nimrod replacement that led to the Nimrod MRA.4.
SR (A) 909	Acoustic processor upgrade for Nimrod.
SR (A) 910	Tactical control system upgrade for Nimrod
USW.111	Improved air-launched, anti-submarine homing weapon. See AST.1186.

RADARS

ASV.13	Air-to-Surface Vessel radar fitted to RAF Shackleton MR.1 but finally replaced by ASV.21 on the Shackleton MR.3.
ASV.19	Thorn-EMI Maritime Reconnaissance System fitted to Gannet and Seamew.
ASV.21	Air-to-Surface Vessel radar initially intended for the Buccaneer S.1 as target acquisition for Green Cheese, it was ultimately fitted to Shackleton MR.3 and Nimrod MR.1.
Searchwater	ASV radar fitted to Nimrod MR.2 and used as airborne early warning radar in Royal Navy Sea King AEW.2 Sea Kings.
Searchwater 2000	Thales Sensors ASV radar fitted to Nimrod MRA.4 and Sea King AS&C.7

ASW EQUIPMENT

APOJI	Automatic processing of Jezebel information. Signal processor used on OR.357 type with Jezebel sonobuoy.
ASQ.901	Marconi Acoustic processing system fitted on RAF Nimrod MR.1.
ASWEPS	ASW environmental prediction system that measured and output the sea's surface temperature, wave height and bathymetric temperature.
Autolycus	Exhaust gas detector to meet OR.3562 fitted to Shackleton and Nimrods to sniff out diesel electric submarines.
Backscratch	Cover name for UK terminals of USN SOSUS system (Sound Surveillance System). Cables come ashore at RAF Brawdy and Scatsta on the Shetlands. Sites in Denmark, Norway, Iceland (2), Nova Scotia, Newfoundland, USA, Bermuda, W Indies (5), Azores, Spain, Italy, Turkey (E.Med and Black Sea).
Barra	Australian-designed SSQ.801 sonobuoy used by RAF Nimrods. UK version is the Plessey SSQ.981 which is designed for cheaper manufacture while maintaining the same capability
CAMBS	Command Active Multi-beam Sonobuoy Active air-dropped sonobuoy used by RAF Nimrods
Clinker	An Admiralty study from 1962 examining the detection of submarines using Infrared that was possibly a follow-on to Yellow Duckling.
DIFAR	Directional Frequency Analysis and Recording Sonobuoy AN/SSQ 53 air-dropped sonobuoy used by RAF Nimrods
Jezebel	Air dropped Size A Sonobuoy.
Jezebel Miniature	A smaller, Size F, version of Jezebel air-dropped Sonobuoy.
Nangana	Australian electronics miniaturisation project that led to the *Barra* sonobuoy
Tandem	Sonobuoy system For Nimrod MR.1
Yellow Duckling	infrared submarine detection system. PbTe sensors to detect the wake of submerged submarines. Ultimately led to Infrared Linescan systems.

ENGINES

Compound-Griffon	Rolls-Royce RGC.30.SM – Griffon development for improved efficiency that used compounding technique similar to the Wright R-3550.
Turbo-Griffon	Rolls-Royce RGT.30.SM – Griffon development for improved efficiency and higher power. Turbo-charged version of the Griffon for flying boats and maritime patrol aircraft.
E.125	Napier Nomad I compound diesel. A 'flat 12' 2-stroke diesel with centrifugal and axial compressor in series with crankshaft driving one half of contraprop and turbine driving the other.

E.145	Napier Nomad II with modifications including linking the turbine to the crankshaft via a Beier gear and driving a single propeller.
Falcon	Putative Rolls-Royce H-24 piston engine of 100 litre capacity proposed for the R.2/48 flying boat. A much-enlarged Rolls-Royce Eagle. However, the Air Staff referred to the Turbo-Griffon being the Falcon.

ASW WEAPONS

Dealer	Mk.30 18in (46cm) surface-launched acoustic torpedo. Dealer-A lacked control surfaces, with lateral control by two side-by-side propellers and depth control by moving the battery fore and aft. Dealer-B had single propeller and eight control surfaces and entered service 1954.
Green Cheese	Fairey anti-ship missile based on Blue Boar airframe with an EMI-developed Red Dean radar seeker intended for use by Blackburn Buccaneer.
Lulu	Mk.101 nuclear depth bomb used by Shackletons
Martel	AJ.168 TV-guided, air-launched, anti-ship missile carried by Blackburn Buccaneer and, initially Hawker Siddeley Nimrod.
Mk.57 bomb	Possibly either a US Mk.57 Project E bomb, or a US Mk.57 Project N bomb. Status uncertain but probably supplied to replace the aged Mk.101 NDB used by RAF maritime aircraft.
Mk.90 Betty	Bomb, Aircraft, AS 2,500lb MC US Mk.90 Project N bomb for Shackleton MR.2 and MR.3.
Mk.101	Lulu NDB Bomb, Aircraft, AS, 1,200lb, MC. US Mk.101 Project N bomb for Shackleton MR.2 and MR.3 and possibly also carried by Nimrod.
P3T	Development designation for the BAe Sea Eagle turbojet-powered anti-ship missile that entered service on RAF Tornados.
Pentane	Vickers/Whitehead Mk.21 air-launched passive homing anti-submarine torpedo, also known as Zeta.

Project 7511	Marconi Stingray air-launched acoustic homing torpedo to replace Mk.44 and Mk.46. Carried by RN Lynx and RAF Nimrods.
Project N (Nuclear)	Cover name for US Navy dual-key nuclear weapons assigned to RAF maritime aircraft. Covered Mk.90 Betty and Mk.101 Lulu nuclear depth bombs (NDBs) assigned to RAF Shackleton MR.2 and MR.3 and Nimrods, and possibly other weapons. Similar custody arrangements as with Project E weapons for the RAF.
Red Angel	Unguided anti-ship rocket for use against Soviet cruisers in the Sverdlov class.
Sea Eagle	BAe Dynamics sea-skimming ASM developed from Martel.
Stingray	Marconi air-launched lightweight torpedo developed from Project 7511 and OR.1163 to replace Mk.44 and Mk.46 torpedoes. Homing torpedo designed to attack the fastest, deepest-diving nuclear submarines. A key weapon in neutralising Soviet ballistic missile submarines.
WE.177	Improved Kiloton Bomb to replace Red Beard. Nuclear weapon in three models, A, B and C. Variable-yield, tactical, boosted-fission gravity bomb and NDB (Nuclear Depth Bomb).
Yellow Sand	EMI anti-ship weapon with its guidance system developed by Smiths. Described as a homing bomb.
Zeta	See Pentane
Zoster	Air-launched homing torpedo. Launched at about 10nm (18.5km) to enter the water at 3nm (5.6km).

A Lockheed P-3C Orion of the *Deutsche Marine* MFG-3 resplendent in 50th anniversary markings. The *Marine* replaced its Breguet Atlantics with former Dutch *Marineluchtvaartdienst* Orions. The Air Staff had dismissed the Orion on a regular basis, but it soldiers on in service around the world. ©*Graham Wheatley 2014*

'Edinburgh Rescue from Rescue Zero One. Rescue helicopter One Three Seven is searching the area to the west of the rig: Rescue One Three One is investigating survivors in the water to the north-east of the rig. Helicopters Bristow Five Zero Yankee, Tango India Golf Bravo, Tango India Golf Oscar and Yankee Bravo are en route to the scene or holding on oil rigs. They will take over search when other helicopters have to refuel. The rig is now believed to be emitting hydrogen sulphide and there is a high probability of an underwater explosion. There is heavy smoke and flames spreading to the north of the rig. This is Rescue Zero One.' Nimrod MR.2 XV228 providing on-scene co-ordination and communications relay, Piper Alpha, 6/7th July 1988. Reproduced, with permission, from 'Fire in the Night' Stephen McGinty

Coastal Command and, later, No.18 Group Strike Command were tasked with the same anti-submarine and maritime patrol roles that they had carried out since the Great War. During World War Two, air/sea rescue became a major role while from the mid-Seventies patrolling the UK's Exclusive Economic Zone also became important. It is through these two roles that the British public became aware of the existence of the maritime patrol force, with photos of a Nimrod overflying an oil installation having similar influence on the public as those of Lightnings and Phantoms intercepting Soviet Tupolev Tu-95 *Bears*. In short these missions became the public face of the Nimrod and that is why these roles require examination.

Nimrod MR.2 XV228 of 201 Sqn on a Tapestry mission flies past the British National Oil Corporation's Beatrice Alpha platform in the Moray Firth. Note the lack of refuelling probe and wingtip ESM pods that identify this Nimrod as an early MR.2. XV228 would play a pivotal role in July 1988 during the Piper Alpha disaster by co-ordinating the SAR operations and is now preserved at Bruntingthorpe. *MoD via Terry Panopalis*

Rescue Zero One: Search and Rescue

You, the reader, might remember where you were on hearing President Kennedy had been shot or to a younger generation, the attacks on the World Trade Centre. British oil workers of a certain age remember where they were when they heard of the Piper Alpha disaster. This author remembers distinctly: a land rig in the Niger Delta, having caught the tail-end of the BBC World Service news on a trusty Sony short-wave radio. Orbiting above the inferno for more than eight hours, *Rescue Zero One*, Nimrod MR.2P XV228 acted as an airborne command post for the operation as rescue craft attempted to save the 226 workers on the Piper Alpha. Sixty one survived.

The RAF has been fishing people; aircrew, seafarers and civilians, out of 'the drink' since its formation in 1918, an aspect of its operations that was honed in World War Two with the need to rescue downed crewmen to fight another day. High-speed launches were the main means of rescue, with seaplanes such as Supermarine Walrus providing more reach in the early years of the war when the main trade was Fighter Command pilots shot down in the English Channel. As the extent of the war widened, the UK-based units of the RAF, in the shape of Coastal Command, became tasked with helping the survival, and if possible, the rescue of aircrew and seamen in the North Sea, Bay of Biscay and Atlantic Ocean. Flying boats such as Shorts Sunderlands and Consolidated Catalinas would be the obvious choice for such a task, alighting on the sea, picking up survivors and whisking them to safety. Unfortunately this, air/sea rescue (ASR), was easier said than done and the oceanic sea-state was not always suitable for landing, despite the valiant efforts of the seaplane pilots.

As the number of flying boats in the RAF inventory diminished and landplanes such as the Lancaster, Liberator, Wellington, Warwick and ultimately Shackleton, took on the maritime patrol task, the ASR role was taken up by these aircraft. The basic premise being that if an object as small as a submarine periscope could be detected by an aircraft, survivors in liferafts and lifeboats could also be found. Note the use of the words *liferafts and lifeboats*, large objects are easier to find than small and finding a single person lost at sea is extremely difficult. Sea survival courses, such as those for offshore oil workers, recommend that the primary aim is to either get out of the water into a liferaft, or make a 'huddle' with fellow survivors to maximise their physical size and make them easier to see from the air. While making the survivors easier to detect, these actions also help reduce the main killer of people in the sea: hypothermia.

Until January 1941, when the Directorate of Air/Sea Rescue (DASR) was formed to co-ordinate the activities of high-speed rescue boats and Coastal Command aircraft, there was no organisation dedicated to saving downed aircrew. DASR only got into its stride from September 1942 and as offensive operations against the Reich escalated, its role became increasingly important.

Throughout the Second World War, all aircraft were fitted with some form of liferaft, pilots of single-seat fighters had a Type K or KS (made from rubberised silk, lighter and thus more compact) 'dinghy pack' that doubled as a seat cushion. Multi-engined aircraft had the larger seven-man J-Type liferaft that could be deployed and boarded once the aircraft had ditched but bomber crews also used the Type K or KS, and took to these on baling out of a doomed aircraft. Once in a liferaft, the airmen (or seamen if they abandoned ship in a liferaft) were at the mercy of the wind and current, the raft having no propulsion apart from rudimentary paddles if the crew were lucky. Deployment of a sea-anchor could reduce the amount of wind-driven drift, but could not prevent the raft moving towards an enemy-held coast on currents. So, aside from the humanitarian act of saving the ditched crews lives, having invested heavily in their training there was a great incentive to recover the personnel rather than have them drift into captivity or be swamped in heavy seas.

The answer was to help the survivors save themselves by delivering the means of their own salvation and the difference between a liferaft and a lifeboat is propulsion. A lifeboat is a rigid, fairly sturdy vessel that can be steered and sailed whereas a liferaft, while keeping its occupants out of the water, is at the mercy of wind and currents. Suffice to say, given the choice, always go for the boat. Uffa Fox was a very successful builder of sailing craft, particularly in the smaller classes of lightweight dinghies. A prolific designer of small boats, Fox had served his time with Sam Saunders the boat builder and of course, builder of flying boats. By 1920 Fox had left Saunders and had established his own boat building business which by 1939 was well-respected in the world of sailing dinghies.

Down in the drink. A Whitley crew takes to their J-Type dinghy after ditching. Even in this somewhat benign sea-state, conditions in a life raft are miserable. The airborne lifeboat might not sound much better but is in fact a great improvement that increased a crew's chances of living to fight another day. *Author's collection*

Warfare drives innovation in many fields, but also brings opportunities for innovators such as Uffa Fox. Well aware of the problems associated with being in a small boat on open seas, Fox could appreciate the difficulties of being adrift in a liferaft and sought to provide the means for aircrew to make their way back to friendly territory, ease their predicament or buy them time until rescue by ship. Fox applied his expertise in designing robust but lightweight boats to develop a lifeboat that could be air-dropped by parachute. The basic design was for a lifeboat that could be sailed under engine power or, after stepping a mast and setting sail, could steer a course for friendly shores. The lifeboat would also carry rations, first aid and survival kits, navigation equipment, radio transmitter, inflatable 'exposure suits' and morale-boosting cigarettes.

Fox's 1942 design entered Coastal Command service as the Airborne Lifeboat Mk.I and was 32ft (10m) long and constructed in marine plywood. Propulsion was a pair of motorcycle engines, with the ultimate design being a Vincent 4hp (3kW) two-stroke outboard motor, two of which gave a speed of 4kt (7km/h) with ranges of up to 1,000nm (1,852km) quoted for some models. In the event of engine failure the lifeboats were equipped with mast and sail plus a layman's guide to sailing included in the survival kit. These boats were also fitted fore-and-aft with inflatable buoyancy chambers that conferred a self-righting capability, useful if the boat landed badly or upturned through mishap or bad weather. After a period in service, lessons learned from operational use of the Mk.I were applied to the Mk.II, which was slightly smaller at 25ft (7.6m) long and the outboard

motors were replaced by a single Austin 8hp (6kW) inboard engine.

The Mk.I was carried by Lockheed Hudsons of No.279 Sqn from April 1942 until February 1943, when the Mk.1A came into service, which along with the Mk.II, was fitted to the Vickers Warwick ASR.I. Postwar, the Warwicks soldiered on until 1946, when they were replaced by Avro Lancaster ASR.III, a modified GR.III that was based on the Lancaster B.III (essentially a B.I fitted with Packard-built Merlin engines). The ASR.IIIs were stripped of most of their defensive armament and carried an Airborne Lifeboat Mk.II in the bomb bay.

As the Avro Shackleton MR.1s came into service, replacing the Lancaster GR.IIIs, the aim was to use standard Shackletons to carry a new Mk.3 airborne lifeboat that could be fitted in

Yachtsman Uffa Fox designed the Airborne Lifeboat to enable more aircrew to be recovered after ditching. The lifeboats were fitted with engines, sails, survival suits, rations and a radio. Many aircrew lived to fight another day thanks to Fox's innovation. This model is on display in the RAF Museum, Hendon.

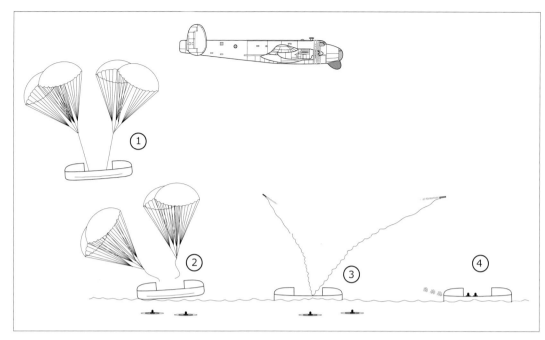

The MK.3 Airborne Lifeboat in action.
1) Boat descends on four parachutes and self-righting chambers inflate.
2) Parachutes release as soon as boat lands on sea.
3) Rockets fire, dragging floating lines to either side of the boat.
4) Survivors in the water pull themselves along the ropes to board and then secure the lifeboat prior to setting sail towards friendly forces.

A Shackleton MR.1 with a Saunders-Roe Mk.3 Airborne Lifeboat in its bomb bay. Such boats became less relevant as the capability of helicopters and other SAR equipment carried by aircraft improved, allowing survivors to be located more quickly and hoisted to safety. *MoD*

around thirty minutes. The Mk.3 was designed and built by Saunders-Roe, its 31ft (9m) hull crafted in aluminium and fitted with the Vincent engine described above, and carrying fuel for a range of 1,080nm (2,000km). The Mk.3 could sustain ten people for 14 days and as well as a radio, was equipped with a radar transponder to aid search aircraft and ships. The Mk.3 used four 42ft (12.8m) diameter parachutes rather than the six of the Mk.II.

In operation the Mk.II airborne lifeboat's deployment comprised a sequence of separate, but co-ordinated actions. The delivery was controlled by the bomb-aimer using the aircraft bomb sight as the aircraft ran in at 700ft (213m) and released the boat to place it upwind of the survivors. A static line pulled on a pilot drogue that slowed the boat before deploying the six

parachutes used for its descent, during which the buoyancy chambers inflated to ensure the boat was upright as soon as it landed. On touchdown the parachutes were released to ensure the canopy and shrouds did not foul the boat and hamper embarkation by the survivors. As soon as the parachutes were released, a sea anchor deployed and a pair of rockets fired, pulling 300ft (91m) floating lifelines either side of the boat. These floating ropes could then be used by survivors to pull themselves to the lifeboat. Once in the boat the crew made it secure, treated injuries and donned the inflatable exposure suits before posting a lookout and starting the engine. Stepping the mast and setting the sails would also be a good idea, as that would make the boat more visible to rescue ships and, if the wind was fair, take the strain off the engine.

Glider Lifeboat

The success of the airborne lifeboat spawned a rather remarkable project, from the home of flying boat: Saunders-Roe's P.108 lifeboat glider. The Admiralty, having seen how the RAF could pluck crews from under the enemy's nose, wanted a similar capability from their carriers. Fairey had in summer 1945 fitted a smaller lifeboat, 17ft 9in (5.4m) long, under a Fairey Barracuda, but that type's limited payload and the need to operate from carriers made these much less capable than the Coastal Command equivalent, only having sufficient fuel for 104nm (193km). This was fine for the North Sea or the English Channel but by mid-1945 the Navy's offensive operations were going to be in the Pacific.

During World War Two, the usual method for carrying heavy or outsize loads that a transport aircraft could not accommodate was to put it in a glider, so Saunders-Roe in January 1951 took the obvious route to allow a well-equipped airborne lifeboat to be launched off an aircraft carrier. The P.108.1 project had originally been to meet a Coastal Command requirement but when members of the Admiralty's Directorate of Air Weapons (DAW) attended a P.108 briefing on a possible replacement for the Supermarine Sea Otter, they were sold on it. The piston-powered Sea Otter was used for ASR, but as the Navy strove to rid itself of gasoline-fuelled types, the Sea Otter's days were numbered. DAW wanted the P.108.1 to operate from carriers, but it was too big and would need to be reduced in overall size to allow it to be towed by the forthcoming GR.17/45 aircraft (which became the Fairey Gannet). This smaller type was designated the P.108.2 to meet Admiralty requirement AW.259, a 30ft (9.1m) long lifeboat that carried two crew and was to be towed behind a Gannet to the survivors' location up to 350nm (648km) from the carrier. Lift and control was provided by 'jettisonable auxiliary flight surfaces' – wings and a butterfly tail were selected – and these were to be of simple construction. With an AUW of 5,000 lb (2,268kg) the lifeboat glider's constant chord wings had a span of 51ft (15.8m) with the chord being 9ft (2.7m) and the overall length was 42ft 9in (13m). As a lifeboat, P.108.2 was powered by a pair of Vincent HRD T5 two-stroke engines and with 500 lb (227kg) of petrol (equates to 68.6 Imperial gallons/312 litres) that could provide a range of 760nm (1,408km) at 6kt (11km/h), which could be augmented or replaced by sails.

The DAW wanted the Gannet/P.108.2 combination to be catapult-launched but as the DAW's H G Jones said in October 1951 'We do not think this will be a practicable proposition.' While the Director of R+D Projects, A S Crouch, considered it 'technically possible but hazardous.' The next option was snatch pick-up by an airborne Gannet, but trials in the latter years of the war using Waco Hadrian gliders and Douglas Dakotas had shown that a winch within the towing aircraft was required to absorb the shock on snatching the tow-line. Although this system would work, such a winch could not be fitted in every Gannet in the fleet. Eventually the Admiralty settled for using rocket-assisted take-off gear (RATOG) on the Gannet to provide the shock-free acceleration for the ensemble to take-off in 450ft (137m) with a 22kt (41km/h) wind-over-deck.

By February 1952 the Admiralty were becoming interested in more diverse uses for the P.108.2, specifically to deliver and extract 'raiding troops' and Saro updated the P.108.2 into the P.147.1 Sea Raider that could accommodate two 'Cockle' kayaks and four Royal Marines Commandoes. The glider lifeboat soon fell from favour as the long-range Shackletons with their Mk.3 lifeboats came into service and the hazards and complexity of operating such equipment were recognised as too much of a risk for little reward. As new, more capable helicopters came into service in the late Fifties, they now provided a more suitable platform for rescue and troop insertion.

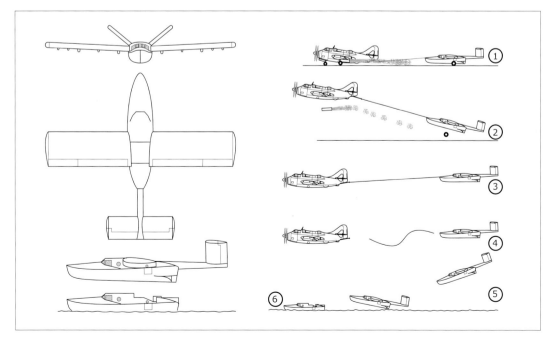

The Saro P.108.2 glider lifeboat.
1) Glider and Gannet on carrier deck. Gannet uses RATO to take-off.
2) Gannet and P.108 climb away, jettisoning the RATO gear.
3) Gannet tows P.108 to area where survivors are in the sea.
4) Gannet releases P.108.
5) P.108 glides in to alight on the sea.
6) P.108 sheds wings and tail, picks up survivors and motors back to safety.

Nimrod SAR

By the 1970s the arrival of the long-range SAR helicopters such as the Westland Wessex HAR.2 and later, the Sea King HAR.3 had revolutionised the UK's SAR capability. The existing SAR procedures had been developed around the Shackleton's '…operating speed and navigational limitations…' and as such it was '…considered inadvisable to apply existing SAR procedures to the Nimrod…', so new methods were developed. Crews soon discovered that the SARBE (search and rescue beacon equipment) homer that had been transferred from Shacktons was unsatisfactory so a new Burndept homing system was installed that improved the use of personal location beacons (PLB) if they were available. One critical aspect in flying SAR missions is accurate navigation as the search patterns depended on accurate positioning and allowance for drift of survivors from an incident site, the 'datum point', and the Nimrod's navigation system improved that no end.

Once the longer-ranged helicopters became available, Coastal Command's role in a SAR operation became one of first response, followed by support in a co-ordination and on-site command role, as shown in the heading. By first response, the Nimrods were normally first on the scene of any incident, be it the 1979 Fastnet Race, the *Alexander L. Keilland* in 1980 or Piper Alpha in 1988, to provide immediate assistance to survivors in the water by dropping Lindholme Gear.

The Fastnet disaster of August 1979 saw eighteen people dead in what was, until then, the largest peacetime search and rescue operation undertaken in the UK SAR area. The bi-annual Fastnet yacht race was hit by storm force winds and 25 yachts were wrecked. Fifteen sailors and three rescue personnel were killed, with many more yachts upturned and their crews in need of rescue. Then in March 1980, the semi-submersible drilling rig, *Alexander L. Keilland*, which was being used as an accommodation vessel for the Edda platform in the Ekofisk field, suffered a structural failure in one of its five legs. It capsized in 23 minutes killing 123 people. However, the event that brought the search and rescue activities of the Nimrod into the public eye was the July 1988 Piper Alpha disaster. The Piper Alpha platform was one of the biggest oil and gas producers in the North Sea and on 6th July 1988 two explosions and subsequent fires killed 167 personnel. The Cullen Inquiry investigated the causes of the disaster and Lord Cullen's report led to a major reassessment of oil and gas operations in the North Sea.

In the event of such disasters, a Nimrod and its crew that had been held at readiness, would be scrambled and despatched to the scene of the incident. On the arrival of rescue helicopters or other ships, the Nimrod would act as a communication relay for the operation, passing information back and forth from Rescue Coordination Centre Edinburgh at RAF Petrevie or for the southern areas of responsibility, Plymouth. In 1997, these were combined to form the Aeronautical Rescue Coordination Centre (ARCC) at RAF Kinloss, which since July 2012 and the end of RAF maritime operations has been Kinloss Barracks.

During the process that led to the HS.801 being adopted, OR.381 had listed the equipment for the search and rescue role which included receivers for the search and rescue beacons such as SARAH and the containers for Lindholme Gear. During World War Two a number of maritime patrol/air sea rescue stations developed their own sets of air-droppable rescue equipment, but eventually a standard set of kit was developed. Called Lindholme Gear it comprised five (later reduced to three) containers that could be fitted to the bomb shackles in the aircraft's bomb bay. The middle container held a nine-man liferaft to which were attached

The United Kingdom area of search and rescue responsibility extends westwards to the middle of the Atlantic, northwards to the Faroes and southwards beyond the Southwest Approaches and into the Atlantic. Until 2010, RAF Nimrods could provide rapid response and support to SAR operations in these areas.

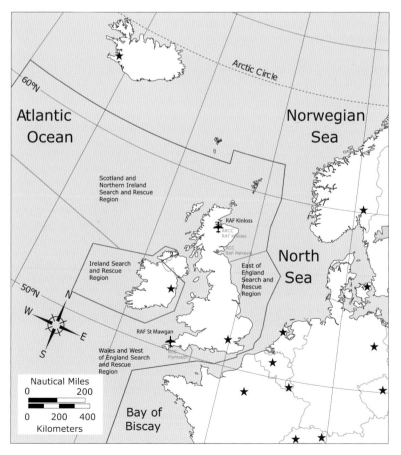

Nimrod XV248 of 201 Sqn overflies a semi-submersible drilling rig similar to the *Alexander L. Keilland*. The *Keilland* lost one of its five legs and capsized in 23 minutes killing 123 people. This was the Nimrod's first major incident in the North Sea and conducted search and co-ordination operations. Many of the lessons learned would contribute to future SAR operations. *Terry Panopalis Collection*

four (later two) containers holding survival equipment including rations and drinking water. Delivered upwind of the survivors and normal to the run of the sea, the containers were dropped as a stick so that the floating rope that joined the containers spread out in a line. The liferaft deployed on impact and drifted downwind to the survivors who could use the floating rope between containers to pull themselves to the liferaft. Once in the raft, they could pull the other containers to the raft.

While the Shackletons had dropped Lindholme Gear to improve the chances of survivors until a ship arrived, new helicopters allowed rescue within the hour. The sensors: radar, infrared and radio beacons, developed to find submarines were ideal for finding survivors at sea and from 1970 this became a very important task for the RAF. The increase in oil exploration activity in the North Sea also saw the Nimrods becoming a fast response SAR asset with a loiter capability to allow on-site co-ordination and it is in this role that the Nimrod entered the public eye. This was the aspect of RAF maritime operations that most exercised the media critics over the scrapping of the RAF's maritime capability, not its anti-submarine or maritime patrol role. Nimrods were seen as a sign of imminent rescue for those in peril on the sea, the sight of a Nimrod circling over a liferaft or people in the sea being a sure sign that rescue, by a boat or helicopter, was on its way.

By 2013 a contract for provision of SAR around the UK had been signed with Bristow Helicopters and by 2017 the RAF's bright yellow Sea Kings would be retired and Sikorsky S-92s and AgustaWestland AW.189s would operate in the SAR role. While not as rushed as the ending of the maritime reconnaissance role, it was the end of yet another era for the RAF.

New Role for Nimrod: Tapestry

The 1970s saw most nation states with a coastline establish Exclusive Economic Zones (EEZ) that since 1982 have been deemed to be an area of sea out to 200nm (370km) from their coastline. A country can exploit natural resources such

BAe's Coastguarder demonstrator G-BCDZ on a demo flight over the Irish Sea. The Coastguarder typified the middle-weight maritime patrol type fitting between the light EEZ patrol aircraft and the full-blown anti-submarine platform such as the Nimrod. Coastguarder was fitted with Searchwater radar and could carry ASW weapons. *via Avro Heritage*

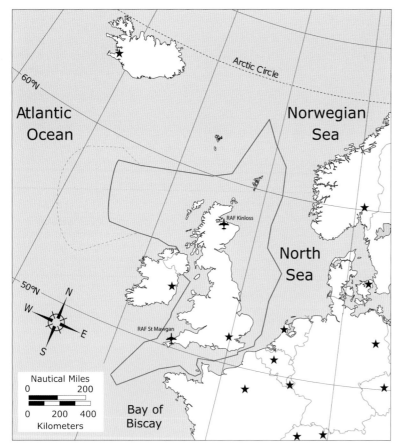

Atlantic
Ocean

Norwegian
Sea

North
Sea

Arctic Circle

60°N

50°N

RAF Kinloss

RAF St Mawgan

Bay of
Biscay

Nautical Miles
0   200

0   200   400
Kilometers

Britain's interest in Exclusive Economic Zones (EEZ) became much more intense after the extent of the oil and gas deposits on the continental shelf became apparent. Note the extension due to the '200-mile' limit extending from St Kilda, with the further, pre-1997 claim (dashed line), extending from Rockall.

BAe utilised the advances made in anti-submarine warfare during the 1980s to develop the P.132, ASW version of the ATP airliner. It was to carry a MEL Marec radar, AQS.902 acoustic suite and a FLIR turret. No sales were forthcoming.
BAE Systems

as oil or fish, within their EEZ with other nations being required to acquire a licence to operate in another's EEZ. The most famous EEZ is that of Iceland, whose declaration of an EEZ beyond its territorial waters (a zone extending 12nm/22.2km from the coast) sparked off the Cod Wars. The adoption of EEZs produced a flurry of

activity in aircraft manufacturer design offices, with companies eager to cash in on the need for aircraft to patrol the seas and guard the nations' assets. These did not require the full anti-submarine capability of a Nimrod or an Orion, but from that point the maritime patrol market could be split into full-blown ASW platforms and small EEZ surveillance aircraft.

During the 1970s at Woodford, in parallel with the NGMR studies, the design teams were taking a different view of the maritime patrol market, particularly that burgeoning market for 'light MR' aircraft for EEZ surveillance, fishery protection and anti-smuggling operations. These roles did not require the endurance and weapons fit of a Nimrod or Orion, but could take advantage of the developments in radar and other sensors that the full-blown ASW platforms were now carrying. In 1978 BAe Woodford built the Coastguarder, a maritime patrol type based on the HS.748 Series 2 and aimed squarely at the EEZ patrol market, but by 1982 BAe was looking at a more capable ASW version, the ASW Coastguarder. This was fitted with a MEL Sea Searcher ASV radar, GEC AQS-902 acoustic processor from the Westland Sea King HAS.5 and four underwing hard points for weapons, including Stingray torpedoes and Sea Eagle ASMs. The Coastguarder did not attract any orders, possibly due to being seen as an old design, effectively an update of the HS.748MR of 1961.

By 1987 BAe began to examine their Advanced Turboprop (ATP) feederliner as a maritime patrol type in what appears to have been

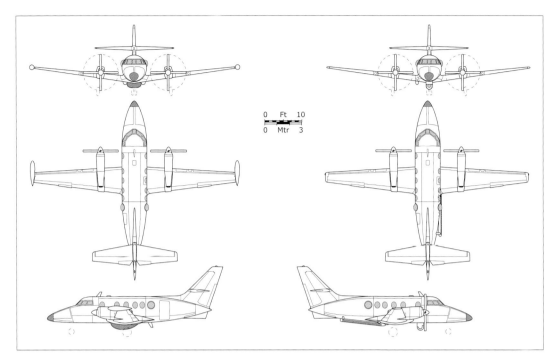

0 Ft 10
0 Mtr 3

Two variations on the Jetstream EZ theme. The left-hand variant is the Jetstream 31EZ with wingtip fuel tanks and ventral ASR 360 radar while the right-hand variant is the Jetstream 31 Coastal Defender fitted with a SLAR on the rear fuselage and a FLIR turret under the forward fuselage. The EEZ patrol role was filled by a variety of commuter types fitted with a wide variation of equipment.

a Warton project as its designation, P.132, continues the Warton projects series. While it could be described as *Nimrod-lite,* it used the AQS.902 system and APS.504(V)5 surveillance radar as used in the later AirTech CN-235 Persuader patrol aircraft. Weapons were to be carried on four underwing pylons and a FLIR was fitted in a chin turret. With its new PW120 turboprops and greater endurance, the P.132 might have been aimed at a market that did not really exist – too good for the 'light MR' EEZ patrol market and not good enough for the full-blown ASW role.

Meanwhile, at the former Scottish Aviation plant at Prestwick, the design teams were looking at the light EEZ patrol market that was being taken over by the Embraer Bandeirante and Britten-Norman Maritime Defender. The smallest airliner in the BAe stable was the Jetstream 31, a 19-seat feederliner that was eminently suitable for the role and so the Prestwick team embarked on design studies. The first was fitted with a ventral radome for an ASV radar such as the Bendix-King RDR-1400C and wingtip fuel tanks, one of which could be fitted with a searchlight. The other variant, the Jetstream 31 Coastal

BAe produced a few studies for the EEZ patrol role, including the Jetstream 31EEZ. This was based on the Jetstream 31. No customers were forthcoming, possibly due to the alternatives being cheaper, particularly the variants of the Embraer Bandeirante. *BAE Systems*

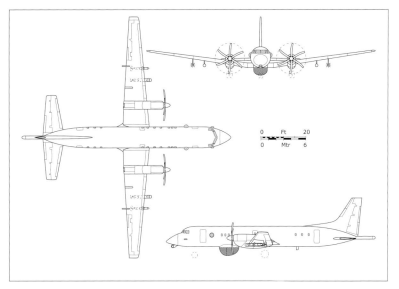

This general arrangement drawing shows the P.132 equipped for ASW and anti-ship missions with Stingray torpedoes and Sea Skua ASMs. The type could also carry Sea Eagle, Harpoon or Exocet ASMs on its underwing hardpoints.

The last flying Vulcan, XH558, is a B.2MRR and conducted Maritime Radar Reconnaissance operations with 27 Sqn from 1973. The type's H2S Mk.9 radar could perform wide area surveillance from high altitude to provide a 'surface picture' of the North Atlantic. They also monitored the North Sea, which soon became a very hectic area with the increase in oil-related operations. These aircraft could also be fitted with air sampling pods.
©Graham Wheatley

Defender, is more interesting as it was to carry a SLAR under the starboard rear fuselage and a FLIR turret on the forward fuselage centreline. Neither version was armed, their role being SAR and offshore surveillance, with observation blisters fitted to the rear fuselage. As with the Coastguarder and the P.132, neither attracted any orders and both were quietly forgotten.

National Treasures

The top-end maritime patrol aircraft were essentially wasted in the EEZ patrol role, but despite this, most countries that possessed anti-submarine maritime patrol aircraft used them for EEZ security. The RAF had a name for such missions – Tapestry.

Britain has been described as an island of coal surrounded by a sea full of fish. Both brought prosperity and development to the country and until the 1970s, there were no real economic threats to these valuable assets. Then, with the opening of markets and economic co-operation with Europe, the seas around Britain were soon being fished by foreign boats. Fishery protection vessels attempted to police this new order, but with limited resources and an ever-increasing workload, the fisheries patrol ships were becoming stretched. The Fishery Protection Squadron had become involved in the Cold War during 1958 when Russian ships interfered with British trawlers, but their operations were further expanded during the three 'Cod Wars'. Icelandic waters had been a rich source of cod for the UK deep sea trawler fleet, but Icelandic concerns about overfishing prompted the Icelandic Government to extend their restricted zone initially from 4nm (7.4km) to 12nm (22km), then to 50nm (97km) and ultimately to 200nm (370km). The claim to a 200nm (370km) limit by maritime nations dates back to 1947, but the extension of a country's sovereignty over maritime resources to 200nm became only became formally adopted after the UN Convention on the Law of the Sea in 1982. As a result of this, the United Kingdom gained the fifth largest Exclusive Economic Zone on earth and within that zone still lie great reserves of oil, gas and fish.

The Royal Navy and RAF became involved in the protection of British trawlers in the late Fifties with Shackletons operating during the First Cod War (1958) and Nimrods during the Second (1972-73) and Third (1975-76). At one point in the Third Cod War, a number of other RAF types were involved including Handley

Page Hastings, which sported mission tallies in the form of stylised fish. These Hastings were T.5 trainers that were normally used to train Bomber Command Navigator/Plotters on the Navigation and Bombing System (NBS) used on the Victor and Vulcan. Fortunately the H2S Mk.9 radar could also be used as an ASV radar and as shown in Chapter Two, was the radar intended for the R.2/48 flying boat. Thus equipped, the Hastings had since 1974 been utilised for Tapestry duties.

After the Cod Wars had been settled in 1976 with UK acceptance of Iceland's 200 nautical mile EEZ, RAF Nimrods continued to monitor the UK fisheries as part of what would become 'Tapestry', which is described below. However, using a full-blown ASW asset for fisheries patrol was inefficient and so the role eventually passed to aircraft operated by the Ministry of Agriculture and Fisheries. As of early 2014 these were under the direction of the Marine Fisheries Agency in England and Wales and Marine Scotland, with both organisations operating radar-equipped Reims Cessna F406 aircraft in the role.

The Black, Black Oil

By 1963 the true extent of the Groningen onshore gas field in The Netherlands was becoming apparent and petroleum geologists had advised that it extended out under the North Sea. Her Majesty's Government issued licences to explore for gas in 1964 and the first gas accumulations were discovered in 1965. Exploration drilling in the British sector had been limited despite the granting of more exploration licences in 1965 for the waters further north. That soon changed with the discovery in 1969 of the massive Ekofisk field on the western edge of the Norwegian sector of the North Sea which prompted renewed exploration drilling in the British sector. In October 1970 BP announced the Forties field in the Central North Sea 97nm (180km) east of Aberdeen and then in 1971 Shell Expro discovered the Brent field in the East Shetland Basin, 270nm (500km) north east of Aberdeen. There now followed a rush by the 'Seven Sisters' as the major oil companies were known, to get a piece of the North Sea oil action, with Occidental finding the Piper and Claymore fields, Mobil discovering the Beryl field and numerous other fields showing 'good pay'. Aberdeen boomed and the clatter of Sikorsky S-61s and later, Aérospatiale Super Pumas, filled the airspace around the city.

Of course, the oil and gas were only useful if they could be produced and in the 1970s that meant a large structure to support the drilling and production equipment that brought the hydrocarbons to surface, then processed and sent them ashore for refining. All this meant that by the late 1970s there were a large number of production platforms in the North Sea and these valuable assets needed protection. The initial effort was to route destroyers and frigates through the areas around the installations, while the RAF operated '...regular (twice a week) routine patrols of the installations using Nimrods, Shackletons, Vulcans and other aircraft.' As for protecting the installations and guarding them from 'possible peacetime threats', a contingency plan was required. Operation *Oilsafe*, was drawn up whereby the Armed Services would be '...acting in support of the civil power.'

Nimrod XV246 flies past the semi-submersible drilling rig *Sea Quest*. This photo shows the future: The Nimrod MR.1 in its new Tapestry role and BP's *Sea Quest*, the rig that drilled the Forties discovery well in late 1970. The largest field in the North Sea, the Forties Field continues to produce oil and contributes greatly to the British economy. *Author's collection*

The Soviet Navy's fleet of intelligence gathering ships ranged from small trawlers, the archetypical spy vessel, to the large purpose-built ship such as the *Primor'ye*-class or the *Balzam*-class shown here. Incidents involving these vessels and UK oil installations prompted the UK Government to formalise Tapestry operations.
US Navy

Prior to the emergence of North Sea oil and the need to patrol the EEZ for foreign fishing vessels, Coastal Command was well practiced in shadowing fishing vessels of a different class: Soviet Bloc intelligence-gathering ships, such as the *Primor'ye* and *Balzam*-class, or the famous Russian spy trawler. These missions against the 'spyships' had a nomenclature of their own that included *Snare* – 'regular air surveillance operation in areas of security importance to ourselves' and *Solitude* – 'Re-locating by aircraft of a known intelligence-gathering vessel.' These also had rules of engagement that specified the height at which the approach was to be made, the distance from the target and the number of passes. On a *Snare* or *Solitude* mission, these were 400ft (122m), 0.25nm (0.5km) and three passes. *Confine* involved 'Special reconnaissance by RAF special aircraft.' and these had to maintain a height of 2,000ft (610m) and all aircraft in the area had to withdraw to 100nm (185km) during a *Confine* operation. Special reconnaissance was an official euphemism for electronic intelligence gathering (Elint) by 51 Sqn and its Comet 2Rs and Nimrod R.1s.

Tapestry, also known as Offshore Tapestry, has its origins in a 1972 Admiralty study of '...the activity on and beneath the offshore waters to see how best it could be safeguarded and organised for the benefit of the country.' This was viewed by the Admiralty as an extension of the fisheries protection activities that the Navy already carried out. At this time the patrol force comprised HMS *Jura*, described as a '30-year-old tug saved from the scrapyard' and some fishery protection vessels, neither of which were up to the new task, while the newspapers described how Russian ships and

submarines harassed oil and gas installations.

In February 1975 Minister of State for Defence, Bill Rodgers announced that five new vessels supported by four RAF aircraft, Nimrods, would be available by 1977. In June 1975 the nations bordering the North Sea held a conference in The Hague on how best to protect the oil and gas installations and '...combat malicious or accidental damage.' Sabotage had been viewed the main threat and *Hansard* in May 1975 reported that twelve bomb threats had been made in an eighteen month period by '...so-called Scottish patriot groups.' Not that the five new ships boosted confidence, they were '...more like deep-sea trawlers than gunboats, having only 16 knots and a 40mm gun.' In short there were two ships to protect 34 installations spread from the Leman gas field east of Cromer to the Thistle field north of the Shetlands and Her Majesty's Most Loyal Opposition demanded action to protect such a valuable asset. Some suggested armed hydrofoils, others proposed helicopters carrying Royal Marines Commandos based in north east Scotland. The latter did materialise as Commacchio Group, becoming the 43 Commando Fleet Protection Group Royal Marines, charged with security of the UK maritime nuclear weapons, ships and offshore installations. It had also been suggested by the NATO Council that these platforms could be fitted with surveillance systems such as air defence radars to close gaps in the radar coverage in the North Sea. An article in the October/November 1974 edition of *NATO's Fifteen Nations* suggested that the platforms should be capable of defending themselves and fitted with air defence radars, SAMs and even fighters! They could also be fitted with anti-submarine defences such as sonar and torpedoes! Such speculation was applicable only for a fully-blown war situation and eventually sense prevailed: this was a cold, not hot, war and none of these plans received any support. In fact some in the ministries viewed the NATO Council's proposals as an attempt to gain access to a share of Britain's oil wealth.

These patrols would cost money and before the oil revenues flowed in, cash was not available as throughout the mid-1970s the armed forces were subject to swingeing cuts. The four aircraft were to come from fleets that were being cut and Nimrod, Andover, Argosy and Hercules were available. All, apart from the Nimrod, would need to be fitted with suitable kit such as radar, cameras and searchlights.

At this point a reasoned voice is heard in the furore, the Head of Defence Secretariat (DS) 8, Brian Robson, in a letter to his colleague Head of DS5. 'I think we would all like to be relieved of the silly agitation which starts up every time a Russian ship approaches an oil or gas installation.' Robson points out that the rigs and platforms are in international waters and the 500m (547 yards) exclusion zone (known to installations across the North Sea as 'Our 500') was a marine safety rather than security zone. Robson also points out that 'A) an attack upon an offshore installation is by no means an easy business; B) proper contingency plans to deal with an attack when it occurs have already been drawn up.' Item C) on Robson's demolition of the security scares is that 'the dedicated patrolling force will be backed up by the full might of the Royal Navy, the Royal Air Force and, if necessary, the Army.'

A month later the size and composition of this patrol force was being queried by the Air Force Board, with the VCAS Air Marshal Sir Ruthven Wade: stating that '…there were two basic issues for consideration: whether the Nimrod was the right choice for the offshore patrol task and, how precisely the Nimrod effort required for the task should be presented.' The DUS (Air) John Brynmor then explained that the monies available for the patrol task, some £8.5m, could also be used to convert three Mk.1 Nimrods to Mk.2 or even create a fleet of maritime patrol versions of the HS.780 Andover C.1, many of which were becoming surplus. The DUS (Air) took the view that the Nimrod was the best option as it allowed the maximum flexibility of the maritime patrol force as a whole, but also warned that 'It would be prudent to avoid using the term "dedicated force" in this context since in practice the aircraft would be interchangeable with the surveillance aircraft.' The board concluded that the preferred option was to use the existing maritime patrol force for the role and that another two Nimrods should be added, but the size of the offshore patrol force should be re-examined when the Nimrod refit programme was finished and with the benefit of experience in the role.

So, four Nimrods were tasked to perform surveillance flights, Operation *Tapestry*, around the North Sea and these flights continued until *Tapestry* officially ended in August 1985 but the Nimrods conducted surveillance flights around the UK until withdrawal in 2010. The RAF's maritime patrol aircraft had gone from protecting the military lifeblood of a nation at war to safeguarding the economic lifeblood of a modern economy.

Surplus elsewhere, Armstrong Whitworth AW.660 Argosies such as XN814 were considered for Tapestry missions. As noted in Chapter Four, AW had proposed a maritime type derived from the Argosy back in the late fifties. Fortunately the Nimrod MR.1 fleet was also being reduced due to the withdrawal from Malta and so Nimrods became available for Tapestry sorties. *Tony Buttler Collection*

Appendix 3
The Blind Alley

'It is debatable whether the Rotodyne truly represents a promising and original line of development for the British aircraft industry or whether it is a freak which will lead the industry up a blind alley.' Brief for the Secretary of State for Air, 1st November 1961

'Normal development in A/S detection equipment allied with the development of the Rotodyne could make this an exceptional anti-submarine aircraft.'
Dr G S Hislop, Chief Engineer, Fairey Aviation, March 1959

The anti-submarine helicopter is generally associated with the helideck at the stern of one of Her Majesty's warships rather than the concrete ramp of a Coastal Command airfield. The RAF did have a short dalliance with helicopters in the maritime role, although much of this was mere curiosity about new ideas and technology rather than hard and fast requirements. RAF helicopters since 1952 have been used in the transport and SAR role while the FAA has been tasked with ASW from ships or land bases such as RNAS Culdrose or HMS *Gannet* at RNAS Prestwick. Search and rescue operations from these bases was an adjunct to the

FAA's anti-submarine task, covering south west England and south west Scotland. HMS *Gannet* relinquished its ASW role in November 2001 and became a dedicated SAR station.

Helicopters in 1950 were beginning to look like they could perform some useful tasks in the ASW field. The Royal and US Navies had tested helicopters from ships, but until the early Fifties, helicopters could barely lift a pilot and a passenger never mind ASW sensors or weapons.

The Air Staff had by September 1951 begun looking at helicopters to fill the short-range ASW role with the additional possibility of using these aircraft for Air-Sea Rescue (ASR), as they should have a faster response time than the ASR boats that were based around the coast. As ever the Air Staff were watching developments across the Atlantic and the Americans had developed a lightweight dipping sonar. Such systems were not new; having been developed for use by flying boats, a method that Saunders-Roe had a particular interest in. However, as has been indicated in Chapter Three, flying boats had to alight on the sea to deploy their dipping sonars and this was dependent on sea state. Helicopters on the other hand could operate a dipping sonar in higher sea states, assuming weather conditions made hovering possible. For littoral, that is, close

Anti-submarine helicopters have come a long way since the RAF conducted trials in the Fifties. This Royal Navy Merlin HM.1 of 820 Naval Air Squadron shows the many sensors associated with modern ASW helicopters. Most obvious is the radome for the Blue Kestrel radar, turret for the MX-15 electro-optical turret and the 'hole' through which the HELRAS dipping sonar is deployed. Weapons are mounted on pylons attached to the side of the cabin.

to shore operations, the flying boat was too expensive to operate, which made helicopters with dipping sonar more attractive.

As shown in Chapter Two, sonobuoys dropped by maritime patrol aircraft were also expensive as they had to be deployed in patterns to triangulate the target and had a life measured in hours. Dipping sonar on the other hand could be used for initial detection then, by retrieving the array and moving to other locations, could triangulate the submarine and provide a continuous track on the target as it moved. This was the rationale behind Hawker Siddeley's HS.828 in the 1970s, although HSA addressed the need for oceanic coverage by deploying the HS.828 from land-based aircraft.

Discussions on helicopter-borne ASW led to the issue in September 1951 of ASR.304, calling for a shore-based helicopter to evaluate the potential of the helicopter in the role. The requirement stated that 'A flight equipped with Bristol 171 helicopters is to be formed as soon as possible in Coastal Command.' The Bristol 171 would be the RAF's first British-built helicopter when it entered service as the Sycamore HR.14 in 1953, but in 1951 it was still in the development stage, having flown in July 1947. The object of the new unit was to evaluate the use of helicopters in the SAR and ASW in the 'Visual sighting' and 'Dipping Sonar' roles and then evolve '…operating techniques and procedures.' The helicopters were to operate as part of the Coastal Command Development Flight.

For this evaluation, the Bristol 171s were to be equipped with either a power-operated hoist, as specified in OR.884, or the American AN/AQS.5 dipping sonar, the 'either/or' being due to payload restrictions. In the SAR role the crew would be two pilots and a hoist operator, while for the ASW task it would carry a sonar operator and the dipping sonar equipment. Unfortunately, by the end of September 1951, there was some confusion about the dipping sonar, with Mr A Clements, R+D, Helicopters at the Air Ministry, advising that '…it cannot be established that the installation of the Dipping Sonar AN/AQS.5 equipment will be satisfactory, since no-one can at present tell us what the equipment is…' In reply, Wing commander J D Hughes, DDOR1, advised that the American sonar might not be available at all and that until the dipping sonar question was answered, no action should be taken on this.

October 1951 saw the rescue hoist being queried as well, with its drive system under discussion. Bristols proposed driving its hydraulic pump from a spur gear on the tail rotor drive

shaft, whereas the Air Staff wanted the pump driven via the main rotor shaft. Bristols were adamant this was not possible. However, by the year's end, there had been a decision on the rescue hoist drive system but no dipping sonar available and therefore the Air Ministry decreed that the first two aircraft should be delivered for evaluation without sonar kit, which was accepted this on 5th January 1952 and all mention of the AN/AQS.5 sonar was deleted from the amended requirement.

Eventually, the Sycamore in its HR.12 form (Type 171 Mk.3A) was evaluated as a SAR type, with the HR.14 (Type 171 Mk.4) entering service in this role, while the Royal Australian Navy used the Type 171 Mk.50 and Mk.51 at sea for plane guard and rescue duties on their carriers. The SAR Sycamore HR.14s were used until replaced by the much more capable Westland Whirlwind HAR.2 from 1955. Dipping sonar became standard kit on the Royal Navy's Westland Whirlwind HAS.7 ASW helicopters from 1957, but was never fitted to the Sycamore. Interestingly the Navy operated its Whirlwind HAS.7s in hunter/killer pairs as the type could carry either a dipping sonar or a torpedo, but not both at the same time.

What the OR.304 story demonstrated was the limitations of the 1950s helicopter. This was appreciated by all involved and indeed by anyone involved with seeking new applications for helicopters, which in the early Fifties had yet to find their ideal role. Helicopters lacked speed, range and most of all payload but with the application of the gas turbine to the helicopter, some if not all of these were addressed, but not to the extent that the advocates of the helicopter had promised. What was desired was an aircraft with the speed, range and payload of an

The RAF used the Bristol 171 Sycamore for search and rescue, but its limited range and lifting capacity limited its utility. When evaluated for ASW, Coastal Command were unconvinced by the land-based anti-submarine helicopter and left that niche to the Royal Navy.
Terry Panopalis Collection

219

aeroplane with the VTOL and hovering capability of the helicopter. An aircraft that can make the transition from vertical take-off with a helicopter rotor to wing-borne flight and land vertically at the end of its journey is generally referred to as a convertiplane.

The convertiplane has been the goal of aircraft designers since the Wright brothers and even after the passage of 100 years only one type that can be described as a convertiplane, the Bell/Boeing OV-22 Osprey, has entered service. The Patent Offices of the world are littered with designs for convertiplanes: tilt-rotor like the OV-22 whose engine pods and rotors translate to become propellers, tilt-wing like the Canadair CL-84 or the Ling-Temco-Vought (LTV) XC-142 whose entire wing moves through 90° to change from vertical to horizontal flight. The CL-84 was evaluated in the ASW role and was touted to the UK armed forces during the 1970s, with the RAF evaluating the type during tri-national (US Navy/Marines, RAF and RCAF) trials. Then there was the Rotodyne.

Most of the world's aerospace concerns, Britain's aircraft companies and especially the

experimental establishments had a fascination for the convertiplane, for no other reason than their being an interesting technology exercise. The Fairey Rotodyne wasn't a convertiplane *per se*, a better description would be a compound gyroplane, but was in the same class of high-speed VTOL aircraft. Rotodyne was a victim of politics and early environmental concerns: the politicians could not see its long-term possibilities and the environmental problems would have been solved in time. Interestingly the concerns were on noise, as were those about Concorde, but as a consequence of these worries, current 'big fan' airliners have become incredibly quiet in comparison. While developments in engineering have been shown to address environmental concerns, such as those about jet noise, there is no foreseeable solution to politicians' short-termism.

The Rotodyne addressed the three perceived weaknesses of helicopters: it was fast, had long range and carried a decent payload. The Aeronautical Research Council's Helicopter Committee in March 1959 examined the Rotodyne with a view to its use as an anti-submarine platform. The committee's report was scathing, and this was included in a letter that found its way to Fairey Aviation at Hayes and in its turn prompted a broadside from Dr G S Hislop, Chief Engineer (Aircraft) of Fairey Aviation. Hislop politely informed the committee secretary that the committee was not in full possession of the facts, but the sentence that aggrieved Fairey most was 'The Fairey Rotodyne is unsuitable for the anti-submarine role because of its low hovering efficiency and high noise level.'

Dr Hislop, suitably piqued, began with the statement 'For example, nowhere is it stated that these remarks apply to the present prototype – which is designed for an altogether different purpose.' Hislop addressed the comments on noise first: 'Sufficient work has been done to confirm our knowledge that the external noise from the tip jets can and will be reduced to an acceptable level for commercial operation – which is more exacting than the naval requirement.' Hislop then cited Mr Roy Greatrex of Rolls-Royce who had praised Fairey's work on jet noise reduction. Greatrex was a pioneer of noise reduction on commercial jet aircraft and had set the standard for such work in the 1950s.

Hislop continued his 'strafing' of the Helicopter Committee's conclusions with 'The statement that the sonar operator would be unable to hear the echo within the aircraft is utter nonsense. The fact is that the major noise

Below: The Bell-Boeing V-22 is the first tilt-rotor aircraft to enter service, in the assault transport role with the US Marines and special operations with the USAF. The type could be used for ASW by the US Navy, who also examined the Osprey for search and rescue and carrier on-board delivery roles. *USAF*

Left: The Fairey Rotodyne was seen as the way forward in high-speed VTOL transport in the late 1950s and as such could combine payload and range with a hover capability. Fairey Aviation considered these attributes ideal for ASW and proposed the Rotodyne for the role. The Aeronautical Research Council's Helicopter Committee thought otherwise, prompting a rather angry response from Fairey. *Fairey Aviation via Phil Butler*

inside the present unfurnished aircraft comes from the powerplant and propeller – not the tip jets, which cannot be heard inside the aircraft.' Note the use of the word 'unfurnished' as Hislop points out that when sound-proofing was applied, the noise levels within the aircraft were '…down to present fixed wing standards which are much better than current helicopters.'

The second criticism, 'low hovering efficiency' now gets Hislop's attention, as he quite obviously thinks this is a nonsense term. 'What is meant by this? It cannot be in terms of h.p. required/lb weight, and presumably can only refer to fuel used.' Hislop then points out that '…this takes no account of developments envisaged or possible within the timescale for the high-speed A/S aircraft.' Not content with demolishing the Committee's arguments on these two items, points out that '…The Rotodyne is the *only* VTOL aircraft with proved high performance that can fly at over 200mph *now*, even in its prototype form.' In concluding his destruction of the Helicopter committee's points, Hislop concluded with 'Normal development in A/S detection equipment allied with the development of the Rotodyne could make this an exceptional anti-submarine aircraft.'

It fell to the Helicopter Committee's Chairman, Mr Herbert Squire, to pour oil on troubled waters in a letter to Hislop. Squire had been employed by the RAE in the immediate postwar period, mainly involved in helicopter research but by 1955 had moved to Imperial College. Squire maintained that the Committee had been misquoted in the letter that Hislop had seen. The sentence that had incensed Hislop so much (about the Rotodyne's unsuitability for ASW) '…implies and is meant to imply that the committee had not reached any firm conclusion about the matter.' Squire then turns to the 'hovering efficiency' comments and asked Hislop to supply '…data on the propulsion system together with a comparison between this system and others.' Adding that the committee would be happy to '…discuss any such data as you can supply.' Squire concluded with an apology, stating that '…if the letter has misrepresented the Rotodyne and assure you that this was not intended.'

The Rotodyne prototype continued to fly into 1962 and as Hislop said the noise from the tipjets was being addressed, but British European Airways (BEA), the main customer for the airliner with an order for six, had lost interest due to the noise level and the RAF had shown little interest in the Tyne-powered Rotodyne Z. This was a larger version to meet OR.344 and capable of carrying 8 tons (8 tonnes) of cargo, including vehicles loaded through the clamshell doors that formed the rear fuselage. The OR.377 requirement was for 100 tons (102 tonnes) of stores per day to be airlifted over a distance of 200 miles (320km). The RAF was to receive twelve Rotodyne Z, but the requirement was eventually filled by the HS.780 Andover. Westland Helicopters took over Fairey's aviation interests in May 1960 and essentially put the Rotodyne on the back-burner. Government funding had been withdrawn in late 1959, oddly enough mostly from the Napier Eland engine that powered the Rotodyne, and Westland were disinclined to take over the project, there being little incentive to continue.

Circular ASW

One of the oddest ASW aircraft proposed was based on yet another idea for a convertiplane. Mr W Stewart of the RAE's Naval Aircraft Dept. at Bedford in December 1958 produced a Tech-

Opposite, bottom: The Canadair CL-84 tilt-wing was used for ASW trials by the Canadian and US Navies. The '84' was also evaluated by the RAF to assess the suitability of the type in a variety of roles. Tilt-wing and tilt-rotor aircraft come into their own when long range or higher speed are required as they fly like a fixed-wing aircraft. Four CL-84s were built and two survive in museums. *Terry Panopalis Collection*

nical Memo for the Admiralty, and its content gives an insight into the innovative thinking on high-speed VTOL. The memorandum comes with a caveat – this is Stewart's personal view and did not reflect the views of the RAE. Stewart pointed out that in ASW the '…primary requirement is for an endurance in hovering of two to three hours.' And that the helicopter is the only solution to this but lacks the transit speed required to reach search area or move between 'dipping' sites. The limitation on speed was caused by the helicopter rotor so Stewart suggested stopping the rotor in flight and '…using the blades as fixed surfaces or retracting them and using a wing for orthodox fixed-wing flight.'

Stewart proposed an aircraft with and an AUW of 20,000 lb (9,072kg) capable of carrying 2,000 lb (907kg) of 'search equipment and weapons' and powered by a pair of turboshafts, such as the Napier N.Ga.4 Gazelle, rated at 1,400shp (1,044kW), as used on early Westland Wessex HAS.1 and 3. The aircraft Stewart sketched out was a fuselage surmounted by a circular wing of around 40ft (12.2m) in diameter with four 17ft 6in x 2ft (5.2m x 0.6m) blades at the periphery giving a total rotor diameter of 75ft (22.9m). The upper surface of the wing was continuous and fixed, while the outer section of the underside, which also carried the blades, rotated around the fixed inner section. This fixed portion of the wing was to carry the fuselage and a pylon-mounted Gazelle in a pod at the furthest outboard point of the fixed undersurface. These engines were to be fitted with tractor propellers providing

thrust in forward flight and '…torque reaction compensation in helicopter flight…' through differential thrust. The Gazelles would also turn the 'rotor' via a ring gear driven by a transmission and clutch system connected to a vertical drive shaft.

As for Stewart's rotor blade retraction system, he described two possible systems. The first was to retract them '…along their radial axis, i.e. by pulling them straight inboard.' However, this meant that the radius of the fixed wing had to be greater than the blade length and drag hinges could not be used. Drag hinges are used in rotor blades to allow them to move back and forth in the horizontal plane to cope with the acceleration and deceleration of the blade as it turns. Stewart then suggested that by including drag hinges, and using these to mount the blades on the rotating portion of the wing, the blades could be swung into the outer section of the wing by folding them forward about the drag hinge. If a rotor brake were applied the blades' momentum would pivot them forward about the hinge and they would swing to slots in the rotating portion as the rotation stopped.

The circular wing needed to be large enough to hold blades of sufficient length to allow the aircraft to hover, but this produced an aircraft with wing-loading too low for efficient high-speed flight. One problem with the smaller wing was reduced effectiveness of the propellers in the anti-torque and differential thrust control, so a ducted fan in the tail fin was proposed. One solution to this blade length/wing diameter/anti-torque problem that Stewart discussed was fitting a coaxial rotor, which would remove the torque reaction. This would involve what are now called 'closely-coupled co-axial rotors'. Stewart ends his memorandum with 'If the rotor blades can be designed without the need for flapping hinges, a fairly close arrangement is possible. Separate forward propulsion would be necessary. A neat aerodynamic layout is possible but mechanical design may be difficult.'

The Rotodyne and Stewart's concept show just how difficult it is to develop a viable high-speed VTOL aircraft. All such aircraft, convertiplanes of whatever stripe, are complicated bits of kit, and as with any aircraft complication means weight – the enemy of all aircraft designers. It is perhaps telling that the only practicable convertiplane in service, the OV-22 Osprey, is used as an assault transport. Perhaps the convertiplane isn't a blind alley after all, just a very, very long tunnel.

The circular wing rotor ASW aircraft was an interesting concept for a high-speed aircraft that could perform the ASW task with dipping sonar. The rotor blades 'flip out' from the rotating annular section of the circular wing, shown here in red. For wing-borne flight, the blades flip back into the circular wing. The rotor is driven via a clutch and gearing from the two underwing Napier Gazelle turboprops.

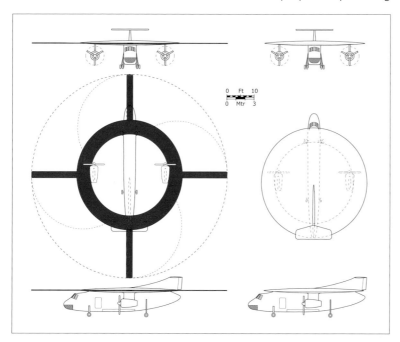

Index

AIRCRAFT
A.33, Saro 55
Albatross, Beriev Be-40P 170, 171, 195
Andover, HS.780 217, 221
Anson, Avro 73, 81, 91, 120
Argus, Canadair CL-28 26, 39, 62, 89, 94, 113, 116, 118, 120, 136, 141, 145, 153, 158, 175, 179, 189, 194
AS.69, Airspeed 35, 77, 78, 90
Atlantic, Breguet Br.1150 9, 100, 103, 105, 106, 117, 121, 159
ATP, BAe 212
Aurora, CP-140 113, 194
AW.189, AgustaWestland 211
BAC One-Eleven 151, 158, 161, 162, 167, 168, 169, 182, 187, 188, 189
BAC Ten-Eleven 22, 143, 167, 168, 169, 187
BAC.208 127
BAe.146 184, 185
Bandeirante, EMB-111A 181, 213
Bear, Tupolev Tu-95/142 13, 45, 205
Belfast, Shorts SC.5 128, 145
Beverley, Blackburn 64, 68, 155
Boeing 707 113
Boeing 737 181, 182, 183
Boeing 777 198
Br.1250, Breguet 61, 103
Brabazon, Bristol 77, 82, 90
Brigand, Bristol 75
Britannia, Bristol Type 175 43, 52, 62, 89, 90, 91, 92, 94, 99, 145
Britannic, Shorts 145
Buccaneer, Blackburn 20, 41, 42, 43, 45, 134, 202, 203, 204
C-121, Lockheed 24
C295ASW, Airbus Military 198, 199
Canberra, English Electric 44, 49, 53, 165
Catalina, Consolidated 7, 35, 49, 58, 61, 82, 202
CL-215, Canadair 63
CL-33, Canadair 53
CL-84, Canadair 220, 221
Coastguarder, BAe 211, 212, 214
Comet, DH.106 5, 52, 67, 123, 158, 160, 162, 163, 164, 165, 169, 170, 171, 174, 175, 176, 177, 179, 181, 182, 183, 185, 187, 189, 191, 193, 194, 195, 196, 216
Condor, Focke-Wulf Fw200 48
Constellation, Lockheed 24, 66, 82
Coronado, Consolidated PB2Y-3B 61
DC-7, Douglas 97, 98, 99
DC-8, Douglas 141, 183, 189
Defender, Britten-Norman 213, 214
DH.121 14, 112, 125, 126, 127, 137, 138, 141, 150, 151, 154, 155, 157, 158, 160, 161, 162, 163, 164, 165, 166, 167, 168, 182, 192, 199
DH.130, de Havilland 111
Do 24ATT, Dornier 64
Duchess, Saro P.131 61, 62
Emily, Kawanishi H8K 63, 65
Export Nimrod 163
F-111K, General Dynamics 156, 164
F.28, Fokker 147
F406, Reims Cessna 9, 215
Flanker, Sukhoi Su-33 18
Fortress, Boeing B-17 81, 172
Freighter, Bristol Type 170 77, 78
Frogfoot, Sukhoi Su-25 18
Fulcrum, MiG-29 18
Galaxy, Lockheed C-5A 145
Gannet, Fairey 37, 38, 42, 43, 72, 73, 94, 100, 103, 203, 209, 218
Globemaster, Boeing C-17 190
Hastings, Handley Page 24, 26, 47, 74, 75, 88, 155, 185, 215
Helix, Kamov Ka-27 18
Hercules, Lockheed C-130 9, 120, 139, 146, 156, 164, 190, 197, 198, 199
Hercules, Lockheed C-130K 156, 164
Hormone, Kamov Ka-25 223
Horsa, Airspeed 77
HP.117, Handley Page 134, 138, 148
HS.1011, Hawker Siddeley 4, 111, 112, 125, 126, 138, 147
HS.1023, Hawker Siddeley 125, 126, 147
HS.748, Hawker Siddeley 80, 159, 212
HS.800, Hawker Siddeley 35, 161, 165, 166, 173
Hudson, Lockheed 7, 207
Jaguar, SEPECAT 35, 159, 163, 183
JetStar, Lockheed 169
Jetstream 31, BAe 213
Knight Hawk, Lockheed MH-60S 27
LARC, Large Aircraft Replacement Concept 32, 39
Liberator, Consolidated 40, 83, 86, 98, 203, 206
Madcap, Antonov An-71 18
Mail, Beriev Be-12 63, 195
Mariner, Martin PBM-5 50, 60, 78
Mars, Martin JRM-3 63, 64
May, Ilyushin IL-38 45
Mercure, Dassault 182, 183, 184
Mosquito, de Havilland 75
MRSA, Multi-Role Support Aircraft 32, 190, 191
Neptune, Lockheed 4, 31, 38, 47, 61, 78, 79, 80, 91, 96, 97, 188
Nimrod 2000 195
Nimrod MR.1, Hawker Siddeley 4, 24, 45, 51, 61, 79, 80, 81, 83, 84, 86, 87, 89, 92, 93, 96, 118, 151, 152, 154, 155, 156, 161, 165, 169, 173, 174, 175, 176, 177, 178, 180, 193, 202, 203, 207, 208, 215, 217
Nimrod MR.2, Hawker Siddeley 4, 9, 19, 20, 43, 45, 46, 62, 73, 82, 83, 84, 86, 87, 88, 92, 93, 94, 116, 118, 135, 136, 152, 153, 154, 156, 157, 158, 161, 165, 166, 168, 170, 174, 176, 179, 180, 181, 187, 189, 190, 193, 194, 195, 196, 197, 202, 203, 204, 205, 206
Nimrod MRA.4, BAe 46, 175, 176, 192, 195, 196, 203
Osprey, Bell/Boeing OV-22 170, 220, 222
Oxford, Airspeed 72, 73, 77
P.104, Saro 4, 29, 58, 59, 60, 69
P.1154, Hawker 153, 155, 156, 165, 170, 176
P.132, BAe 213, 214
P.155, Piaggio 104, 105
P.162, Saro 28, 31
P.208, Saro 28, 31, 100, 101
P.340, Dornier 101, 102
PD.2, Shorts 50, 51, 57, 58, 62, 64, 66, 203
Percival Prince 73
Phantom, McDonnell Douglas F-4K/M 156, 158, 160, 165
Poseidon, Boeing P-8 5, 197, 198, 199
Princess, Saro SR.45 52, 70
Project 83, Fairey 100
R100, Vickers 77
Rotodyne, Fairey 129, 218, 220, 221, 222
S-92, Sikorsky 211
Sabre, Canadair 79
Sandringham, Short S.25 223
Sea Fury, Hawker 64
Seaford, Shorts PD.3 53, 54, 55, 56, 57, 59, 60
SeaMaster, Martin P6M 52, 61
Seamew, Shorts 35, 71, 73, 88, 202, 203
Sentinel R.1, Raytheon 198
Sentry, Boeing E-3D and E-3F 183, 186, 197, 199
SH-5, Harbin 63
Shackleton, Avro 4, 8, 9, 15, 19, 20, 22, 24, 26, 32, 34, 35, 37, 38, 39, 42, 43, 44, 47, 50, 51, 52, 58, 59, 60, 62, 63, 64, 66, 72, 74, 75, 79, 80, 81, 82, 83, 84, 85, 86, 87, 88, 89, 90, 91, 92, 93, 94, 95, 99, 104, 105, 107, 112, 114, 115, 116, 117, 118, 119, 120, 122, 123, 125, 127, 129, 131, 135, 136, 139, 140, 145, 150, 153, 155, 156, 157, 158, 162, 163, 164, 165, 168, 169, 170, 171, 173, 174, 177, 178, 186, 192, 196, 201, 202, 203, 204, 206, 207, 208, 210
Shetland, Shorts S.35 47, 48, 53, 55, 56, 57, 58, 64, 69, 215
Shin Meiwa PS-1 63
Solent, Shorts S.45 53, 61, 64
Southampton, Supermarine 78
Spectre, Lockheed AC-130A 25
Sperrin, Shorts 49
Stratocruiser, Boeing 82
Sturgeon, Shorts 35, 37
Sunderland, Shorts 4, 8, 9, 47, 48, 49, 50, 53, 54, 60, 61, 62, 63, 64, 74, 82, 88, 94, 201
Swordfish, Fairey 5, 40, 197, 199
Sycamore, Bristol Type 171 219
Tempest, Hawker 64, 171
Tiger Moth, de Havilland 72
Tornado, Panavia 45, 171, 188, 190
Tradewind, Consolidated R3Y 50
TriStar, Lockheed L-1011 145
Triton, Northrop Grumman MQ-4C 198
TSR.2, BAC 107, 153, 155, 164
Tudor, Avro 67, 82, 93
Type 175MR 43, 52, 62, 89, 90, 91, 92, 145
Type 189, Bristol 43, 52, 91, 145
Type 191, Bristol 37, 202
Type 192, Bristol 37
Type 206, Bristol 99
Type 207, Bristol 99
Type 524, Supermarine 58, 59, 69
Type 716, Avro 62, 92, 94
Type 719, Avro 66, 92, 93, 94
Typhoon, Hawker 64
Valetta, Vickers 47
Valiant, Vickers 42, 61, 76, 78, 109, 169, 206
Vanguard, Vickers 29, 79, 121, 130, 131, 132, 133, 135, 138, 145, 153, 158, 161, 162, 167, 169, 192
Varsity, Vickers 35, 37, 74, 75, 76, 77, 78, 80, 90
VC10, Vickers 4, 22, 108, 109, 113, 118, 119, 129, 130, 135, 138, 140, 141, 142, 143, 145, 148, 150, 151, 155, 157, 161, 164, 167, 168, 169, 182, 187, 188, 189, 201, 202
Victor, Handley Page 20, 21, 61, 179, 185, 186, 215
Viscount, Vickers 67, 71, 76, 77, 78, 80, 90
Vulcan, Avro 4, 20, 61, 92, 114, 118, 120, 123, 124, 134, 155, 170, 185, 186, 201, 214, 215
Warwick, Vickers 81, 172, 206, 207
Wasp, Westland 34, 35, 45, 49
Wessex, Westland 34, 35, 45, 210, 222
Whirlwind, Westland 219
Wyvern, Westland 66
Yak-44, Yakovlev 18

COMPANIES
Airbus Industrie 189, 196
AWA, Armstrong Whitworth Aviation 104, 127, 149, 200
BAC, British Aircraft Corporation 6, 22, 108, 109, 112, 127, 138, 139, 140, 141, 142, 143, 144, 145, 146, 147, 148, 149, 150, 151, 153, 156, 157, 158, 161, 162, 165, 167, 168, 169, 182, 184, 187, 188, 189, 190, 200
BESL, Bristol Siddeley Engines Ltd 183, 200
Bombardier 47, 199
BP, British Petroleum 70, 215
Breguet 9, 61, 64, 99, 100, 103, 105, 106, 107, 117, 119, 120, 121, 122, 123, 138, 139, 145, 157, 158, 159, 160, 178, 182, 183, 195, 199, 202, 204
Bristol Aeroplane Company 99, 153
Bristol Engines 64, 65, 68, 70, 76, 146
Dassault 97, 182, 183, 194, 195, 198
Dornier Flugzeugwerke 101, 102
Dunlop 53
English Electric 42, 68, 97, 98, 99, 200
Fairey 20, 37, 38, 40, 42, 43, 44, 66, 72, 73, 83, 91, 94, 100, 103, 129, 204, 208, 209, 218, 220, 221
Flight Refuelling Ltd 83, 196
Fokker 147, 159
Gulf Oil 20
Handley Page Ltd 133, 148
HSA, Hawker Siddeley Aviation 32, 111, 112, 120, 125, 126, 135, 140, 151, 154, 159, 160, 161, 164, 165, 166, 171, 174, 181, 182, 190, 201, 219
Marconi 31, 38, 39, 44, 68, 187, 194, 200, 203, 204
Marshalls 194, 199
Mobil 215
Napier, D. Napier and Sons 49, 53, 54, 57, 60, 61, 64, 65, 66, 67, 68, 69, 70, 77, 85, 86, 89, 91, 92, 93, 97, 98, 99, 101, 145, 148, 200, 203, 204, 221, 222
Nord 45, 103, 105, 116
Occidental 215
Rolls-Royce 6, 43, 54, 61, 64, 66, 68, 69, 70, 77, 81, 82, 83, 85, 86, 97, 99, 101, 102, 103, 104, 105, 111, 112, 120, 122, 127, 129, 133, 134, 139, 141, 143, 145, 151, 152, 153, 161, 163, 166, 168, 184, 187, 188, 195, 203, 204, 220
SAAB 199
Saro, Saunders-Roe 28, 29, 31, 48, 49, 50, 51, 52, 53, 54, 55, 57, 58, 59, 60, 61, 62, 64, 66, 69, 70, 100, 101, 203, 208, 209, 218
Selex 199
Shell Expro 215
Supermarine 50, 53, 55, 57, 58, 59, 66, 69, 201, 203, 206, 209
Ultra Electronics 29, 30, 196, 199
Vickers 6, 22, 29, 37, 40, 42, 44, 47, 53, 55, 57, 71, 74, 75, 76, 77, 78, 81, 90, 104, 108, 109, 110, 111, 112, 113, 116, 119, 121, 127, 129, 130, 131, 133, 135, 140, 143, 145, 147, 148, 150, 151, 153, 157, 162, 167, 169, 172, 189, 201, 202, 204, 207

ENGINES AND MOTORS
15-KS-1000, Aerojet 40
Adour, Rolls-Royce Turbomeca 163, 187, 188
Avon, Rolls-Royce AJ.65 61, 62, 163
BE.53, Bristol 99, 100
BS.75, Bristol Siddeley 147
BS.98, Bristol Siddeley 147
BS.106, Bristol Siddeley 143
Centaurus, Bristol 49, 50, 54, 57, 58, 59, 60, 62, 64, 65, 68, 69, 70, 76, 77, 90, 91, 92, 93
Compound-Griffon, Rolls-Royce 54, 58, 65, 66, 203
Conway, Rolls-Royce RB.80 111, 112, 125, 132, 141, 143, 144, 148, 167
Corgi, rocket motor 42
Crecy, Rolls-Royce 4, 6, 66
D30KPV, Soloviev 195
Eagle, Rolls-Royce 45, 65, 66, 187, 189, 199, 204, 212, 214
Eland, Napier 77, 85, 86, 98, 99, 101, 221
Falcon, Rolls-Royce 65, 66, 198, 204
Griffon, Rolls-Royce 51, 54, 58, 59, 64, 65, 66, 68, 69, 70, 83, 85, 86, 88, 93, 122, 138, 203, 204
Hercules, Bristol 9, 53, 54, 64, 65, 74, 76, 77, 88, 120, 138, 139, 146, 156, 190, 197, 198, 199, 217
JT8D, Pratt & Whitney 182
Mamba, Armstrong Siddeley 73, 100, 103, 104
Medway, Rolls-Royce 127, 128, 144, 148, 151, 152, 161
Nomad, Napier 49, 50, 53, 54, 57, 58, 59, 60, 61, 62, 64, 65, 66, 67, 68, 69, 70, 86, 89, 91, 92, 93, 94, 145, 148, 203, 204
NPT171, Noel Penny Turbines 40
Pegasus, Bristol Siddeley 127, 153
Proteus, Bristol 54, 57, 58, 59, 64, 65, 68, 69, 70, 77, 90, 91, 145, 146, 153
R-3350, Pratt & Whitney 65, 66, 145
RB.211, Rolls-Royce 145
RB.401, Rolls-Royce 184, 185
RB.410, Rolls-Royce 185, 189
RB.415, Rolls-Royce 183, 184, 185
RD36-35, Kolesov 195
Sabre, Napier 64
Smokey Joe, rocket motor 42
Soar, Roll-Royce RB.93 43
Spey, Rolls-Royce 120, 121, 122, 132, 144, 163, 168, 180, 187, 188, 195
Super Conway, Rolls-Royce RB.178 112, 125, 141, 148
T56, Allison 101, 102, 139, 146
T78, Allison 146
TF39, General Electric 145
Theseus, Bristol 68, 146
Turbo-Griffon, Rolls-Royce 51, 54, 58, 59, 65, 66, 68, 69, 203, 204
Twin Wasp, Pratt & Whitney 49
Tyne, Rolls-Royce RB.109 68, 70, 77, 97, 99, 100, 101, 102, 103, 104, 105, 106, 120, 122, 125, 126, 128, 130, 131, 133, 138, 139, 145, 146, 158, 162, 221

GENERAL
Alexander L. Keilland 210, 211
Backscratch 13, 203
Big Three 156, 158, 161, 164, 165, 169, 176
Fastnet Race 210
Have Garden 24, 25
Lindholme Gear 75, 116, 197, 210, 211
Nangana 30, 203
NBMR.2 4, 6, 20, 96, 97, 99, 100, 101, 103, 104, 105, 107, 114, 117, 119, 120, 127, 136, 169
Paraquat, Op. 186
Paraquet, Op. 186
Pave Pronto 25
Piper Alpha 181, 205, 206, 210
Poffler 109, 202
Porosint 148, 149
Project N 34, 204
Ranch Hand 23
Tapestry 4, 5, 186, 188, 205, 207, 209, 211, 213, 214, 215, 216, 217
Three-in-One 107, 108, 109, 110, 111, 112, 113, 114, 116, 117, 123, 126, 128, 129, 134, 136
Trinity 5, 34, 96, 97, 99, 101, 103, 105, 107, 108, 109, 110, 111, 113, 114, 115, 117, 119, 121, 123, 125, 127, 129, 131, 133, 135
ULTRA 19, 21, 28, 29, 30, 196, 199

ORGANISATIONS
A&AEE, Aircraft and Armament Evaluation Establishment 44, 84, 86, 118, 179, 200
Admiralty Research Laboratory (ARL) 23
Admiralty, The 7, 17, 23, 25, 26, 27, 32, 38, 42, 43, 60, 62, 71, 72, 79, 98, 117, 208, 209, 216, 222
Air Inter 182
Air Staff, The 8, 13, 14, 27, 29, 32, 33, 34, 37, 38, 41, 42, 44, 45, 47, 48, 49, 50, 51, 52, 53, 54, 55, 56, 57, 58, 60, 61, 62, 63, 64, 65, 66, 67, 68, 69, 70, 71, 72, 73, 75, 77, 78, 80, 84, 85, 86, 88, 89, 91, 92, 94, 95, 97, 98, 101, 105, 106, 107, 108, 110, 111, 112, 113, 114, 115, 116, 117, 118, 120, 121, 122, 126, 127, 128, 129, 133, 134, 135, 136, 137, 138, 139, 140,

141, 145, 148, 150, 151, 153, 154, 155, 156, 157, 158, 159, 160, 161, 162, 164, 167, 168, 172, 175, 176, 177, 178, 180, 182, 185, 187, 189, 192, 193, 194, 195, 198, 199, 200, 201, 202, 203, 204, 218, 219
an tAerchór (Irish Air Corps) 199
Armée de l'Air 183
BAE Systems 6, 33, 45, 144, 175, 195, 199, 200, 212, 213
BEA, British European Airways 77, 171, 221
BOAC, British Overseas Airways Corporation 53, 59, 82, 97, 108, 140, 142, 164, 167
Bomber Command 8, 19, 20, 33, 140, 173, 185, 215
Coastal Command 4, 7, 8, 10, 15, 17, 32, 33, 35, 37, 40, 41, 44, 47, 49, 50, 59, 61, 62, 67, 68, 71, 73, 74, 75, 76, 78, 79, 80, 81, 83, 84, 86, 88, 89, 90, 91, 94, 96, 114, 115, 116, 117, 118, 119, 129, 133, 134, 135, 140, 141, 153, 156, 157, 158, 164, 165, 170, 171, 173, 174, 177, 179, 205, 206, 207, 208, 209, 210, 216, 218, 219
Imperial Japanese Navy 10, 21
Kriegsmarine 7, 10, 39
MinTech, Ministry of Technology 175
MoA, Ministry of Aviation 8, 59, 106, 107, 108, 114, 117, 119, 121, 126, 130, 134, 135, 136, 138, 139, 140, 150, 151, 153, 154, 155, 158, 159, 160, 161, 164, 165, 168, 171, 176, 201
MoD, Ministry of Defence 8, 31, 35, 39, 106, 107, 159, 165, 171, 173, 181, 185, 186, 189, 194, 196, 197, 201, 205, 208
MoS, Ministry of Supply 50, 51, 52, 60, 61, 65, 66, 68, 69, 70, 72, 73, 78, 81, 88, 89, 92, 99, 201
NATO 5, 11, 12, 13, 14, 15, 16, 17, 20, 22, 33, 45, 64, 79, 95, 96, 97, 99, 100, 101, 102, 103, 105, 106, 107, 109, 111, 113, 114, 117, 119, 137, 171, 173, 186, 192, 195, 216, 217
RAAF, Royal Australian Air Force 63, 175, 177, 198
RCAF, Royal Canadian Air Force 39, 89, 104, 113, 118, 120, 175, 177, 220
RNZAF, Royal New Zealand Air Force 63, 118, 138, 157, 175, 177, 198
Royal Aircraft Establishment (RAE) 25, 27, 42, 86, 149, 159, 160, 164, 177, 178, 201, 221, 222
Royal Marines 180, 186, 209, 216
Royal Navy 11, 16, 21, 24, 28, 33, 35, 42, 45, 73, 170, 178, 192, 197, 199, 201, 203, 214, 217, 218, 219
Russian Navy 9, 18, 192, 195, 199
SECBAT 106
Signals Command 24
Soviet Navy 4, 11, 13, 14, 15, 16, 17, 18, 40, 42, 116, 216
Strike Command 17, 179, 185, 205
Telecommunications Research Establishment (TRE) 25, 55, 201
Tiger Force 82
Torpedo Experimental Establishment (TEE) 38
Transport Command 141, 160, 164, 167, 171
US Navy 5, 10, 11, 13, 14, 15, 16, 17, 18, 21, 24, 27, 35, 46, 50, 61, 78, 79, 96, 98, 130, 139, 140, 145, 173, 193, 194, 197, 198, 204, 216, 220

PEOPLE
Amery, Julian 117, 159
Attlee, Clement 10
Bancroft, Ian 68
Bateman, Air Vice-Marshal R N 134
Beaugeard, Wing Commander G G 170, 171, 199
Benwell, Wing Commander R H 152
Bovenizer, Vernon 106
Broad, Group Captain H P 73, 83, 90, 98, 101, 111, 151, 198
Brotherhood, Air Commodore W R 108
Brynmor, John 217
Burnell, Brian 34
Burnett, Air Marshal Sir Brian 173, 174
Burwell, Air Commodore R H C 114, 169
Butler, R A (Rab) 6, 47, 67, 68, 69, 71, 83, 86, 221
Case, Air Vice-Marshal A A 171
Chadwick, Roy 82
Chamant, Jean 117
Chandler, Squadron Leader G J 75, 76, 89, 94
Chilton, Air Marshal Sir Edward 47, 53, 63
Clements, A 219
Cochrane, Air Chief Marshal Ralph 51, 60, 94
Cockburn, Dr Robert 44
Coleman, Dr W S 149, 150
Cooper, Frank 159
Crouch, A S 78, 89, 209
Cullen, Lord 210
Cunningham, Group Captain John 150, 171
Davis, Air Marshal Sir John 170
Dickson, Marshal of the RAF Sir William 62
Diesel, Rudolf 22
Dobson, G T 42
Duke of Edinburgh, the 68
Duke of Wellington, the 140, 162
Edelsten, Admiral J N 98
Edwards, George 76, 140
Emson, Air Vice-Marshal R H E 140, 158, 165, 170
Esplin, Air Commodore I G 96
Fedden, Roy 64
Fletcher, Air Commodore P C 119, 138, 139, 140, 150, 153
Foster, Air Vice-Marshal Robert 50
Fox, Uffa 206, 207
Fraser, Hugh 156, 158, 159
Gauss, Carl 20
Gray, W E 149
Greatrex, Roy 220
Haddon-Cave, Charles 180, 181, 196
Hall, Sir Arnold 159, 160, 164, 197
Harding, Field Marshal Sir John 63
Hardman, Sir Henry 108, 159, 160, 164
Hartley, Air Marshal C H 44, 164, 165, 170, 171, 173
Harvey, Nick 197
Herbecq, J E 107
Hill, Rear Admiral Richard 32
Hislop, Dr G S 218, 220, 221
Hodges, Air Vice-Marshal Lewis M 174
Hodgkinson, Air Vice-Marshal W D 171
Holder, Air Marshal Paul 170, 171, 172, 173
Hudleston, Air Chief Marshal Sir Edmund C 108
Hughes, Wing Commander J D E 73, 219
Ingrams, RCAF Wing Commander 118
Ivelaw-Chapman, Air Marshal Sir Ronald 52
Jenkins, Roy 164
Johnson, D 149
Knott, Air Commodore R G 151, 156, 157, 161, 165, 171, 172, 173
Lees, Air Marshal Sir Ronald 108, 117, 136, 138, 139, 140
Lord Caldecote 98
Lord de l'Isle and Dudley 53
Luby, Captain M 69
Lugg, Group Captain S 89
Macmillan, Harold 53, 63
McFadyen, Air Vice-Marshal Douglas 74
McGrigor, First Sea Lord, Sir Rhoderick 62
Meili, Ernest H 106
Merton, Professor Sir Thomas 171, 172, 173
Messmer, Pierre 159
Millar, Air Commodore J C 47, 48, 53
Moreton, Group Captain N V 71
Musgrove, Cyril 99
Nelson, Sir George 68, 69
Padmore, Sir Thomas 98, 99
Pearson, D J 171
Pelly, Air Vice-Marshal Claude 74
Pike, Air Commodore Thomas 49, 50, 69, 70, 108, 140
Raban, Squadron Leader M C 57, 58, 59
Ragg, Air Vice-Marshal Robert 78, 79
Rawet, B A 170, 171
Roberts, Group Captain J F 107, 118
Rodgers, Bill 216
Sandys, Duncan 63, 67, 68, 69, 73, 117
Satterly, Air Commodore Harold 72
Selway, Air Marshal A D 4, 164, 165
Serby, J E 70
Shute Norway, Neville 77
Simmons, R W 74
Slatter, Air Marshal Leonard 50
Slessor, Marshal of the RAF Sir John 50, 51, 67, 68, 169, 170
Smallwood, Air Vice-Marshal Denis 161, 170
Sorley, Air Vice-Marshal R S 98
Squire, Herbert 221
Steele, Air Marshal Charles 74
Stevens, Air Marshal Alick 78, 79
Stewart, W 221, 222
Strath, William 106, 107
Symmons, R W 78
Symonds, Captain G O 38
Tall, Commander Jeff 19
Thorneycroft, Peter 159, 165
Tiltman, Hessel 77
Truman, President Harry S 79
Tuttle, Air Vice-Marshal Geoffrey W 37, 47, 50, 52, 53, 65
Ubee, Group Captain S R 75, 77, 84
Vacquier, Victor 21
Wade, Air Marshal Sir Ruthven 217
Wardle, Group Captain A R 98
Way, Sir Richard 'Sam' 159
Whittan, Wing Commander R 119
Whittock, G S 51
Williams, Group Captain M F D 6, 71
Wilson, Admiral Sir Arthur 6, 19, 160, 175
Woodhouse, Monty 117
Woodward-Nutt, A E 81, 88
Zuckerman, Solly 164

PLACES
Afghanistan 157, 179, 180, 181
Atlantic Ocean 4, 7, 8, 9, 13, 14, 16, 17, 20, 22, 26, 32, 35, 46, 48, 49, 50, 51, 52, 53, 61, 62, 71, 74, 79, 81, 83, 89, 90, 97, 98, 100, 103, 104, 105, 106, 107, 116, 117, 119, 120, 121, 122, 123, 133, 138, 139, 145, 155, 156, 157, 158, 159, 160, 161, 164, 165, 166, 171, 172, 173, 178, 180, 181, 182, 183, 186, 192, 193, 194, 195, 202, 206, 210, 212, 214, 218
Ballykelly 34, 81
Baltic Sea 11, 97
Banff 8, 40
Black Sea 11, 173, 203
Boscombe Down 179
Brawdy 13, 203
British Columbia 63
Chadderton 91, 120, 126, 182
Defford 25
Eglin AFB 24
Embakasi 108
Entebbe 108
Farnborough 25, 67, 86
Filton 90, 99, 110, 112, 141, 145, 153, 189
Freidrichshafen 64
Greenland 14, 115
Greenock 38
Groningen 215
Hatfield 110, 111, 123, 154, 161, 162, 184
Hayes 220
Hurn 196
Iceland 13, 14, 115, 203, 212, 215
Indian Ocean 50, 60, 90, 97
Kinshasa 70
Lake Constance 64, 101
Liverpool 7
London 7, 67, 118
Mediterranean Sea 11, 26, 52, 61, 62, 97, 173
Mojave Desert 24
Moray Firth 9, 18, 205
N'jdili Airport 70
North Sea 9, 40, 75, 77, †80, 88, 97, 192, 206, 208, 210, 211, 214, 215, 216, 217
Pacific Ocean 13, 64, 82, 208
Petrevie 210
Portsmouth 9, 18, 77
Radlett 141
RAF Changi 24
RAF Kinloss 34, 78, 79, 86, 174, 210
RAF Manston 197
RAF Seletar 63
RAF Silloth 80
RAF St Mawgan 34, 79, 81, 174, 176, 182
RAF Topcliffe 79
RNAS Culdrose 218
RNAS Prestwick 74, 213, 218
RRE Pershore 24
Scatsta 13, 203
Singapore Island 24
South Georgia 186
The Hague 216
Toulon 165
Toulouse 106
Viet Nam 23, 24, 25
Weybridge 44, 108, 110, 130, 133, 140, 141, 144, 145, 147, 148, 151, 161, 167, 187, 189
Woodford 6, 28, 31, 32, 39, 45, 80, 84, 91, 105, 117, 120, 121, 123, 126, 135, 153, 154, 155, 159, 161, 165, 166, 170, 175, 179, 181, 182, 183, 184, 185, 186, 187, 189, 190, 191, 192, 196, 212

SENSORS
ALMDS 27
AN/APS.20 4, 78, 79, 80, 94
AN/AQS.5 219
ARAR/ARAX 178
ASDIC 28
Ash 24
ASQ.901 19, 31, 39, 203
ASV.13 19, 20, 55, 75, 76, 77, 83, 84, 86, 90, 203
ASV.19 20, 73, 124, 203
ASV.21 20, 43, 80, 87, 90, 92, 94, 109, 116, 117, 123, 125, 127, 130, 131, 132, 133, 154, 162, 163, 167, 168, 177, 178, 203
ASV.VI 19
ASV.VII 19
ASV.VIII 19
ASWEPS 31, 39, 137, 203
Autolycus 13, 22, 23, 24, 25, 26, 116, 167, 178, 189, 203
Barra 28, 30, 31, 32, 177, 203
Black Crow 25
Blue Lagoon 25, 26
Blue Sapphire 26
CAMBS 29, 31, 177, 203
Clinker 26, 116, 117, 162, 203
DIFAR 31, 177, 203
Green Thistle 25
H2S 19, 20, 42, 54, 55, 56, 59, 123, 124, 137, 185, 186, 214, 215
High Tea 29, 30
HS.828, Hawker Siddeley 31, 32, 39, 40, 52, 219
Huff-Duff 19, 21
Jezebel 28, 29, 30, 31, 32, 177, 203
Julie 31, 203
Kielgerät 25
ORADS 27, 178, 191
Orange Harvest 20, 87
Orange Tartan 26
Rasputin 31
Searchwater 19, 20, 45, 177, 178, 179, 180, 183, 184, 185, 187, 189, 196, 199, 202, 203, 211
SOSUS †8, 13, 15, 203
Sperber 25
SQUID 22, 185, 191
Tandem 31, 32, 73, 76, 103, 203
Valkyrie 25, 194, 195
Vampir 25
Violet Banner 25, 26
Yellow Duckling 25, 26, 117, 162, 203
Yellow Gate 180

SHIPS
Admiral Kuznetzov 9, 18
Balzam-class 216
Gannet, HMS 218
Jura, HMS 216
Kiev-class 18
Kirov-class 17
Komar-class 45
Moskva-Class 173
Primor'ye-class 216
RMS Queen Mary 67
Sea Devil, HMS 26
Slava-class 17
SS America 67
Sverdlov-class 17, 41, 42
Wilhelm Gustloff 11
York, HMS 9, 18

SIGNIFICANT REQUIREMENTS
OR.231 49, 50, 53, 89, 94, 202, 203
OR.320 86, 202
OR.347 126, 202
OR.350 5, 8, 26, 29, 30, 35, 44, 113, 114, 115, 116, 117, 118, 119, 120, 121, 122, 123, 124, 125, 126, 127, 128, 129, 130, 131, 132, 133, 134, 135, 136, 137, 138, 145, 148, 153, 162, 192, 202
OR.357 5, 20, 30, 35, 40, 44, 88, 126, 129, 136, 137, 138, 139, 141, 142, 143, 145, 147, 148, 149, 150, 151, 153, 155, 156, 157, 158, 182, 202, 203
OR.381 5, 30, 156, 157, 159, 161, 163, 164, 165, 166, 167, 168, 169, 171, 173, 174, 202, 210
R.112 50, 202, 203
R.2/48 20, 29, 30, 35, 36, 37, 50, 51, 52, 53, 54, 55, 56, 57, 59, 60, 61, 62, 63, 64, 65, 66, 68, 69, 70, 101, 143, 146, 202, 203, 204, 215
R.5/46 29, 30, 83, 85, 86, 202, 203
SR(A).420 195

SUBMARINES
41 for Freedom 13
Akula 11, 15, 16
Alpha 16, 17, 22
Charlie 16
Delta 13, 14
Echo 11, 16
Foxtrot 11, 12, 14, 15
Golf 12
Hotel 12, 13, 176
Juliett 11
Kilo 14, 40, 116
November 12, 13, 14, 16
Oscar 16
Resolution 115
Santa Fe 45
Tango 14
Typhoon 14, 15
Victor 10, 11, 15, 16, 18
Walker 10, 15
Whiskey 11, 12, 37
Yankee 4, 13, 114
Zulu 12, 37
Weapons
3in RP 40, 41, 45, 90
ADCAP 17
AS.12, Nord 44, 45, 46, 202
AS.30, Nord 44, 116, 155
AS.37 44
ASROC 34
Bidder 36
Blue Duck 34
Blue Moon 25, 26
Bootleg, BA.920 37, 202, 203
Bullpup, Martin AGM-12 44, 100, 116, 202
Dealer 32, 33, 35, 37, 38, 74, 75, 80, 89, 202, 204
Fancy 36, 73
Fido, Mk.24 35
Gadget 34
Gecko, SA-N-4 46
Goblet, SA-N-3 46
Grand Slam 33
Green Cheese 20, 42, 43, 44, 203, 204
Harpoon, McDonnell Douglas AGM-84 33, 45, 46, 180, 187, 195, 197, 214
Ikara 34
Lulu, Mk.101 34, 35, 203, 204
Martel, HSD 33, 44, 45, 46, 116, 163, 174, 180, 189, 202, 204
Mk.30 torpedo 38, 115
Mk.31 torpedo 38
Mk.43 torpedo 224
Mk.57 19, 34, 35, 204
Nozzle 41, 127, 201
Pentane 32, 35, 36, 37, 49, 51, 54, 56, 57, 72, 73, 74, 75, 76, 89, 92, 203, 204
Polaris 12, 13, 16, 17, 115, 116
Red Angel 41, 42, 204
Red Dean, Vickers Type 888 42, 43, 204
Red Shoes, English Electric 42
Retrobomb 21, 33, 34
Sandbox, SS-N-12 11, 18
Sark, SS-N-4 12, 15
Sawfly, SS-N-8 13, 115
Scud, SS-N-1 12
Sea Eagle, BAe Dynamics 45, 187, 189, 204, 212, 214
Sea Skimmer, Fairey 91
Serb, SS-N-5 12, 13, 15
Shaddock, SS-N-3 11
Shipwreck, SS-N-19 16
Sidewinder, AIM-9 45, 179, 180
Siren, SS-N-9 16
Skybolt, Douglas AGM-48 107, 112, 121, 122, 123, 127
SS.10/11, Nord 45, 46
Starbright, SS-N-7 16
Starfish, SS-N-15 14, 16
Stingray 19, 33, 36, 38, 39, 40, 180, 189, 195, 202, 203, 204, 212, 214
Sturgeon, SS-N-20 14, 15
Tallboy 33
Tigerfish 17
Uncle Tom 41, 202
W.112, Avro 11
WE.177 34, 35, 130, 204
Z-series 36
Zannet 36
Zeta 36, 37, 55, 203, 204
Zombi 36
Zonal 36
Zorster 224